Colour Chemistry
2nd edition

KU-227-038

Robert M Christie
School of Textiles & Design, Heriot-Watt University, UK and Department of Chemistry, King Abdulaziz University, Saudi Arabia
Email: r.m.christie@hw.ac.uk

THE QUEEN'S AWARDS
FOR ENTERPRISE:
INTERNATIONAL TRADE
2013

Print ISBN: 978-1-84973-328-1

A catalogue record for this book is available from the British Library

© R M Christie 2015

All rights reserved

Apart from fair dealing for the purposes of research for non-commercial purposes or for private study, criticism or review, as permitted under the Copyright, Designs and Patents Act 1988 and the Copyright and Related Rights Regulations 2003, this publication may not be reproduced, stored or transmitted, in any form or by any means, without the prior permission in writing of The Royal Society of Chemistry or the copyright owner, or in the case of reproduction in accordance with the terms of licences issued by the Copyright Licensing Agency in the UK, or in accordance with the terms of the licences issued by the appropriate Reproduction Rights Organization outside the UK. Enquiries concerning reproduction outside the terms stated here should be sent to The Royal Society of Chemistry at the address printed on this page.

The RSC is not responsible for individual opinions expressed in this work.

Published by The Royal Society of Chemistry,
Thomas Graham House, Science Park, Milton Road,
Cambridge CB4 0WF, UK

Registered Charity Number 207890

Visit our website at www.rsc.org/books

Preface

I was pleased to receive the invitation from the RSC to write this second edition of *Colour Chemistry* following the success of the first edition published in 2001. I am also appreciative of the broadly positive reviews that the first edition received and of the favourable comments that I have received from a wide range of individuals. The initial approach to compiling this second edition involved taking stock of the original content, while also assessing the extensive range of developments in colour chemistry that have taken place in the intervening years. While the chemistry of the traditional classes and applications of dyes and pigments is well-established, there have been significant developments in other areas, especially in topics related to functional dyes. The industry associated with the manufacture and application of dyes and pigments has continued to transfer substantially away from Europe and the USA towards the emerging economies in Asia, especially to China and India, and consequently many new developments are emerging from research undertaken in that region. Two important textbooks have been published in the last decade or so. I am honoured to pay a special tribute to the late Heinrich Zollinger, whose third edition of *Color Chemistry* appeared in 2003, and maintained the standard of detail, originality and excellence for which this eminent author was renowned. I also acknowledge the importance of *Chromic Phenomena: Technological Applications of Colour Chemistry*, by Peter Bamfield and Michael Hutchings, the second edition of which appeared in

Colour Chemistry, 2nd edition
By Robert M Christie
© R M Christie 2015
Published by the Royal Society of Chemistry, www.rsc.org

UCB
298103

2010. This excellent textbook, also published by the RSC, adopts an original approach to the subject, organising the topics according to the phenomena giving rise to colour. The experience and knowledge of these authors from an industrial perspective is evident throughout their book.

This second edition of *Colour Chemistry* adopts broadly the original philosophy and structure, retaining a relatively traditional approach to the subject. The content has been significantly revised and expanded throughout, especially to reflect newer developments. The book thus remains aimed at providing an insight into the chemistry of colour, with a particular focus on the most important colorants produced industrially. It is aimed at students or graduates who have knowledge of the principles of chemistry, to provide an illustration of how these principles are applied in producing the range of colours that are all around us. In addition, it is anticipated that professionals who are specialists in colour science, or have some involvement with the diverse range of coloured materials in an industrial or academic environment, will benefit from the overview of the subject that is provided.

The opening chapter provides a historical perspective on how our understanding of colour chemistry has evolved, leading to the development of an innovative global industry. The second chapter provides a general introduction to the physical, chemical and, to a certain extent, biological principles which allow us to perceive colours. This chapter has been expanded in particular to provide a discussion of the recent developments that have taken place in the use of computational methods used to model and predict the properties of colorants by calculation. Chapters 3–6 encompass the essential principles of the structural and synthetic chemistry associated with the most important chemical classes of industrial dyes and pigments. Chapters 7–11 deal with the applications of dyes and pigments, and in particular the chemical principles underlying their technical performance, not only in traditional applications such as textiles, printing inks, coatings and plastics but also in an expanding range of high technology or functional applications. The chapter on functional dyes has been significantly re-written to reflect recent and current developments in, for example, display technologies, solar energy conversion and biomedical applications. A new chapter introduces the chemistry of colour in cosmetics, with particular emphasis on hair dyes, which reflects the continuing growth of a sector of the colour industry that has thus far largely resisted the move from West to East. I express my gratitude to my co-author of this chapter, Olivier Morel, for his contribution. The final chapter provides an account of the most important environmental issues associated with the manufacture and use of colour, which the industry is increasingly required to acknowledge and address.

R. M. Christie

Contents

Colour Chemistry, 2nd edition
By Robert M Christie
© R M Christie 2015
Published by the Royal Society of Chemistry, www.rsc.org

CHAPTER 1

Colour: A Historical Perspective

1.1 INTRODUCTION

We only have to open our eyes and look around to observe how important a part colour plays in our everyday lives. Colour pervades all aspects of our lives, influencing our moods and emotions and generally enhancing the way in which we enjoy our environment. In addition to its literal meaning, we often use the term colour in more abstract ways, for example to describe aspects of music, language and personality. We surround ourselves with the colours we like and which make us feel good. Our experience of colour emanates from a rich diversity of sources, both natural and synthetic. Natural colours are all around us, in the earth, the sky, the sea, animals and birds and in the vegetation, for example in the trees, leaves, grass and flowers. These colours can play important roles in the natural world, for example as sources of attraction and in defence mechanisms associated with camouflage. Plant pigments, especially chlorophyll, the dominant natural green pigment, play a vital role in photosynthesis in plants, and thus may be considered as vital to our existence! Colour is an important aspect in our enjoyment of food. We frequently judge the quality of meat products, fruit and vegetables by the richness of their colour. There is also convincing evidence that colorants present naturally in foods may bring us positive health benefits, for example as anti-oxidants, which are suggested to play a role in protection against cancer. In addition, there is a myriad of examples of synthetic colours, products of the chemical manufacturing industry, which we

Colour Chemistry, 2nd edition
By Robert M Christie
© R M Christie 2015
Published by the Royal Society of Chemistry, www.rsc.org

tend to take so much for granted these days. Synthetic colours often serve a purely decorative or aesthetic purpose in the clothes we wear, in paints, plastic articles, in a wide range of multicoloured printed materials such as posters, magazines and newspapers, in photography, cosmetics, toiletries, ceramics, and on television and film. There are many examples of colours playing pivotal roles in society. In clothing, the desire for fashion sets colour trends, and the symbolism of colours is important in corporate wear and uniforms. Individual nations adopt specific national colours that are reflected, for example, in national flags and as displayed by sports teams. In some cases, colours may be used to convey vital information associated with safety, for example in traffic lights and colour-coded electrical cables.

Colour is introduced into these materials and applications using substances known as *dyes* and *pigments*, or collectively as *colorants*. The essential difference between these two colorant types is that dyes are soluble coloured compounds which are applied mainly to textile materials from solution in water, whereas pigments are insoluble compounds incorporated by a dispersion process into products such as paints, printing inks and plastics. The reader is directed to Chapter 2 of this book for a more detailed discussion of the distinction between dyes and pigments as colouring materials.

1.2   THE EARLY HISTORY OF DYES AND PIGMENTS

The human race has made use of colour since prehistoric times, for example in decorating the body, in colouring the furs and skins worn as clothing and in the paintings that adorned cave dwellings.[1] Of course, in those days the colours used were derived from natural resources. The dyes used to colour clothing were commonly extracted either from botanical sources, including plants, trees, roots, seeds, nuts, fruit skins, berries and lichens, or from animal sources such as crushed insects and molluscs. Pigments for paints were obtained from coloured minerals, such as ochre and haematite which are mostly based on iron oxides, giving yellows, reds and browns, dug from the earth, ground to a fine powder and mixed into a crude binder. Charcoal from burnt wood provided the early forerunners of carbon black pigments. The durability of these natural inorganic pigments, which contrasts with the more fugitive nature of natural dyes, is demonstrated in the remarkably well-preserved Palaeolithic cave paintings found, for example, in Lascaux in France and Altamira in Spain.

Synthetic colorants may also be described as having an ancient history, although this statement applies only to a range of pigments produced from rudimentary applications of inorganic chemistry. These very early synthetic inorganic pigments have been manufactured and used in paints for thousands of years.[2,3] The ancient Egyptians were responsible for the development of probably the earliest synthetic pigment, Egyptian Blue later known as Alexandria blue, a mixed silicate of copper and calcium, which has been identified in murals dating from around 1000 BC. This development added bright blue, a colour not readily available from natural minerals, to the artists' palette. Arguably the oldest synthetic colorant still used significantly today is Prussian blue, the structure of which has been established as iron(III) hexacyanoferrate(II). The manufacture of this blue inorganic pigment is much less ancient, dating originally from the middle of the seventeenth century. However, it is noteworthy that this product pre-dates the origin of synthetic organic dyes and pigments by more than a century.

Synthetic textile dyes are exclusively organic compounds and, in relative historical terms, their origin is much more recent. Textile materials were coloured exclusively with natural dyes until the mid-nineteenth century.[4–9] Since most of nature's dyes are rather unstable, the dyeings produced in the very early days tended to be quite fugitive, for example to washing and light. Over the centuries, however, dyeing procedures, generally quite complex, using a selected range of natural dyes were developed that were capable of giving reasonable quality dyeing on textile fabrics. Since natural dyes generally have little direct affinity for textile materials, they were usually applied together with compounds known as *mordants*, which were effectively 'fixing-agents'. Metal salts, for example of aluminium, iron, tin, chromium or copper, were the most commonly used mordants. They functioned by forming metal complexes of the dyes within the fibre. These complexes were insoluble and hence more resistant to washing processes. These agents not only improved the fastness properties of the dyeings, but also in many instances were essential to develop the intensity and brightness of the colours produced by the natural dyes. Some natural organic materials, such as tannic and tartaric acids, may also be used as mordants. The most important natural blue dye, and arguably the most important natural dye, is indigo, **1.1a**, obtained from certain plants, for example *Indigofera tinctoria* found in India and in other regions of Asia, and woad, *Isatis tinctoria*, a flowering plant that grows in Europe and the USA.[10,11] Natural indigo dyeing – still practised quite widely as a traditional craft process in Asia and North Africa for

textiles and clothing, often to provide textile garments with traditional symbolic status – commonly starts with the fermentation of
extracts of the leaves harvested from the plants to release the indigo
precursors. Dyeing may be carried out directly from a vat where the
fermentation of composted leaves takes place in the presence of alkali
from wood ash or limestone to produce precursors that are oxidised
in air on the fibre to give indigo. Alternatively, the blue pigment may
be isolated and applied by a reduction/oxidation process, in a
'natural' version of vat dyeing (Chapter 7). In these ways, indigo
produces attractive deep blue dyeings of good quality without the
need for a mordant. A chemically-related product is Tyrian purple, the
principal constituent of which is 6,6′-dibromoindigo (**1.1b**). This
colouring material was for many years a fashionable, aristocratic
purple dye extracted from the glands of *Murex brandaris*, a shellfish
found on the Mediterranean and Atlantic coasts.[12,13] It is said to have
required the use of 10 000 shellfish to provide one gram of dye, which
no doubt explains why the luxurious, bright purple fabrics were
available only to the ruling class elite in Mediterranean and Middle
Eastern societies, and also the consequent association of the colour
purple with wealth and nobility. Natural red dyes were derived from
vegetable (madder) or animal (cochineal, kermes and lac insect)
sources. Madder is extracted from the roots of shrubs of the *Rubia*
species, such as *Rubia tinctorum*. The main constituent is alizarin, 1,2-
dihydroxyanthraquinone, **1.2**. Alizarin provides a relevant example of
the use of the mordanting process, since it readily forms metal
complexes within fibres, notably with aluminium. These complexes
show more intense colours and an enhanced set of fastness properties
compared with the uncomplexed dyestuff. The main constituent of
cochineal, obtained from dried parasitic insect species, is carminic
acid, **1.3**, a rather more complex anthraquinone derivative. There is a
wide range of natural yellow dyes of plant origin, one of the best-
known being weld, obtained from flowering plant species such as
Reseda luteola. The main constituents of the dye obtained from these
plants, which also requires mordanting for application to textiles, are
the flavononoids, leuteolin, **1.4a**, and apigenin, **1.4b**. Natural green
textile dyes proved elusive, because pigments such as chlorophyll, **1.5**,
could not be made to fix to natural fibres and also faded rapidly.
Lincoln Green was commonly obtained from weld over-dyed with
indigo. Over the centuries, natural dyes and pigments have also
been used for their medicinal qualities. Logwood is a flowering tree
(*Haematoxylum campechianum*) used as a natural dye source; it still
remains an importance source of haematoxylin, **1.6**, which generates

the chromophoric species by oxidation. Logwood has also been used in histology as a staining agent and extracts also have medical applications. On textiles, the colour developed from logwood varies (black, grey, blue, purple) depending on the mordant used as well as the application pH.

1.1

(**a**): R = H; (**b**) R = Br.

1.2

1.3

1.4 (a) R=OH; (**b**) R=H

1.5

R = CH₃: chlorophyll a
R = CHO: chlorophyll b

1.6

1.3   THE ERA OF THE SYNTHETIC DYE

It may be argued that the first synthetic dye was picric acid, **1.7**, which was first prepared in the laboratory in 1771 by treating indigo with nitric acid. Much later, a more efficient synthetic route to picric acid from phenol as the starting material was developed. Picric acid was found to dye silk a bright greenish-yellow colour but it did not attain any real significance as a practical dye mainly because the dyeings obtained were of poor quality, especially in terms of lightfastness. However, it did find limited use at the time to shade indigo dyeings to give bright greens.

1.7

The foundation of the synthetic dye industry is universally attributed to Sir William Henry Perkin on account of his discovery in 1856 of a purple dye that he originally gave the name Aniline Purple, but which was later to become known as Mauveine.[14–18] Perkin was a young enthusiastic British organic chemist who was carrying out research not aimed initially at dyes but rather at developing a synthetic route to quinine, the antimalarial drug. Malaria was a devastating condition at the time and natural quinine, a product often in short supply and expensive, was the most effective treatment. His objective in one particular set of experiments was to attempt to prepare synthetic quinine from the oxidation of allyltoluidine, but his attempts to this end proved singularly unsuccessful. With hindsight, this is not too surprising in view of our current knowledge of the complex heteroalicyclic structure of quinine. As an extension of this research, he turned his attention to the reaction of the simplest aromatic amine, aniline, with the oxidizing agent, potassium dichromate. This reaction gave a black product which might have seemed rather unpromising to many chemists, but from which Perkin discovered that a low yield of a purple dye could be extracted with organic solvents. An evaluation of the new dye in a silk dyeworks in Perth, Scotland, established that it could be used to dye silk a rich purple colour and that the resulting textile dyeings gave reasonable fastness properties. The positive response and also the technical assistance from an application perspective provided to Perkin by Roger Pullar, the dyer, was probably a decisive feature in what was to follow, since other traditional dyers proved more sceptical towards this revolutionary concept. The particular colour of the dye was significant to its ultimate success. It offered a potentially low cost means to reproduce the rich purple colour that was formerly obtainable from Tyrian purple, the use of which had been more or less discontinued centuries before. The colour was certainly superior to the 'false shellfish purples' of the time, which were extractable from lichens and to the dull purples associated with mixtures of red and blue natural dyes, such as madder and indigo. Perkin showed remarkable foresight in recognising the potential of his discovery. He took out a patent on the product and had the boldness to instigate the development of a large-scale manufacturing process, using his father's life savings, to build a factory at Greenford Green, near London, to manufacture the dye. Since the manufacture required the development of large-scale industrial procedures for the manufacture of aniline from benzene via reduction of nitrobenzene, the real significance of Perkin's venture was as the origin of the organic chemicals industry. This industry has evolved from such a humble beginning to become a dominant feature of the industrial base of many economies worldwide and to influence fundamentally the development of a wide range of indispensible modern products such as pharmaceuticals, agrochemicals, plastics, synthetic fibres, explosives, perfumes and photography. For many years, the structure of Mauveine was reported erroneously as **1.8**. It has been demonstrated from an analytical investigation of an original sample that the dye is a mixture, and that the structures of the principal constituents are in fact compounds **1.9** and **1.10**, with other minor constituents also identified.[19,20] The presence of the methyl groups, which are an essential feature of the product, demonstrate that it was fortuitous that Perkin's crude aniline contained significant quantities of the toluidines. Compound **1.9**, the major component of the dye, is derived from two molecules of aniline, one of *p*-toluidine and one of *o*-toluidine, while compound **1.10** is formed from one molecule of aniline, one of *p*-toluidine and two molecules of *o*-toluidine. It is likely, as the manufacturing process developed, that individual batches of the dye were variable in composition. Mauveine was launched on the market in 1857 and enjoyed rapid commercial success. Through its unique colour, it became highly desirable in the fashion houses of London and Paris. As an example important to the marketing of the product, Queen Victoria wore a mauve dress to her daughter's wedding. Indeed, the introduction of Mauveine, in association with other concurrent developments such as the emergence of department stores, the sewing machine and fashion magazines, arguably initiated the democratisation of fashion that had previously been available only to the wealthy upper classes of society.

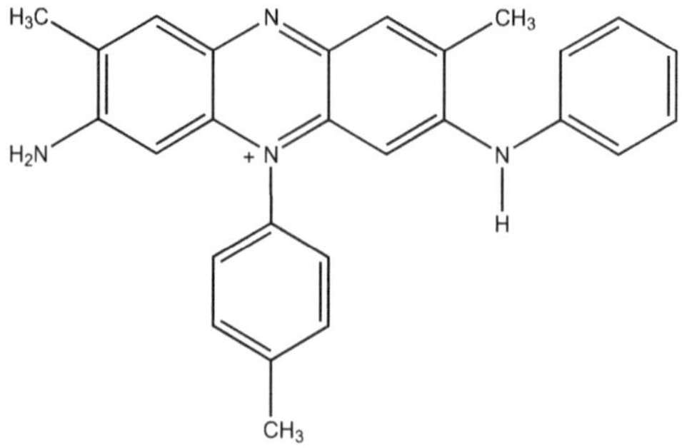

1.8

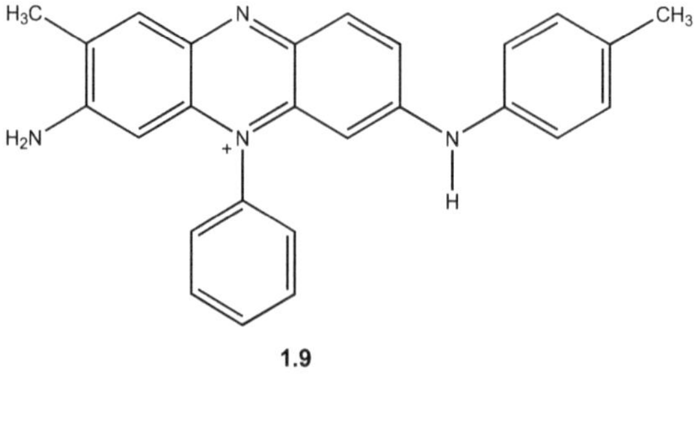

1.9

1.10

During the several years following the discovery of Mauveine, research activity in dye chemistry intensified especially in Britain, Germany and France.[21] For the most part, chemists concentrated on aniline as the starting material, adopting a largely empirical approach to its conversion into coloured compounds, and this resulted in the discovery, within a very short period of time, of several other synthetic textile dyes with commercial potential. In fact the term 'Aniline Dyes' was for many decades synonymous with synthetic dyes.[22] The most notable among the initial discoveries were in the chemical class now known as the arylcarbonium ion or triphenylmethine dyes (Chapter 6). An important commercially successful product that rapidly followed Mauveine was Fuschine, a rich red dye, also to become known as Magenta, which was introduced in 1859.[23,24] Magenta was first prepared by the oxidation of crude aniline (containing variable quantities of the toluidines) with tin(IV) chloride. The dye contains two principal constituents, rosaniline, **1.11** and homorosaniline, **1.12**, the central carbon atom being derived from the methyl group of *p*-toluidine. A structurally-related dye, rosolic acid, had been prepared

1.11

1.12

in the laboratory in 1834 by the oxidation of crude phenol, and therefore may also be considered as one of the earliest synthetic dyes, although its commercial manufacture was not attempted until the 1860s. Structure **1.13** has been suggested for rosolic acid, although it seems likely that other components were present. A range of new dyes, providing a wide range of bright fashion colours, yellows, reds, blues, violets and greens, as well as browns and blacks, soon emerged and these proved ultimately to be superior in properties and more economic compared with Mauveine, the production of which ceased after about ten years.

1.13

Undoubtedly the most significant discovery in colour chemistry in the 'post-Mauveine' period was due to the work of the German chemist Peter Griess, which provided the foundation for the development of the chemistry of azo dyes and pigments (Chapter 3). In 1858, Griess demonstrated that the treatment of a primary aromatic amine with nitrous acid gave rise to an unstable salt (a diazonium salt), which could be used to prepare highly coloured compounds. The earliest azo dyes were prepared by treatment of primary aromatic amines with a half equivalent of nitrous acid, so that half of the amine was diazotised and the remainder acted as the coupling component in the formation of the azo compound. The first commercial azo dye was 4-aminoazobenzene, **1.14**, Aniline Yellow, prepared in this way from aniline, although it proved to have quite poor dyeing properties. A much more successful commercial product was Bismarck Brown (originally named Manchester Brown), which was actually a mixture of compounds, the principal constituent being compound **1.15**. This dye was obtained directly from *m*-phenylenediamine as the starting material and was introduced commercially in 1861. The true value of azo dyes emerged eventually when it was demonstrated that different diazo and coupling components could be used, thus extending dramatically the range of coloured compounds that could be prepared. The first commercial azo dye of this type was chrysoidine, which was derived from reaction of diazotized aniline with *m*-phenylenediamine and was introduced to the market in 1876. This was followed soon after by a series of orange dyes (Orange I, II, III and IV), which were prepared by reacting diazotized sulfanilic acid (4-amino-benzene-1-sulfonic acid) with, respectively, 1-naphthol, 2-naphthol, *N,N*-dimethylaniline and diphenylamine. In 1879, Biebrich Scarlet, **1.16**, the first commercial disazo dye to be prepared from separate diazo and coupling components, was introduced. From this historical beginning, azo colorants have emerged as by far the most important chemical class of dyes and pigments, dominating most applications (Chapter 3). It was becoming apparent that the synthetic textile dyes that were being developed were less expensive, easier to produce on an industrial scale, easier to apply, more versatile, and capable of providing better colour and technical performance than the range of natural dyes applied by traditional methods. As a consequence, within 50 years of Perkin's initial discovery, around 90% of textile dyes were synthetic rather than natural, and azo dyes had emerged as the dominant chemical type.

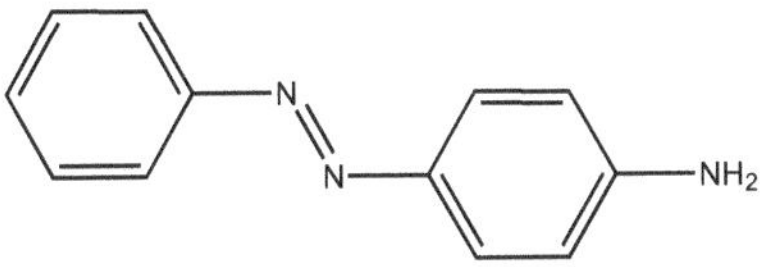

1.14

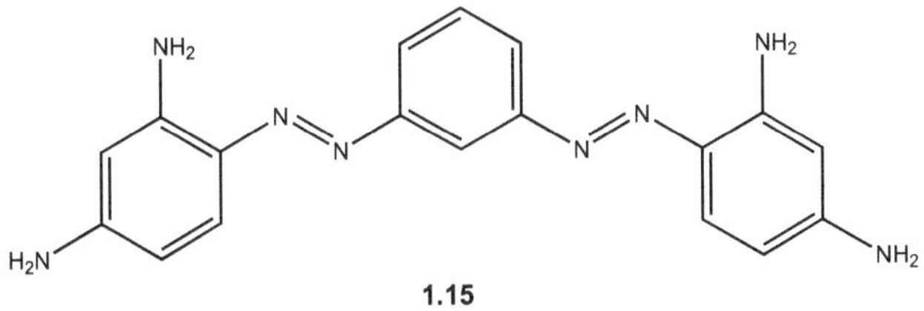

1.15

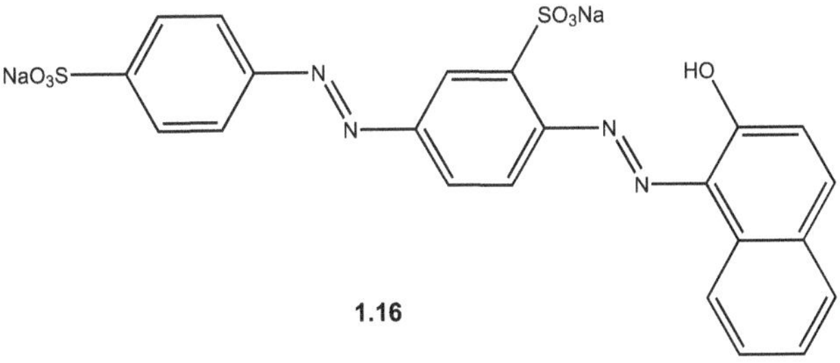

1.16

Towards the end of the nineteenth century, a range of organic pigments was also being developed commercially, particularly for paint applications. Inorganic pigments had been in use for many years, providing excellent durability, but generally rather dull colours. It was well-known that brighter, more intense colours could be provided by products commonly referred to as *lakes*, which were obtained from dyes by precipitation on to inert white powders. The name is derived from the lac insect from which a red colorant related to carminic acid, **1.3**, was derived. An early pigment lake was prepared by precipitation of this colorant on to an inorganic mineral substrate.[25] This technology proved to be readily applicable to the range of established water-soluble synthetic textile dyes, whereby anionic dyes were rendered insoluble by precipitation on to inert colourless inorganic substrates such as alumina and barium sulfate while cationic dyes were treated with tannin or antimony potassium tartrate to give insoluble pigments. Their introduction was followed soon after by the development of a group of yellow and red azo pigments, such as the Hansa Yellows and β-naphthol reds, which did not contain substituents capable of salt formation. Many of these products are still of considerable importance today, and are referred to commonly as the classical azo pigments (Chapter 9).

It is of interest, and in a sense quite remarkable, that at the time of Perkin's discovery of Mauveine chemists had very little understanding of the principles of organic chemistry. As an example, even the structure of benzene, the simplest aromatic compound, was an unknown quantity. Kekulé's proposal concerning the cyclic structure of benzene in 1865 without doubt made one of the most significant contributions to the development of organic chemistry. It is certain that the commercial developments in synthetic colour chemistry which took place from that time onwards owed much to the coming of age of organic chemistry as a science. For example, the structures of some of the more important natural dyes, including indigo, **1.1a**, and alizarin, **2**, were elucidated. In this period, well before the advent of the modern range of instrumental analytical techniques that are now used routinely for structural analysis, these deductions usually arose from painstaking investigations of the chemistry of the dyes, commonly involving a planned series of degradation experiments from which identifiable products could be isolated. Following the elucidation of the chemical structures of these natural dyes, a considerable amount of research effort was devoted to devising efficient synthetic routes to these products. The synthetic routes that were developed for the manufacture of these dyes ultimately proved to be significantly more cost-effective than the traditional methods, which involved extracting the dyes from natural sources, and in addition gave the products more consistently and with better purity. At the same time, by exploring the chemistry of these natural dye systems, chemists were discovering a wide range of structurally-related dyes that could be produced synthetically and had excellent colour properties and technical performance. As a consequence, the field of carbonyl dye chemistry, and the anthraquinones in particular, had opened up. This group of dyes remains for many textile applications the second most important chemical class, after azo dyes, in use today (Chapter 4).

1.4 COLOUR CHEMISTRY IN THE TWENTIETH CENTURY

In the first half of the twentieth century, new chemical classes of organic dyes and pigments continued to be discovered. Probably the most significant discovery was of the phthalocyanines, which have become established as the most important group of blue and green organic pigments.[26] As with virtually every other new type of chromophore developed over the years, the discovery of the phthalocyanines was fortuitous. In 1928, chemists at Scottish Dyes, Grangemouth (later to become part of ICI), observed the presence of a blue impurity in certain batches of phthalimide produced from the reaction of phthalic anhydride with ammonia. They were able to isolate the blue impurity and subsequently its structure was established as iron(ii) phthalocyanine. The source of the iron proved to be the reactor vessel wall, which had become exposed to the reactants as a result of a damaged glass lining. As it turned out, the formation of phthalocyanines had almost certainly been observed previously, although the compounds were not characterised and the significance of the observations was not recognised. Following their industrial discovery in Scotland, the chemistry of formation of phthalocyanines, together with its relationship with their chemical structure and properties, was investigated extensively by Linstead of Imperial College, London.[27] The elucidation of the structure of the phthalocyanine system by Robertson was historically important as one of the first successful applications of X-ray crystallography in the structure determination of organic molecules.[28] Copper phthalocyanine, **1.17**, has emerged as by far the most important product, a blue pigment that is capable of providing a brilliant intense blue colour and excellent technical performance, yet at the same time can be manufactured at low cost in high yield from commodity starting materials (Chapter 5). The discovery of this unique product set new standards for subsequent developments in dye and pigment chemistry. Although copper derivatives provide the most important colorants, complexes of phthalocyanines with an extensive range of other metals are well-established and have other industrial applications, for example as photosensitisers, semiconductors and catalysts.[29]

1.17

As time progressed, the strategies adopted in dye and pigment research evolved from the early approaches based largely on empiricism and involving the synthesis and evaluation of large numbers of products, to a more structured approach involving more fundamental studies of chemical principles. For example, attention turned to the reaction mechanisms involved in the synthesis of dyes and pigments and to the interactions between dye molecules and textile fibres. Probably the most notable advance in textile dyeing in the twentieth century, which arguably emerged from such fundamental investigations, is the process of reactive dyeing. Reactive dyes contain functional groups that, after application of the dyes to certain fibres, can be linked covalently to the polymer molecules that make up the fibres, and this gives rise to dyeings with superior washfastness compared with the more traditional dyeing processes. Dyes that contain the 1,3,5-triazinyl group, discovered by ICI in 1954, were the first successful group of fibre-reactive dyes. The introduction of these products to the market as Procion dyes by ICI in 1956, initially for application to cellulosic fibres such as cotton, proved to be a rapid commercial success.[30,31] The chemistry involved when Procion dyes react with the hydroxyl groups present on cellulosic fibres under alkaline conditions involves the aromatic nucleophilic substitution process outlined in Scheme 1.1, in which the cellulosate anion is the effective nucleophile. The range of industrial reactive dyes developed significantly in the second half of the twentieth century, with the

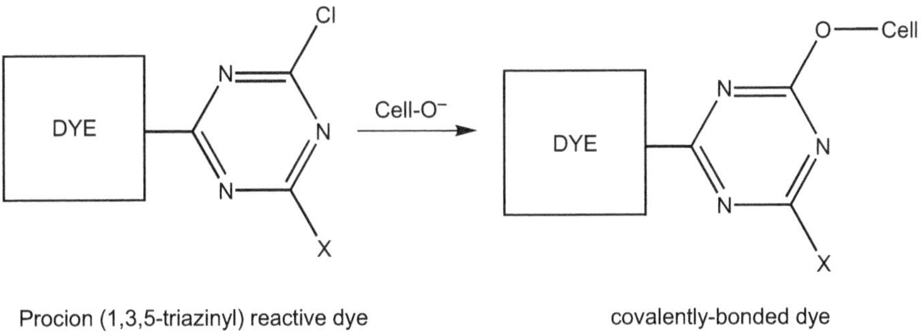

Procion (1,3,5-triazinyl) reactive dye   covalently-bonded dye

Scheme 1.1   Reaction of Procion dyes with cellulosic fibres.

introduction of alternative types of reactive groups and with the aim to address the enhancement of fixation together with a series of related environmental issues. Reactive dyes have become the most popular application class of dyes for cellulosic fibres, and their use has been extended to a certain extent to other types of fibres, notably wool, silk and nylon (Chapter 8).

1.5   RECENT AND CURRENT TRENDS IN COLOUR CHEMISTRY

In the latter part of the twentieth century, new types of dyes and pigments for the traditional applications of textiles, leather, plastics, paints and printing inks continued to be developed and introduced commercially but at a declining rate. Clearly, the colour manufacturing industries considered that a mature range of products existed for these conventional applications. The cost of introducing new products to the market, not only in terms of R&D effort but also in addressing the increasing demands of toxicological evaluation, was becoming increasingly prohibitive. Emphasis transferred towards process and product development and optimisation, and the consolidation of existing product ranges. At the same time, during this period, research in organic colour chemistry developed new directions, as a result of the opportunities presented by the emergence of a range of applications in new technologies, demanding new types of colorant. These colorants have commonly been termed *functional dyes* because the applications require the dyes to perform certain functions beyond simply providing colour.[32,33] The concept of functional dyes was, of course, not new. The role of dyes found in nature almost always extends beyond the need to provide colour, familiar examples of which include solar energy harvesting in photosynthesis and the mechanism of visual perception.[34] Qualitative and quantitative analytical applications have traditionally relied extensively on the use of specific dyes or colour-forming reactions, although modern instrumental analytical techniques have commonly superseded these approaches. Biological applications of dyes, such as staining techniques are also long-established. The applications of the modern range of functional dyes, which have also been referred to as *π-functional materials*, include reprographic techniques, such as electrophotography and digital inkjet printing, electronic applications including optical data storage, display technologies, lasers and solar energy conversion, and a range of medical uses (Chapter 11). In addition, dyes and pigments that change colour when exposed to an external stimulus, especially when that change is controllable and reversible, for example thermochromic and photochromic dyes, have attracted renewed interest in view of their potential applications as 'smart' colouring materials.

For these new applications, but also for traditional uses, the concepts of computer-aided molecular design of dyes for specific properties have become increasingly important. Of particular significance in this respect is the fact that a range of molecular and quantum mechanical calculation methods has developed from their origin and perception as complex academic theoretical exercises into accessible, routine tools that can be applied to the calculation of properties of dyes, facilitated by the dramatic advances in computing technology that have taken place. The Pariser–Pople–Parr (PPP) molecular orbital method initially proved of particular value, largely because of its versatility and modest computing demands, although subsequently a range of more sophisticated methods has become increasingly available. Indeed, methods involving density functional theory (DFT) have become the current methods of choice. From knowledge of the molecular structure of a dye, these methods may be used with a reasonable degree of confidence to predict its colour properties by calculation, including the hue of the dye from its absorption maximum, and its intensity as indicated by its molar extinction coefficient. The availability of these methods allows the potential properties of large numbers of dye molecules to be screened as an aid to the selection of synthetic targets (Chapter 2). These methods also apply to the calculation of the molecular properties of pigments. However, to address further the calculation of the properties of pigments, which commonly exist as discrete nanocrystalline particles, the principles of crystal engineering have been developed. Crystal engineering may be defined as an understanding of intermolecular interactions in the context of crystal packing and the utilisation of such understanding in the design of new materials with desired chemical and physical properties. Structural information obtained from X-ray crystallographic studies of pigments may be used to predict, for example, the morphology of the particles and the effect of additive incorporation at the particle surfaces (Chapter 9).

The diketopyrrolopyrroles (DPP), exemplified by compound **1.18**, represent probably the most successful new chromophoric system introduced in more modern times. These have provided high performance brilliant red pigments that exhibit properties similar to the phthalocyanines. The formation of a DPP molecule was first reported in 1974 as a minor product obtained in low yield from the reaction of benzonitrile with ethyl bromoacetate and zinc.[35] A study by research chemists at Ciba Geigy into the mechanistic pathways involved in the formation of the molecules led to the development of an efficient 'one-pot' synthetic procedure for the manufacture of DPP pigments from readily available starting materials (Chapter 4). The development of DPP pigments has emerged as an outstanding example of the way in which an application of the fundamental principles of synthetic and mechanistic organic chemistry can lead to an important commercial outcome. These products are now manufactured on a large commercial scale with purpose-built facilities in Switzerland and the USA. There are potential lessons for colour chemists. It demonstrates that, well over a century after Perkin's discovery of Mauveine, there may well be unrealised scope for the development of new improved colorants for traditional colour applications.

1.18

In addition, while powerful sophisticated molecular modelling methods are now available to assist in the design of new coloured molecules, the foresight to follow-up and exploit the fortuitous discovery of a coloured compound, perhaps as a trace impurity in a reaction, will remain a vital complementary element in the search for new dyes and pigments.

The increasing public sensitivity towards the environment has had a major impact on the chemical industry in recent years. There is no doubt that one of the most important challenges to industry currently is the requirement to satisfy increasingly demanding toxicological and environmental constraints not only as a consequence of legislation but also of the growing public concern surrounding issues associated with the ecology of the planet. Since the products of the colour industry are designed to enhance our living environment, it may be argued that this industry has a special responsibility to ensure that its products and processes do not have an adverse impact on the environment in its wider sense.[36,37] Recent decades have seen major progressive changes in the colour industry. The structure of the traditional colorant manufacturing industry in Europe changed dramatically as a result of a series of mergers, acquisitions and the development of toll manufacturing arrangements. Long-established entities such as ICI in the UK, Hoechst in Germany and Ciba in Switzerland either disappeared or had little remaining association with colour. Manufacturing capacity declined substantially in the traditional heartlands where the original discoveries and inventions were made and the products and processes were developed, in Europe and in the USA. This has been accompanied by rapid growth of the industry in Asia, mainly in China and India. As well as the influence on the national economies and industrial cultures which is accompanying this trend, there is a consequent transfer of responsibility for dealing with the environmental issues towards those countries who are progressively manufacturing, supplying and applying most of the world's colour. Some of the most important environmental issues associated with colour manufacture and application are discussed in Chapter 12. However, Western Europe and the USA currently retain some dominance in industries dealing with the manufacture of dyes and pigments for cosmetics applications, including hair dyes. This feature may be attributed, to a certain extent, to the cultural association of products branded as cosmetics with the Western fashion and design industry, often linked with specific brand names. The important topic of the chemistry of colour as applied in cosmetics is covered in Chapter 10.

REFERENCES

1. M. V. Orna, *The Chemical History of Color*, Springer, Heidelberg, 2013.
2. H. Skelton, *Rev. Prog. Coloration*, 1999, **29**, 43.
3. F. Brunello, *The Art of Dyeing in the History of Mankind*, Neri Pozza, Vicenza, 1973.
4. E. Knecht, C. Rawson and R. Loewenthal, *A Manual of Dyeing*, Charles Griffiths and Co. Ltd., London, 4th edn, 1917, vol. 1.
5. J. H. Hofenk de Graaff, *The Colourful Past, Origins, Chemistry and Identification of Natural Dyestuffs*, Archetype Publications Ltd., Riggisberg, Switzerland, 2004.
6. D. Cardon, *Natural Dyes: Sources, Tradition, Technology and Science*, Archetype Publications, London, 2007.
7. H. Schweppe, *Handbuch der Naturfarbstoffe: Vorkommen, Verwendung, Nachweis*, Nikol, Hamburg, 1993.
8. A. S. B. Ferreira, A. H. Hulme, H. McNab and A. Quye, *Chem. Soc. Rev.*, 2004, **33**, 329.
9. T. Bechtold and R. Mussak, *Handbook of Natural Colorants*, John Wiley & Sons Ltd., Chichester, 2009.
10. J. Balfour-Paul, *Indigo*, British Museum Press, London, 1998.
11. C. J. Cooksey, *Biotech Histotech.*, 2007, **82**, 105.
12. I. I. Ziderman, *Rev. Prog. Coloration*, 1986, **16**, 46.
13. C. J. Cooksey, *Dyes History Archaeol.*, 1994, **12**, 57.
14. W. H. Perkin, *J. Chem. Soc. Trans.*, 1896, 596.
15. W. H. Perkin, *J. Chem. Soc. Trans.*, 1879, 717.
16. J. Boulton, *J. Soc. Dyers Colour.*, 1957, **73**, 81.
17. S. Garfield, *Mauve: How One Man Invented a Colour that Changed the World*, WW Norton, New York, 2001.
18. I. Holme, *Color. Technol.*, 2006, **122**, 235.
19. O. Methcohn and M. Smith, *J. Chem. Soc., Perkin Trans. I*, 1994, 5.
20. J. Seixas de Melo, S. Takato, M. Sousa, M. J. Melo and A. J. Parola, *Chem. Commun.*, 2007, 2624.
21. M. R. Fox, *Dye-Makers of Great Britain 1856–1976. A History of Chemists, Companies, Products and Changes*, Imperial Chemical Industries plc, Manchester, 1987.
22. W. T. Johnston, *Biotech Histotech.*, 2008, **83**, 83.
23. M. Chastrette, *Actualite Chim.*, 2009, 48.
24. C. Cooksey and A. Dronsfield, *Biotech Histotech.*, 2009, **84**, 179.
25. R. M. Christie and J. L. Mackay, *Color. Technol.*, 2008, **24**, 133.
26. C. C. Leznoff and A. B. P. Lever, *Phthalocyanines: Properties and Applications*, VCH, Weinheim, 1983.

27. R. P. Linstead, *J. Chem. Soc.*, 1934, 1035.
28. J. Robertson, *J. Chem. Soc.*, 1935, 613.
29. N. B. McKeown, *Phthalocyanine Materials: Synthesis, Structure and Function*, Cambridge University Press, Cambridge, UK, 1998.
30. I. D. Rattee, *Rev. Prog. Color*, 1984, **14**, 50.
31. A. H. M. Renfrew and J. A. Taylor, *Rev. Prog. Color.*, 1990, **20**, 1.
32. P. Gregory, *High Technology Applications of Organic Colorants*, Plenum Press, New York, 1991.
33. S-H. Kim (ed.), *Functional Dyes*, Elsevier, Amsterdam, 2006.
34. N. Hampp and A. Silber, *Pure Appl. Chem.*, 1996, **68**, 1361.
35. D. G. Farnum, G. Mehta, G. G. I. Moore and F. P. Siegal, *Tetrahedron Lett.*, 1974, **29**, 2549.
36. A. Reife and H. S. Freeman, *Environmental Chemistry of Dyes and Pigments*, John Wiley & Sons, Inc., New York, 1996.
37. R. M. Christie (ed.), *Environmental Aspects of Textile Dyeing*, Woodhead Publishing, Cambridge, 2007.

CHAPTER 2

The Physical and Chemical Basis of Colour

2.1 INTRODUCTION

It has been said that the presence of colour requires three things: a source of illumination, an object to interact with the light that emanates from this source and a human eye to observe the effect which results. In the absence of any one of these, it may be argued that colour does not exist. Indeed, colour is a perceptual phenomenon rather than a property of an object. A treatment of the basic principles underlying the origin of colour thus requires consideration of each of these three aspects, which brings together concepts arising from three natural science disciplines: chemistry, physics and biology. Although the principal aim of this textbook is to deal with the chemistry of dyes and pigments, a complete appreciation of the science of colour cannot be achieved without some knowledge of the fundamental principles of the physical and biological processes that ultimately give rise to our ability to observe colours. This chapter therefore presents an introduction to the physics of visible light and the way it interacts with materials, together with a brief description of the physiology of the eye and how it responds to stimulation by light, thus giving rise to the sensation of colour. In addition, the chapter contains a discussion of some of the fundamental chemical principles associated with coloured compounds, including a description of how dyes and pigments may be classified, followed by an overview of the

Colour Chemistry, 2nd edition
By Robert M Christie
© R M Christie 2015
Published by the Royal Society of Chemistry, www.rsc.org

ways in which the chemical structure of a molecule influences its colour properties. This section places special emphasis on the principles as applied to azo colorants because of their particular importance in the colour industry. These topics are presented as a prelude to the more detailed discussion of the chemistry of dyes and pigments contained in later chapters of this book.

2.2 VISIBLE LIGHT

Visible light refers to the region of the electromagnetic spectrum to which our eyes are sensitive and corresponds to radiation within the very narrow wavelength range 360–780 nm. Since the sensitivity of the eye to radiation is very low at each of these extremes, in practice the visual spectrum is commonly taken as 380–720 nm. Beyond the extremes of this range are the ultraviolet (UV) region of the spectrum (below 360 nm) and the infrared (IR) region (above 780 nm). Normal white light contains this entire wavelength range, although not necessarily in equal intensities. There are numerous sources of white light, some natural and some artificial in origin. The most familiar natural illumination is daylight, originating from the sun. The visible light from the sun not only allows us to see objects, but it is in fact essential for life since it is the source of energy responsible for photosynthesis, the vital process that allows plants to grow and thus provides us with an essential food source. Normal daylight encompasses the complete visible wavelength range although its exact composition is extremely variable and dependent on various factors such as the geographical location, the prevailing weather conditions, the time of day and the season. Artificial illuminants, such as the tungsten lamps, fluorescent lights and light emitting diodes (LEDs) which are used for interior lighting, are also sources nominally of white light, although the composition of the light from these sources varies markedly depending on the type of lamp in question. For example, tungsten lights appear yellowish as the light they emit is deficient in the blue region of the spectrum. Colours do appear different under different illumination sources, although when the human visual system assesses colours it is capable of making some allowance for the nature of the light source, for example by compensating for some of the deficiencies of particular artificial light sources.

The splitting of white light into its various component colours is a familiar phenomenon. It may be achieved in the laboratory, for example, by passing a beam of white light through a glass prism, or naturally, as in a rainbow where the colours are produced by the

Table 2.1 Complementary colour relationships.

Wavelength range (nm)	Colour	Complementary colour
400–435	Violet	Greenish-yellow
435–480	Blue	Yellow
480–490	Greenish-blue	Orange
490–500	Bluish-green	Red
500–560	Green	Purple
560–580	Yellowish-Green	Violet
580–595	Yellow	Blue
595–605	Orange	Greenish-blue
605–750	Red	Bluish-green

interaction of sunlight with raindrops. The visible spectrum is made up of specific wavelength regions that are recognised by the eye in terms of their characteristic colours. The approximate wavelength ranges of light corresponding to these observed colours are given in the first column of Table 2.1. Fundamental to the specification of colours is an understanding of the laws of colour mixing, the processes by which two or more colours are combined to 'synthesise' new colours. There are two fundamentally different ways in which this may be achieved: *additive* and *subtractive* colour mixing. Additive colour mixing, as the name implies, refers to the mixing of coloured lights, so that the source of illumination is observed directly by the eye. Subtractive colour mixing is involved when colours are observed as a result of reflection from or transmission through an object after it interacts with incident white light. The colours *red*, *green* and *blue* are referred to as the *additive primary colours*. Their particular significance is that they are colours that cannot be obtained by the mixing of lights of other colours, but they may be combined additively in appropriate proportions to produce the other colours. As illustrated in Figure 2.1(a), additive mixing of red and blue produces magenta, blue and green gives cyan, while combining red and green additively gives yellow. When all three primaries are mixed in this way, white light is created since the entire visible spectral range is present. An everyday situation in which additive colour mixing is encountered is in displays, such as those used in colour television and computer monitors. To create multicolour effects, separate red, green and blue (RGB) emitting phosphors are used in the case of traditional cathode-ray tubes, which are rapidly declining in popularity. The flat-screen devices that have largely replaced cathode-ray tubes include plasma, liquid crystal and electroluminescent displays. These use various technologies to create the multicolour effect, for example with emitting phosphors or using coloured microfilters, although in each case

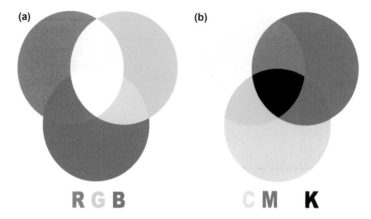

Figure 2.1 (a) Additive colour mixing; (b) subtractive colour mixing.
© Shutterstock.

employing the principles of additive RGB mixing (Chapter 11). When an object absorbs light of a given colour corresponding to its particular wavelength range, it is the complementary colour that is observed. The complementary colour corresponds to the remaining wavelengths of incident light, which are either transmitted or reflected, depending on whether the object is transparent or opaque, and are then detected by the eye. These complementary colour relationships are also given in Table 2.1. For example, an object that absorbs blue light (*i.e.*, in the range 435–480 nm) will appear yellow, because the red and green components are reflected or transmitted. This forms the basis of *subtractive colour mixing*. This type of colour mixing, which is involved when dyes and pigments are mixed in application, is the more familiar of the two processes. The subtractive primary colours are *yellow*, *magenta* and *cyan*. These are the colours, for example, of the three printing inks used commonly to produce the vast quantities of multicolour printed material we encounter in our daily lives, such as magazines, posters, newspapers, *etc.*, and also in the inkjet printers that are linked to computers as used in home and office printing. The principles of subtractive colour mixing are illustrated in Figure 2.1(b).

The colours described in Table 2.1 that are observed as a result of the selective light absorption process are referred to as *chromatic*. If all wavelengths of light are reflected from an object, it appears to the eye as white. If no light is reflected, we recognise it as black. When dyes or pigments of the three subtractive primaries, yellow, magenta and cyan, are mixed together, black is produced as all wavelengths of light are absorbed. If the object absorbs a constant fraction of the incident light throughout the visible region, it appears grey. White, black and grey are therefore referred to as *achromatic* since in those cases there is no selective absorption of light involved.

2.3 THE EYE

Colour is a perception rather than a property of an object. The sensation of colour that we experience arises from the interpretation by the brain of the signals that it receives *via* the optic nerve from the eye in response to stimulation by light. Indeed, we can never know whether the sensation of colour that we experience matches with that of another observer. This section contains a brief description of the components of the eye and an outline of how each of these contributes to the mechanism by which we observe colours. Figure 2.2 shows a cross-section diagram of the eye, indicating some of the more important components.

The eye is enclosed in a white casing known as the *sclera*, or colloquially as the 'white of the eye'. The retina is the photosensitive component and is located at the rear of the eye. It is here that the image is formed by the focusing system. Light enters the eye through the *cornea*, a transparent section of the sclera, which is kept moist and free from dust by the tear ducts and by blinking of the eyelids. The light passes through a transparent flexible lens, the shape of which is determined by muscular control, and which acts to form an inverted image on the retina. The light control mechanism involves the *iris*, an annular shaped opaque layer, the inner diameter of which is

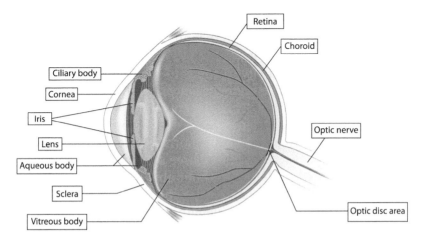

Figure 2.2 Components of the eye.
    © Shutterstock.

controlled by the contraction and expansion of a set of circular and radial muscles. The aperture formed by the iris is termed the *pupil*. Light passes into the eye through the pupil, which normally appears black since little of the light entering the eye is reflected back. The diameter of the pupil is small under high illumination, but expands when illumination is low in order to allow more light to enter.

The retina owes its photosensitivity to a mosaic of light sensitive cells known as *rods* and *cones*, which derive their names from their physical shape. There are about 6 million cone cells, 120 million rod cells and 1 million nerve fibres distributed across the retina. It is the rods and cones that translate the optical image into a pattern of nerve activity that is transmitted to the brain by the fibres in the optic nerve. At low levels of illumination only the rod cells are active and a type of vision known as *scotopic* vision operates, while at medium and high illumination levels only the cone cells are active, and this is gives rise to *photopic* vision. There is only one type of rod-shaped cell present in the eye. The rods provide essentially a monochromatic view, allowing perception only of lightness and darkness. The sensitivity of rods to light depends on the presence of a photosensitive pigment known as *rhodopsin*, which consists chemically of the carotenoid, retinal, bonded to the protein, opsin. Rhodopsin is continuously generated in the eye and is also destroyed by bleaching on exposure to light. At low levels of illumination (night or dark-adapted vision), this rate of bleaching is low and thus there is sufficient rhodopsin present for the rods to be sensitive to the small amount of light available. At high levels of illumination, however, the rate of bleaching is high so that only a small amount of rhodopsin is present and the rods consequently have low sensitivity to light. At these higher levels of illumination, it is only the cone cells that are sensitive. The cones provide us with full colour vision as well as the ability to perceive lightness and darkness. The sensitivity of cones to light depends on the presence of the photosensitive pigment *iodopsin*, which is retained up to high levels of illumination. Thus, in normal daylight when the rods are inactive, vision is provided virtually entirely by the response of the cone cells. Under ideal conditions, a normal observer can distinguish about 10 million separate colours. Three separate types of cone cells have been identified in the eye and our ability to distinguish colours is associated with the fact that each of the three types is sensitive to light of a particular range of wavelengths. The three types of cone cell have been classified as long, medium and short corresponding to the wavelength of maximum response of each type. Short cones are most sensitive to blue light, the maximum response being at a wavelength of about 440 nm. Medium cones are most sensitive to green light, the maximum response being at a wavelength of about 545 nm. Long cones are most sensitive to red light, the maximum response being at a wavelength of about 585 nm. Colour vision is thus trichromatic and we see colours based on the principles of colour mixing. The specific colour sensation perceived by the eye is governed by the response of the three types of cone cells to the particular wavelength profile with which they are interacting.[1]

We do not all perceive colours in the same way. Colour vision deficiency (a term that is preferable to 'colour blindness') is relatively common. Its prevalence may be as high as 8% in males although only around 0.5% in females. The reasons for the gender differences are associated with the fact the most important genes governing colour vision are located in the X chromosome, of which males have only one, while females have two. Colour vision deficiencies are not considered as seriously debilitating conditions, but the individuals concerned are at a distinct disadvantage when performing tasks that require colour discrimination. There are a number of different types of colour vision deficiency which vary in the degree of severity.[1] The mildest and most common form of colour vision deficiency is *anomalous trichromacy*. In the individuals concerned, there is a defect in the colour mixing mechanism so that they accept colour matches that an individual with normal colour vision will not. *Dichromacy* is a more severe form of colour vision deficiency in which one set of cones is not functional. *Monochromacy* is the most severe form of congenital colour vision deficiency in which colour discrimination is absent because at least two sets of cones are dysfunctional and vision is dominated by the rods.

2.4 THE CAUSES OF COLOUR

It is commonly stated that there are 15 specific causes of colour, arising from various physical and chemical mechanisms.[2] These mechanisms may be collected into five broad groupings:

(a) colour from simple excitations: from gas excitation (*e.g.*, vapour lamps, neon signs), and from vibrations and rotations (*e.g.*, ice, halogens);
(b) colour from ligand field effects: from transition metal compounds and from transition metal impurities;
(c) colour from molecular orbitals: from organic compounds and from charge transfer;

    (d) colour from band theory: in metals, in semiconductors, in doped semiconductors and from colour centres;

    (e) colour from geometrical and physical optics: colour from dispersion, scattering, interference and diffraction.

This book is focussed on the industrially important organic dyes and pigments and, to a certain extent, inorganic pigments and thus deals mostly with colour generated by the mechanisms described by group (c), although the reader will find examples of the other mechanisms on occasions also.

2.5   THE INTERACTION OF LIGHT WITH OBJECTS

The most obvious requirement of a dye or pigment to be useful in its applications is that it must have an appropriate colour. Of the many ways in which light can interact with objects, the two most important from the point of view of their influence on colour are *absorption* and *scattering*. Absorption is the process by which radiant energy is utilised to raise molecules in the object to higher energy states. Scattering is the interaction by which light is re-directed as a result of multiple refractions and reflections. In general, if only absorption is involved when light interacts with an object, then the object will appear transparent as the light that is not absorbed is transmitted through the object. If there are scattering centres present, the object will appear either translucent or opaque, depending on the degree of scattering, as light is reflected back to the observer.

Electronic spectroscopy, often referred to as UV/visible spectroscopy, is a useful instrumental technique for characterising the colours of dyes and pigments.[3] These spectra are obtained from appropriate samples using a spectrophotometer operating either in transmission (absorption) or reflection mode. UV/visible absorption spectra of dyes in solution, such as that illustrated in Figure 2.3, are useful analytically, qualitatively to assist the characterisation of the dyes and as a sensitive method of quantitative analysis. They also provide important information to enable relationships between the colour and the molecular structure of the dyes to be developed.

A dye in solution owes its colour to the selective absorption by dye molecules of certain wavelengths of visible light. The remaining wavelengths of light are transmitted, thus giving rise to the observed colour. The absorption of visible light energy by the molecule promotes electrons in the molecule from a low energy state, or *ground state*, to a higher energy state, or *excited state*. The dye molecule is therefore said

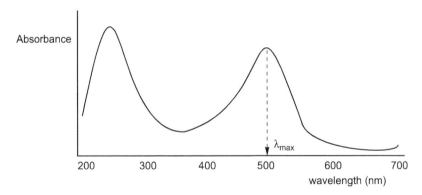

Figure 2.3 UV/visible absorption spectrum of a typical red dye in solution.

to undergo an electronic transition during this excitation process. The energy difference, ΔE, between the electronic ground state and the electronic excited state is given by Planck's relationship:

$$\Delta E = h\nu$$

where h is a constant (Planck's constant) and ν is the frequency of light absorbed. Alternatively, the relationship may be expressed as:

$$\Delta E = hc/\lambda$$

where c is the velocity of light (also a constant) and λ is the wavelength of light absorbed. Thus, there is an inverse relationship between the energy difference between the ground and excited states of the dye and the wavelength of light that it absorbs. As a consequence, for example, a yellow dye, which absorbs short wavelength (blue) light, requires a higher excitation energy than, say, a red dye which absorbs longer wavelength (bluish-green) light (Table 2.1).

There are several ways of describing in scientific terms the characteristics of a particular colour. One method that is especially useful for the purposes of relating the colour of a dye to its UV/visible spectrum in solution is to define the colour in terms of three attributes: hue (or shade), strength (or intensity) and brightness. The hue of a dye is determined essentially by the absorbed wavelengths of light, and so it may be characterised to a reasonable approximation by the wavelength of maximum absorbance (the λ_{max} value) obtained from the UV/visible spectrum, at least in those cases where there is a single visible absorption band. A shift of the absorption band towards longer wavelengths, (*i.e.*, a change of hue in the direction yellow → orange → red → violet → blue → green), for example as a result of a structural change in a dye molecule, is referred to as a *bathochromic* shift. The reverse effect, a shift towards shorter absorbed wavelengths, is described as a *hypsochromic* shift.

A useful measure of the strength or intensity of the colour of a dye is given by the *molar extinction coefficient* (ε) at its λ_{max} value. This quantity may be obtained from the UV/visible absorption spectrum of the dye using the Beer–Lambert Law, *i.e.*:

$$A = \varepsilon c l$$

where A is the absorbance of the dye at a particular wavelength, ε is the molar extinction coefficient at that wavelength, c is the concentration of the dye and l is the path length of the cell (commonly 1 cm) used for measurement of the spectrum. The Beer–Lambert law is obeyed by most dyes in solution at low concentrations, although when dyes show molecular aggregation effects in solution, deviations from the law may be encountered. However, since the colour strength of a dye is more correctly related to the area under the absorption band, it is important to treat its relationship with the molar extinction coefficient as qualitative and dependent to a certain extent on the shape of the absorption curve.

A third attribute of colour is brightness, although this property may be described in various other ways, for example as brilliance, vibrance or vividness. This characteristic of the colour depends on the absence of wavelengths of transmitted light other than those of the hue concerned. Electronic absorption bands of molecular compounds are not infinitely narrow because they are broadened by the superimposition of numerous vibrational energy levels on both the ground and excited electronic states. Brightness of colour is characterised, in terms of the UV/visible spectrum, by the shape of the absorption band. Dyes that exhibit bright colours show narrow absorption bands, whereas broad absorption bands are characteristic of dull colours, such as browns, navy blues and olive greens.

Visible reflectance spectroscopy is used routinely to measure the colour of opaque objects such as textile fabrics, paint films and plastics for purposes such as colour matching and dye and pigment recipe prediction. There is now a wide range of commercially-available reflectance spectrophotometers used industrially as colour measurement devices for such purposes. In many ways, this technique may be considered as complementary to the use of visible absorption spectroscopy for the measurement of transparent dye solutions. Reflectance spectra of typical red, green and blue surfaces

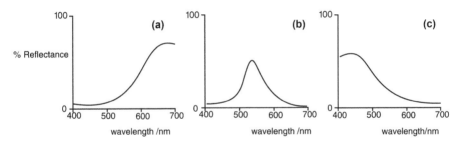

Figure 2.4   Visible reflectance spectra of (a) red, (b) green and (c) blue surfaces.

are shown in Figure 2.4. The spectrum of the red surface, for example, shows low reflectance (high absorption) in the 400–500 nm (blue) and 500–600 nm (green) ranges, and high reflectance of the red wavelengths (600–700 nm).

When colour is assessed on the basis of reflectance measurements, it is common to consider the three relevant attributes of perception of colour as hue, chroma (or saturation) which is the 'colourfulness' or richness of the colour, and lightness, which refers to the amount of reflected light. These three attributes may be described using the concept of colour space, which shows the relationships of colours to one another and which illustrates the three-dimensional nature of colour, as shown in Figure 2.5(a). The hue of a particular colour is represented in a colour circle. The three additive primaries, red, green and blue, are equally spaced around the colour circle. The three subtractive primaries, yellow, magenta and cyan, are located between the pairs of additive primaries from which they are obtained by mixing. The second attribute, chroma, increases with distance from the centre of the circle. The third attribute, lightness, requires a third dimension that is at right angles to the plane of the colour circle. The achromatic colours, white and black, are located at either extreme of the lightness scale.

Mathematical approaches that make use of concepts associated with colour space for colour measurement and specification based on reflectance spectroscopy have become well-established.[4,5] Indeed the calculations involved are normally implemented in the software that is provided within an instrument used for colour measurement. The basis of all colour measurement systems is the CIE system of colour specification, which was agreed in 1931. The *Commission Internationale de l'Eclairage* (CIE) is an international organisation promoting cooperation and information exchange on matters relating to the science and art of lighting. An important approach used for the

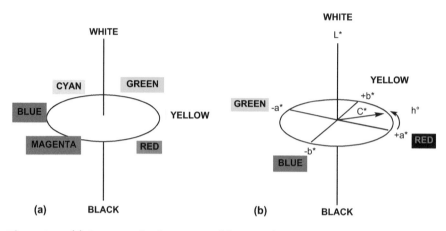

Figure 2.5   (a) Concept of colour space; (b) LAB colour space.

specification of colour involves CIELAB colour space, which makes use of the L^*, a^*, b^* parameters, as illustrated in Figure 2.5(b). Lightness, L^*, and chroma, C^*, are quantified as illustrated on the diagram. The hue is described by the coordinates a^* (redness/greenness) and b^* (yellowness/blueness), and by the hue angle, $h°$. Any specific single colour may be represented as a point in colour space. An important use of CIELAB colour space is in the calculation of the difference between two colours. In this approach, colour differences are calculated as ΔE values, potentially confusingly since this same term is used to describe energy differences, as the reader will find elsewhere in this chapter. Colour difference equations are generally based on the principle of calculating the distance in three-dimensional colour space between two colours, for example between a sample and a standard. Most commonly, this is achieved by application of the Pythagoras theorem. These ΔE values have particular significance in instrumental colour matching, which has become a vital feature of those industries, such as textiles, paint, plastics, printing inks, cosmetics, *etc.*, where there is a requirement to produce articles to strict colour specifications and reproducibly over a period of time. For a perfect colour match, the ΔE value should be as low as possible. Historically, a ΔE value of less than 1 has been regarded as an acceptable commercial match, although currently more demanding matching criteria often set smaller tolerance values (0.5–0.75). Colour measurement, and its mathematical basis, generally referred to as colour physics, is a well-developed and well-documented science and is not considered in further detail here.[4–7]

2.6 FLUORESCENCE AND PHOSPHORESCENCE

Most dyes and pigments owe their colour to the selective absorption of incident light. In some compounds, colour can also be observed as a result of the emission of visible light of specific wavelengths. These compounds are referred to as *luminescent*. The most commonly encountered luminescent effects are fluorescence and phosphorescence.[8-11] The energy transitions that occur in a molecule exhibiting fluorescence are illustrated in Figure 2.6.

When a molecule absorbs light, it does so in discrete amounts of energy, termed *quanta*. It is thus excited from the lowest vibrational level in its ground state (S_0) to a range of vibrational levels in the first excited singlet state (S_1^*). In the case of fluorescent dyes, this is generally a $\pi \rightarrow \pi^*$ electronic transition which occurs very rapidly, in about 10^{-15} s (femtoseconds). During the time the molecule spends in the excited state, energy is dissipated from the higher vibrational levels (*vibrational relaxation* or *internal conversion*) until the lowest vibrational level is attained, typically within a timeframe of 10^{-10} to 10^{-12} s (picoseconds). Fluorescence occurs if the molecule then emits light as it reverts from this level to various vibrational levels in the ground state ($S_1^* \rightarrow S_0$), which occurs within a timeframe of 10^{-7} to 10^{-9} s (nanoseconds). Non-radiative processes also cause dissipation of energy from the excited state, either as heat or by transfer of energy in collision with other molecules, the latter effect giving rise to the phenomenon of *quenching*. The extent to which these processes

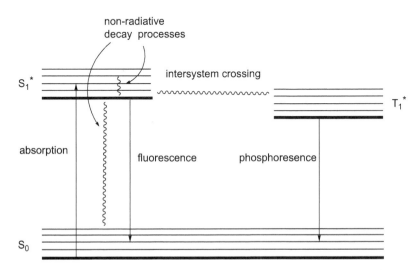

Figure 2.6 Energy transitions in fluorescent and phosphorescent molecules.

compete with fluorescence determines its intensity. Because vibrational energy is lost in the excited state before fluorescence occurs, and in accordance with the inverse relationship between electronic transition energy (ΔE) and wavelength (λ) as described by Planck's relationship ($\Delta E = hc/\lambda$), emission inevitably occurs at longer wavelengths than absorption. The difference between the emission and absorption wavelength maxima is referred to as the *Stokes shift*. In reality, most dyes do not exhibit significant fluorescence. The quenching phenomenon means that that the fluorescence intensity exhibited by a dye is influenced profoundly by its environment. Factors that influence fluorescence quenching include the chemical nature of the solvent or matrix into which the dye is incorporated, its concentration and purity, and the presence of other species. For example, molecular oxygen and certain ions can cause significant fluorescence quenching. The intensity of light emission from a fluorescent dye is characterised by its *quantum yield* (η), which is defined as the ratio of the emitted quanta to the absorbed quanta of light. Thus, a non-fluorescent material has a quantum yield of 0, while a perfect fluorescer would give a quantum yield of 1. With some fluorescent dyes, and in ideal circumstances, the quantum yield may approach unity. Another process that may occur in this photoinitiated process is *intersystem crossing* from the singlet excited state (S_1^*) to an excited triplet state (T_1^*). Subsequent emission of light from the triplet state ($T_1^* \rightarrow S_0$) is termed *phosphorescence*, and is a much longer lasting (glow in the dark) light emission phenomenon (10^{-1}–10^2 s). The visual effect provided by a fluorescent dye, whether in solution or incorporated into a textile fabric, plastic material or surface coating, is due to a combination of two effects: the base colour due to the wavelengths of light transmitted or reflected as a result of selective light absorption, supplemented by the colour due to the wavelengths of emitted light by fluorescence.[10] The traditional applications of fluorescent colorants are associated with their ability to attract attention, because of the remarkable vivid brilliance of the colours that results from the extra glow of emitted fluorescent light.[12] Fluorescent dyes may be used on textiles for specific aesthetic and fashion purposes, the eye-catching bright colours providing unique creative opportunities for the textile and garment designer. Wearing garments with fluorescent colours is also desirable when safety is paramount, for example as used by police, fire-fighters and construction workers who require visibility in their work surroundings. Daylight fluorescent pigments consist of fluorescent dyes dissolved in a transparent, colourless solid polymer in a fine particle size form.[13] They are used for their high visual impact, in printing inks for advertising, posters, magazines and supermarket packaging, coatings especially where safety is an important feature such as emergency and security vehicles and in a range of plastic products such as toys and bottles. Fluorescent dyes are also used in a range of applications that specifically exploit their intense light emission properties. These functional uses of the dyes include chemical and biochemical analytical techniques, non-destructive flaw detection, dye lasers, solar collectors and electronic display technologies (Chapter 11). Another type of compound that makes use of light-emitting properties is fluorescent brightening agents (FBAs).[14,15] These are compounds, structurally closely related to fluorescent dyes, that absorb light in the UV region of the spectrum and re-emit the energy at the lower (blue) end of the visible spectrum. When incorporated into a white substrate, such as a textile fabric or a plastic article, FBAs provide a particularly appealing bluish cast. One of the most important uses of FBAs is in washing powders to impart a bluish whiteness to washed fabrics.

2.7 DYES AND PIGMENTS

Colour may be introduced into manufactured articles, for example textiles and plastics, or into a range of other application media, for example paints and printing inks, for various reasons, but most commonly the purpose is to enhance the appearance and attractiveness of a product and improve its market appeal. Indeed it is often the colour that first attracts us to a particular article. The desired colour is generally achieved by the incorporation into the product of coloured compounds referred to as dyes and pigments. The term *colorant* is frequently used to encompass both types of colouring materials. For an appreciation of the chemistry of colour application, it is of fundamental importance that the distinction between dyes and pigments as quite different types of colouring materials is made. Dyes and pigments are both commonly supplied by the manufacturers as coloured powders. Indeed, as the discussion of their molecular structures contained in subsequent chapters of this book will illustrate, the two groups of colouring materials may often be quite similar chemically. However, they are distinctly different in their properties and especially in the way they are used. Dyes and pigments are distinguished on the basis of their solubility characteristics: essentially, dyes are soluble, pigments are insoluble.

The traditional use of dyes is in the coloration of textiles, a topic covered in considerable depth in Chapters 7 and 8. Dyes are almost invariably applied to the textile materials from an aqueous medium, so that they are generally required to dissolve in water.[16-18] Frequently, as is the case for example with acid dyes, direct dyes, cationic dyes and reactive dyes, they dissolve completely and very readily in water. This is not true, however, of every application class of textile dye. Disperse dyes for polyester fibres, for example, are only sparingly soluble in water and are applied as a fine aqueous dispersion. Vat dyes, an important application class of dyes for cellulosic fibres, are completely insoluble materials but they are converted by a chemical reduction process into a water-soluble form that may then be applied to the fibre. There is also a wide range of non-textile applications of dyes, many of which have emerged in recent years as a result of developments in the electronic and reprographic industries (Chapter 11). For many of these applications, solubility in specific organic solvents rather than in water may be of importance.

In contrast, pigments are colouring materials that are required to be completely insoluble in the medium into which they are incorporated. The principal traditional applications of pigments are in paints, printing inks and plastics, although they are also used more widely, for example, in the coloration of building materials, such as concrete and cement, and in ceramics and glass. The chemistry of pigments and their application is discussed in more detail in Chapter 9. Pigments are applied into a medium by a dispersion process, which reduces the clusters of solid particles into a more finely-divided form, but they do not dissolve in the medium. They remain as solid particles held in place mechanically, usually in a matrix of a solid polymer. A further distinction between dyes and pigments is that while dye molecules are designed to be attracted strongly to the polymer molecules which make up the textile fibre, pigment molecules are not required to exhibit such affinity for their medium. Pigment molecules are, however, attracted strongly to one another in their solid crystal lattice structure in order to resist dissolving in solvents.[19-21]

2.8 CLASSIFICATION OF COLORANTS

Colorants may be classified usefully in two separate ways, either according to their chemical structure or according to the method of application. The most important reference work dealing with the classification of dyes and pigments is the *Colour Index*, a publication produced by the Society of Dyers and Colourists, Bradford, England, which provides a comprehensive listing of the known commercial dyes and pigments and is up-dated on a regular basis. The first two editions of the Index were published as hard copies in several volumes, the third edition was produced on a CD-ROM, and the current edition is now web-based.[22] In the Index, each colorant is given a CI Generic Name, which incorporates its application class, its hue, and a number that simply reflects the chronological order in which the colorants were introduced commercially. It is extremely useful that this system of nomenclature for dyes and pigments is more or less universally accepted by all those involved in their manufacture and application, and so it is frequently used throughout this book as a means of identifying specific dyes and pigments. The *Colour Index* provides useful information for each dye and pigment on the methods of application and on the range of fastness properties that may be expected. The Index also lists the companies that manufacture each of the products, together with trade names, and gives the appropriate chemical constitutions where these have been disclosed by the manufacturer.

In the chemical classification, colorants are grouped according to certain common chemical structural features. The most important organic dyes and pigments, in roughly decreasing order of importance, belong to the azo (–N=N–), carbonyl (C=O) (including anthraquinones), phthalocyanine, arylcarbonium ion (including triphenylmethines), sulfur, polymethine and nitro chemical classes. A discussion of the principal structural characteristics of each of these chemical classes, together with a discussion of the major synthetic strategies used in the manufacture of these chemical groups of organic colorants, is found in Chapters 3–6 of this book. Added to this, Chapter 9 contains a discussion of the most important inorganic pigments, a group of colorants that has no counterpart in dye chemistry

To the textile dyer whose role it is to apply colour to a particular textile fibre, the classification of dyes according to the method of application is arguably of greater interest than the chemical classification. Dye and pigment molecules are carefully designed to ensure that they have a set of properties, which are appropriate to their particular application. Obvious requirements for both types of colorant are that they must possess the desired colours, in terms of hue, strength and brightness, and an appropriate range of fastness properties. Fastness properties refer to the ability of a dye or pigment to resist colour change when exposed to certain conditions, such as to light, weathering, heat, washing, solvents or to chemical agencies such as acids and alkalis. For textile applications, dye molecules are designed so that they are attracted strongly to the molecules of the fibre to which they are applied. The chemical and physical nature of the different types of textile fibres, both natural and synthetic, require that the dyes used, in each case, have an appropriate set of chemical features to promote affinity for the particular fibre concerned. The chemical principles of the most important application classes of textile dyes are described in Chapters 7 and 8. The application classes discussed include acid dyes, mordant dyes and premetallised dyes for protein fibres, direct dyes, reactive dyes and vat dyes for cellulosic fibres, disperse dyes for polyester and basic (cationic) dyes for acrylic fibres. In contrast to textile dyes, pigments tend to be versatile colouring materials requiring less tailoring to individual applications. With pigments therefore, classification according to application is relatively unimportant. Pigments are simply designed to resist dissolving in solvents with which they may come into contact in paint, printing ink or plastics applications.

There is a third method proposed for classifying colorants which is in terms of the mechanism of the electronic excitation process.[23] According to this method, organic colorants may be classified as donor/acceptor, polyene, cyanine or n-π^* chromogens. While this method of classification is undoubtedly of importance theoretically, it is arguably of lesser practical importance. Most commercial organic dyes and pigments for traditional applications belong to the donor/acceptor category. The mechanism of the electronic excitation process, an understanding of which is fundamental in establishing relationships between the colour and the molecular constitution of dyes, is discussed in the following sections of this chapter.

2.9   COLOUR AND MOLECULAR STRUCTURE

Ever since the discovery of the first synthetic dyes in the mid-nineteenth century, chemists have been intrigued by the relationship between the colour of a dye and its molecular structure, a topic often referred to as *colour and constitution*.[23,24] Since these early days, the subject has been of special academic interest to those fascinated by the origin of colour in organic molecules. In addition, an understanding of colour/structure relationships has always been of critical importance in the design of new dyes. In the early days of synthetic colour chemistry little was known about the structures of organic molecules. However, following Kekulé's proposal concerning the structure of benzene in 1865, organic chemistry made significant and rapid progress as a science and, almost immediately, theories concerning the influence of organic structures on the colour of the molecules began to appear in the literature. One of the earliest observations of relevance was due to Graebe and Liebermann who, in 1867, noted that treatment of the dyes known at the time with reducing agents caused a rapid destruction of their colour. They concluded, with some justification, that the dyes were unsaturated compounds and that this unsaturation was destroyed by reduction.[25]

Perhaps the most notable early contribution to the science of colour/structure relationships was due to Witt who, in 1876, proposed that dyes contain two types of group that are responsible for their colour.[26] The first of these is referred to as the *chromophore*, which is defined as a group of atoms that is principally responsible for the colour of the dye. Secondly, there are the *auxochromes*, which he suggested were 'salt-forming' groups of atoms whose role, rather more loosely defined, was to provide an essential 'enhancement' of the colour. This terminology is still used to a certain extent today to provide a simple explanation of colour, although Witt's original suggestion that auxochromes were also essential for dyeing properties was quickly recognised as having less validity. A further notable contribution was made by Hewitt and Mitchell, who first proposed in 1907 that conjugation is essential for the colour of a dye molecule.[27] In 1928, this concept was incorporated by Dilthey and Wizinger in their refinement of Witt's theory of chromophores and auxochromes.[28] They recognised that the chromophore is commonly an electron-withdrawing group, that auxochromes are usually electron-releasing groups and that they are linked to one another through a conjugated system. In essence, the concept of the donor/acceptor chromogen was born. Furthermore, it was observed that a bathochromic shift of the colour, *i.e.*, a shift of the absorption band to longer wavelength, might be obtained by increasing the electron-withdrawing power of the chromophore, by increasing the electron-releasing power of the auxochromes and by extending the length of the conjugation.

The chromophore and auxochrome theory, which was first proposed more than 100 years ago, still retains some merit today as a simple method for explaining the origin of colour in dye molecules, although it lacks rigorous theoretical justification. The most important chromophores as defined in this way are the azo ($-N{=}N-$), carbonyl ($C{=}O$), methine ($-CH{=}$) and nitro (NO_2) groups. Commonly-encountered auxochromes, groups that normally increase the intensity of the colour and shift the absorption to longer wavelengths of light, include hydroxyl (OH) and amino (NR_2) groups. The numerous examples of chemical structures that follow in later sections of this book will illustrate the many ways in which chromophores, auxochromes and conjugated aromatic systems, together with other structural features designed to confer particular application properties, are incorporated into dye and pigment molecules. The concept may be applied to most chemical classes of dye, including azo, carbonyl, methine and nitro dyes, but for some classes which are not of the donor/acceptor type, for example the phthalocyanines, it is less appropriate. Nowadays, modern theories of chemical bonding, based on either the valence bond or the molecular orbital approaches, are capable of providing a much more sophisticated account of colour/structure relationships.

2.9.1 Valence-Bond Approach to Colour/Structure Relationships

The valence-bond (or resonance) approach to bonding in organic molecules is a particularly useful traditional approach to explaining the molecular structure and properties of aromatic compounds. The approach involves postulating a series of organic structures representing a particular compound in each of which the electrons are localised in bonds between atoms. These structures are referred to as canonical, or resonance forms. The individual structures do not have a separate existence, but rather each makes a contribution to the overall structure of the molecule, which is considered to be a resonance hybrid of the contributing forms. Arguably the simplest example of resonance is benzene, which, to a first approximation, may be considered as a resonance hybrid of the two Kekulé structures as shown in Figure 2.7. It was Bury who, in 1935, first highlighted the relationship between resonance and the colour of a dye, noting that the more resonance structures of comparable energy that could be drawn for a particular dye, the more bathochromic were the dyes.[29]

The valence bond approach may be used to provide a qualitative account of the λ_{max} values, and hence the hues, of many dyes, particularly those of the donor–acceptor chromogen type. The use of this approach to rationalise differences in colour is illustrated in this section with reference to a series of dyes which may be envisaged as being derived from azobenzene, although in principle the method may be used to account for the colours of a much wider range of chemical classes of dye including anthraquinones (Chapter 4), polymethines and nitro dyes.[30]

Figure 2.7 Valence-bond (resonance) approach to the structure of benzene.

The valence-bond approach to colour/structure relationships requires that certain assumptions be made concerning the structures of the electronic ground state and of the electronic first excited state of the dye molecules. Invariably, a dye molecule may be represented as a resonance hybrid of a large number of resonance forms, some of which are 'neutral' or normal Kekulé-type structures, and some of which involve charge separation, particularly involving electron release from the donor through to the acceptor groups. For the purpose of explaining the colour of dyes, a first assumption is made that the ground electronic state of the dye most closely resembles the most stable resonance forms, the normal Kekulé-type structures. A second assumption is that the first excited state of the dye more closely resembles the less stable, charge-separated forms. The nature of these assumptions will be clarified in the consideration of the examples that follow. As a consequence of Planck's relationship ($\Delta E = hc/\lambda$), the wavelength at which the dye absorbs increases (a bathochromic shift) as the difference in energy between the ground state and the first excited state decreases. Essentially, in its interpretation of colour/ structure relationships in dye molecules, the valence bond approach is used to account for these energy differences. Structural factors, both electronic and steric, which either stabilise or destabilise the first excited state relative to the ground state, are analysed to provide a qualitative explanation of colour. The assumptions that the approach makes concerning the structures of the ground and the first excited states are clearly approximations and cannot be rigorously justified. Nevertheless, evidence that the approximations are reasonable is provided by the fact that the method works well, at least in qualitative terms, in such a large number of cases. There are, however, numerous examples where the approach fails, no doubt due to the inadequacy of the assumptions for those cases.

Table 2.2 shows the λ_{max} values obtained from the UV/visible spectra, recorded in ethanol as the common solvent, of a series of substituted azobenzenes, which may be considered as the basis of the

Table 2.2 The λ_{max} values for a series of substituted azobenzenes, **2.1a–f**.

Compound	X	Y	λ_{max} (nm) in C_2H_5OH
2.1a	H	H	320
2.1b	NO_2	H	332
2.1c	H	NH_2	385
2.1d	H	NMe_2	407
2.1e	H	NEt_2	415
2.1f	NO_2	NEt_2	486

simplest azo dyes. Several observations may be made from the data given in Table 2.2. Azobenzene, **2.1a**, the simplest aromatic azo compound, gives a λ_{max} value of 320 nm. This compound is only weakly coloured because its principal absorption is in the UV region. An electron-withdrawing group, such as the nitro group, on its own produces only a weakly bathochromic shift, so that nitroazobenzene, **2.1b** (λ_{max} 332 nm) is still only weakly coloured. In contrast, an electron-releasing group, such as the amino group, in the *p*-position leads to a pronounced bathochromic shift, the magnitude of which increases with increasing electron-releasing power of the group in question. The absorption bands of compounds **2.1c–e** are thus shifted into the visible region (λ_{max} 385, 407 and 415 nm, respectively) and the compounds are reasonably intense yellows. Even larger bathochromic shifts are provided when there is an electron-releasing group in one aromatic ring and an electron-withdrawing group in the other, *i.e.*, a typical donor–acceptor chromogen, as in the case of compound **2.1f**, which is orange-red (λ_{max} 486 nm).

2.1

Valence bond representations consisting of the most relevant canonical forms contributing towards the structures of compounds **2.1a–f** are illustrated in Figure 2.8. The arguments that follow illustrate how the valence bond approach may be used to explain colour by rationalising the trends in λ_{max} values for the compounds given in Table 2.2. Using the approach, it is assumed that the ground state of azobenzene, **2.1a**, the parent compound, most closely resembles the more stable traditional Kekulé resonance forms, such as structure I. Further, it is assumed that charge-separated resonance forms, such as structure II, make a major contribution to the first excited state of azobenzene. In structure II, the negative charge is accommodated on the electronegative nitrogen atom of the azo group, a reasonably stable situation, but the structure also contains a carbocationic centre, a source of instability. The first excited state of azobenzene is therefore rather unstable, *i.e.*, of high energy. In the case of 4-nitroazobenzene, **2.1b**, the carbocationic centre is still present as a destabilising feature in the first excited state (structures IV and V). There is, however, a marginal stabilisation of the first excited state in compound **2.1b** as a result of the additional contribution to the first

(I)　　　　　　　　　　　　　　　　　(II)

2.1a

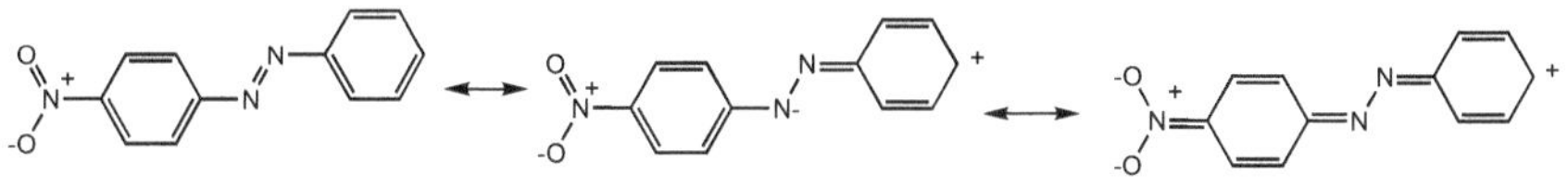

(III)　　　　　　　　　(IV)　　　　　　　　　(V)

2.1b

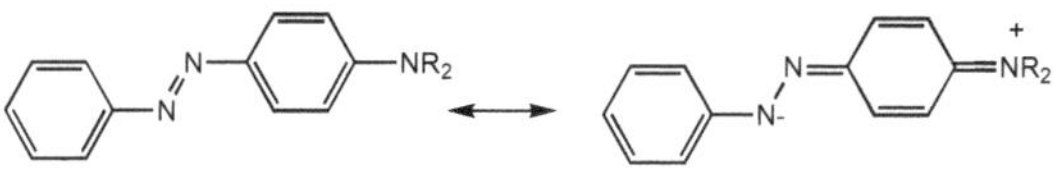

(VI)　　　　　　　　　　　　　　　　　(VII)

2.1c-e

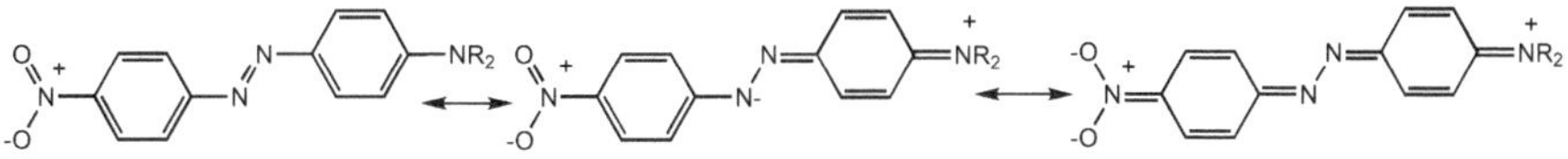

(VIII)　　　　　　　　　(IX)　　　　　　　　　(X)

2.1f

Figure 2.8　Valence-bond (resonance) approach to the structure of azobenzenes **2.1a–f**.

excited state from canonical forms such as structure V, in which charge is delocalised on to the nitro group. Consequently, the energy difference between the ground and first excited states in compound **2.1b** becomes smaller than in azobenzene, **2.1a**, and a small bathochromic shift is observed as a result of the inverse relationship between energy difference and the absorbed wavelength.

The situation is significantly different with the 4-aminoazobenzenes **2.1c–e**. In the charge-separated structures that make the major contribution to the first excited states of these compounds (structure VII), donation of the lone pair from the amino nitrogen atom removes the carbocationic centre, which has a sextet of electrons in its valence shell, and the positive charge becomes accommodated on the nitrogen atom, deriving much higher stability from its full octet of electrons. There is thus a marked stabilisation of the first excited state, lowering its energy and leading to a pronounced bathochromic shift. The electron-releasing inductive effect of the

Table 2.3 The λ_{max} values for a series of donor–acceptor substituted azobenzenes, 2.2a–f.

Compound	Acceptor substituents (A)	λ_{max} (nm) in C_2H_5OH
2.2a	o-CN	462
2.2b	m-CN	446
2.2c	p-CN	466
2.2d	o-,o'-,p-tricyano	562
2.2e	o-NO$_2$	462
2.2f	p-NO$_2$	486

N-alkyl groups in compounds **2.1d** and **2.1e** serves to increase the electron-donor power of the lone pair on the amino nitrogen atom compared with compound **2.1c**, thus further stabilising the first excited state and causing a more pronounced bathochromic shift, the magnitude of which increases with increasing electron-releasing power of the alkyl groups. In the case of compound **2.1f**, typical of most aminoazobenzene dyes in which one aromatic ring contains electron-donor groups and the other electron-acceptor groups, the strong bathochromicity is explained by a first excited state in which there is further stabilisation by charge delocalisation on to the nitro group, as a result of a contribution from structure X.

The spectral data for a further group of donor–acceptor aminoazobenzenes are given in Table 2.3. The valence bond approach may be used to provide a good qualitative account of the data in the table. Some relevant resonance structures, which may be used to explain the λ_{max} values for cyano compounds **2.2a–d**, are shown in Figure 2.9.

2.2

Inspection of the spectral data for the isomeric cyano compounds **2.2a–c** demonstrates that the bathochromicity is most pronounced when the donor and acceptor groups are conjugated with one another, *i.e.*, *ortho*- or *para*- to the azo group allowing the full mesomeric interaction between the amino and cyano groups in the first excited states to operate, as illustrated for compounds **2.2a** and **2.2c** in structures XI and XII respectively. In contrast, the presence of a cyano group in the *meta* position gives rise to a less pronounced bathochromic effect (compound **2.2b**) as only the inductive effect of the cyano group is in operation. Increasing the number of

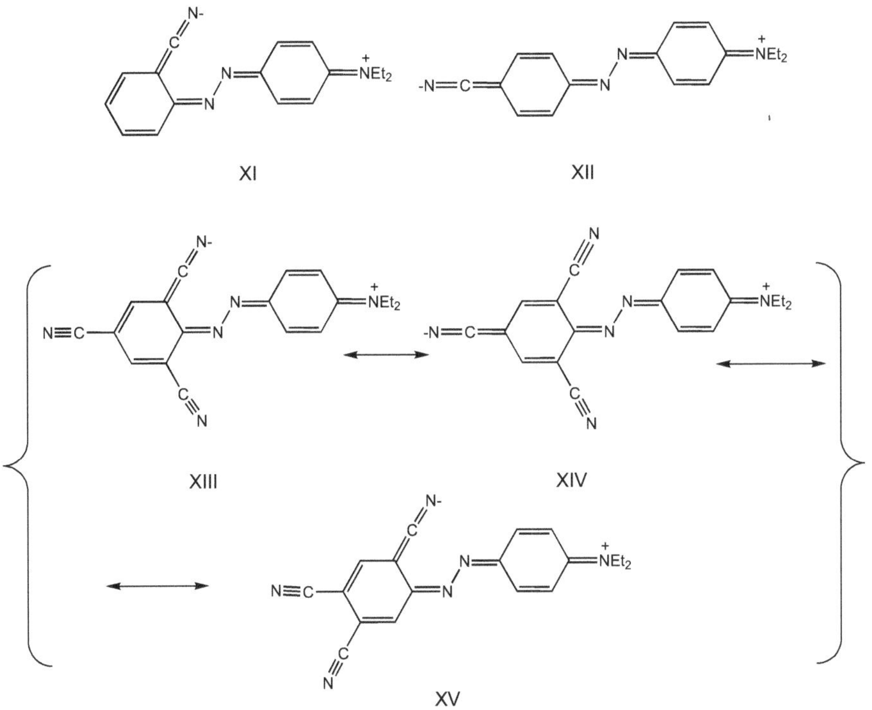

Figure 2.9 Resonance forms that make a major contribution to the first excited state of azobenzenes **2.2a**, **2.2c** and **2.2d**.

electron-withdrawing groups in the acceptor ring increases the bathochromicity, as illustrated by the tricyano compound **2.2d**. The bathochromicity of this dye (λ_{max} 562 nm) is explained by the extensive stabilisation of the first excited state as a result of resonance involving structures XIII–XV in which the negative charge is accommodated on the nitrogen atoms of each of the three cyano groups in turn.

The colour of dyes may be affected by steric as well as electronic effects. For example, a comparison of isomers **2.2e** and **2.2f** shows that an *o*-nitro group produces a significantly smaller bathochromic shift than a *p*-nitro group. On the basis of an argument based purely on electronic effects, delocalisation of charge on to the oxygen atom of the *o*-nitro group, as illustrated by the valence bond representation of the structure of compound **2.2e** given in Figure 2.10, might be expected to give rise to a bathochromic shift similar to that given by a *p*-nitro group. The reason for the differences arises from the fact that while compound **2.2f** is a planar molecule, steric congestion forces

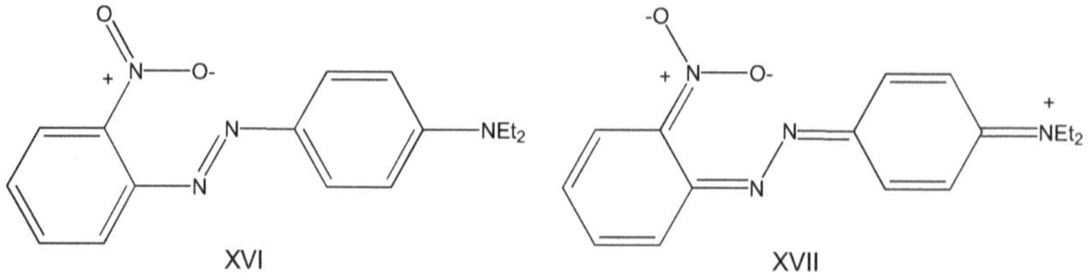

Figure 2.10 Valence-bond (resonance) approach to the structure of compound **2.2e**.

compound **2.2e** to adopt a non-planar conformation. This happens because the *o*-nitro group clashes sterically with the lone pair of electrons on one of the azo nitrogen atoms and is thus forced to rotate out of a planar conformation. It may be envisaged that this rotation takes place about the bond between the carbon atom of the aromatic ring and the nitrogen atom of the NO_2 group. As Figure 2.10 shows, the C–N bond in question is of higher bond order in the first excited state (represented approximately by structure XVII), but of lower bond order in the ground state (similarly represented by structure XVI). Since rotation about a double bond requires more energy than rotation about a single bond, the first excited state of compound **2.2e** is destabilised relative to its ground state, with a consequent reduction in the bathochromic effect. When it is desirable to have electron-withdrawing substituents in the positions *ortho* to the azo group to maximise the bathochromicity, cyano groups are commonly preferred. Their linear, rod-like shape minimises the steric clash with the lone pair on the azo nitrogen atoms and allows a more planar conformation to be adopted.

The absorption band of aminoazobenzenes may be shifted to very long wavelengths to produce blue dyes by incorporating a number of electron-withdrawing groups into the acceptor ring and a number of electron-releasing groups into the donor ring. An example of such a dye is compound **2.3**, which gives a λ_{max} of 608 nm in ethanol as a result of the three electron-withdrawing groups (two nitro, one bromo) in one ring and the three electron-releasing groups (dialkylamino, acylamino and methoxy) in the other. Alternatively, particularly bathochromic azo dyes, which are structurally analogous to the aminoazobenzenes, may be obtained by replacement of the carbocyclic acceptor ring with a five-membered aromatic heterocyclic ring. Examples of such dyes are provided by CI Disperse Blue 339, **2.4**, a bright blue dye that contains a thiazole ring and gives a λ_{max} of 590 nm and thiophene derivative **2.5** which gives a λ_{max} of 614 nm.

2.3

2.4

2.5

Several reasons for the particular bathochromicity of heterocyclic azo dyes of this type may be postulated from the valence bond approach. Some important resonance forms which may contribute to the structure of the first electronic excited state of dye **2.4** are illustrated in Figure 2.11. The first reason involves a consideration of aromatic character. Evidently, when the structures that are considered to make a major contribution to the first excited states of azo dyes are compared with those of the respective ground states, electronic excitation causes a loss of aromatic character, *i.e.*, a loss of resonance stabilisation energy. It is generally accepted that five-membered ring heterocyclic systems of this type are less aromatic than benzene derivatives. The loss of resonance stabilisation energy on electronic excitation of dyes such as **2.4** is therefore less than in the corresponding carbocyclic systems and, as a consequence, the difference in energy between the ground and excited states is less.

Steric effects also play a part in the explanation. For example, the contribution to the first excited state from structure XX (Figure 2.11) illustrates that the heterocyclic nitrogen atom can effectively act as an electron-withdrawing *o*-substituent with no associated steric effect that might otherwise tend to reduce the bathochromicity. In addition,

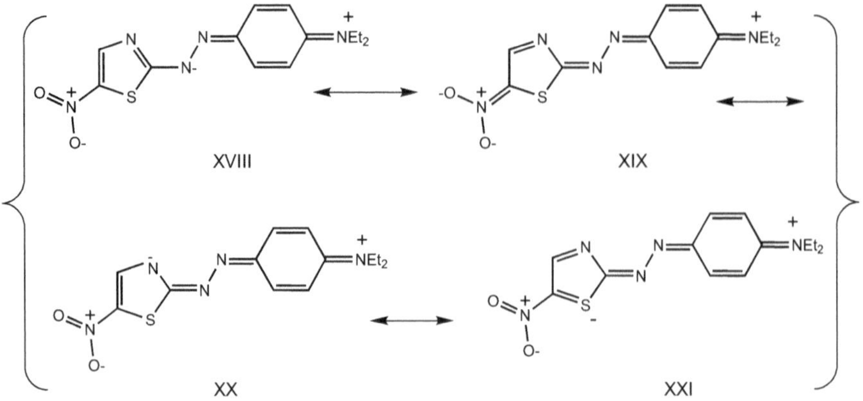

Figure 2.11   Resonance forms making a major contribution to the first excited state of heterocyclic azo dye **2.4**.

in the case of five-membered heterocyclic rings that contain *o*-substituents, such as the *o*-nitro group in compound **2.5**, the steric interaction between the substituent and the lone pair on the azo nitrogen atoms is less than in the case of six-membered rings, as in dye **2.2e**, allowing the group to adopt a more coplanar arrangement and hence maximise the bathochromic effect. This may be explained by a comparison of the geometrical arrangements. As illustrated in Figure 2.12, the azo group makes an angle of 126° with a five-membered ring, compared with a corresponding angle of 120° in the case of a six-membered ring, thus reducing the steric congestion between the oxygen atom of the nitro group and the lone pair on the azo nitrogen atom. A further explanation has been put forward to explain why sulfur heterocycles, in particular, appear in general to provide a pronounced bathochromic effect. It has been argued, although it has to be said that there is incomplete consensus on the argument, that there is a contribution to the first excited state from structures such as XXI (Figure 2.11) in which there is valence shell expansion at sulfur. This is possible in principle as a consequence of the availability of vacant 3d-orbitals into which valence electrons may be donated, a situation that is not available in the case of nitrogen and oxygen heterocycles.

Compound **2.6** is noteworthy in that it was the first azo dye reported whose absorption band is shifted beyond the visible region into the near-infrared region of the spectrum, showing a λ_{max} of 778 nm in dichloromethane.[31] The extreme bathochromicity of this dye may be explained by a combination of the effects discussed throughout this

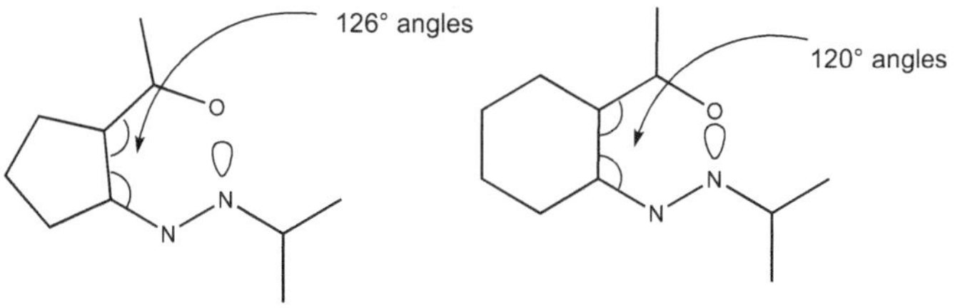

Figure 2.12 Reduced steric congestion with substituents ortho to the azo group in the case of a five-membered ring compared with a six-membered ring.

section, including the extended conjugation, the influence of the thiazole ring and the maximising of both the electron donor (dialkylamino, acylamino, methoxy) and electron accepting (cyano, sulfone, chloro) effects in appropriate parts of the molecule.

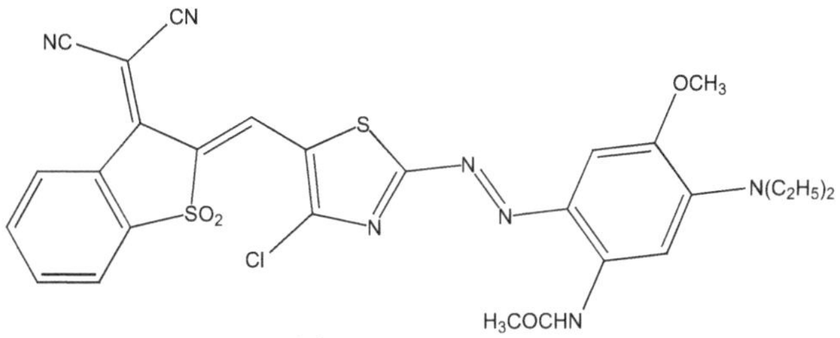

2.6

It can be seen that the valence-bond (resonance) approach may be used to provide a reasonable qualitative explanation of the λ_{max} values of a wide variety of azo dyes. While the use of the approach has been exemplified in this chapter for a series of azo dyes, the argument may be extended successfully to a range of donor–acceptor type chromogens such as carbonyl (see Chapter 4 for an illustration of its application to some anthraquinone dyes, for example), nitro and methine dyes. Nevertheless, the approach does have deficiencies. There are several examples where wrong predictions are made. For example, the order of bathochromicities of the o-, m- and p-aminoazobenzenes is wrongly predicted (see next section for a further discussion of this observation). Secondly, the approach cannot readily be used to account, even qualitatively, for the intensity of colour by addressing trends in molar extinction coefficients. Finally, and arguably most importantly, the method cannot at the present time be used quantitatively. Quantitative treatment of the light absorption properties of dyes has, however, been made possible by developments in molecular orbital methods, as discussed in the next section.

2.9.2 Molecular Orbital Approach to Colour/Structure Relationships

For several decades, there has been intense interest in the ability to predict the colour of dyes based on theoretical calculations. An early approach towards this aim involved the use of tables of empirical data representing the effects of auxochromes on λ_{max} values in particular, established based on experimental spectral data for dyes from the important chemical classes. However, an inherent limitation in this approach is the inability to take into account new auxochromes and chromophoric systems. Subsequently, the application of quantitative molecular orbital methods has contributed immensely to developments in colour chemistry, providing a more rigorous theoretical understanding of the light absorption properties of colorants in relation to their molecular structure, and enabling the calculation of many of the physical properties of dyes, including colour properties, from a knowledge of their chemical structure, with the aid of a computer. Thus, the characteristics of the colour of a dye whose structure may be drawn on paper and visualised on a computer screen may be predicted, with some expectation of accuracy, without the need to resort to devising a method for the synthesis of the dye in the laboratory. The value to the chemist whose aim is the synthesis of new dyes with specific properties, perhaps for new applications, is obvious. The properties of a large number of structures may be predicted in a short period of time using computational methods, and specific compounds for which interesting properties are predicted may be selected for synthesis and an evaluation of application performance. In the last half century or more, the major advances in computer technology that have taken place and which will no doubt continue into the future, in terms of both software and hardware, have meant that increasingly sophisticated methods have become accessible as routine tools for the colour chemist. The concepts and mathematical basis of molecular theory, commonly referred to as quantum mechanics, are well documented, and this particular text makes little attempt to address the detail of these.[32–34] The section that follows provides an outline of the basis of the molecular orbital approach to bonding in organic molecules, an overview of the methods that have historically proved valuable to the colour chemist, and those aspects of colour properties that the methods can address.

In molecular orbital theory, electrons are considered as a form of electromagnetic radiation, *i.e.*, in terms of their wave nature rather than their particulate nature. Electrons, rather than being treated as localised in individual bonds between atoms, are considered to be moving under the influence of all of the nuclei in the molecule. Of fundamental importance to the theory is the *quantum principle*, which states that the electron can only exist in a fixed series of discrete energy states. An essential concept in quantum theory that is of relevance to colour chemistry is that electrons are contained in regions of high probability referred to as *orbitals*. The mathematics underlying molecular orbital theory was first formulated in 1925 by Schrödinger, the solution of whose equation gives a fixed number of values of E, the energy states available to the electrons in a particular atom or molecule. The complexity of the mathematics means that, even with the computing power currently available, the equation may not be solved exactly for the complex molecular frameworks of coloured organic molecules. However, numerous quantum mechanical approaches have been devised using various approximations to provide solutions to the equation. These methods range from the simplest approaches based on empirically or semi-empirically derived parameters, which were formerly the methods of choice for the colour chemist, through to the *ab initio* approaches, such as those based on time-dependent density functional theory (TD-DFT) which now predominate, and the discussion in this section reflects these historical developments.

Molecular orbitals are considered to be generated by overlap of atomic orbitals. There are two types of overlap. Direct or 'end-on' overlap gives rise to σ-orbitals, either bonding, the low energy orbitals that are occupied by two electrons in the ground state of a molecule, and the high energy antibonding (σ*) orbitals that remain unoccupied, while π-molecular orbitals are obtained by indirect or 'sideways' overlap, for example, from overlap of two singly occupied $2p_z$ atomic orbitals. Dyes are invariably organic molecules with extended conjugation containing a framework of σ-bonds and an associated π-system. The lowest energy electronic transition occurs when an electron is promoted from the highest occupied molecular π-orbital (HOMO) to the lowest unoccupied π* molecular orbital (LUMO). It is these π–π* transitions (rather than σ–σ*, which are of much higher energy) that give rise to the absorption of most organic dyes and pigments in the UV and visible regions of the spectrum.

The *Hückel molecular orbital* (HMO) method is one of the earliest and simplest molecular orbital methods that proved to be appropriate for calculations on conjugated molecules in the pioneering years of quantum chemistry, before computers became routinely available.[35–37] In the HMO method it is assumed that the set of molecular orbitals (described by the molecular orbital wave function, φ) of a particular structure may be expressed as a linear combination (or weighted sum) of the atomic orbitals (described by the atomic orbital wave function, χ) in the molecular system (the LCAO approximation). Essentially this represents a mathematical expression of the assumption that molecular orbitals are derived from overlap of atomic orbitals. The HMO method also assumes that, in conjugated systems, only the π-electrons are involved in the electronic transitions that take place when light is absorbed and that these transitions are unaffected by the framework of σ-bonds in the molecule (the σ–π separation principle). This would at first sight appear to be a reasonable first approximation as in conjugated molecules it is generally assumed that the low energy π–π* transitions are principally responsible for UV and visible light absorption. By calculating the energies of each molecular orbital, the HMO method may be used to provide electronic transition energies for the promotion of an electron from an occupied molecular orbital to a higher energy unoccupied orbital. These energies are obtained in terms of β, the bond resonance integral, which is treated as an empirical parameter and given a value by comparison of the calculated values with the experimental values obtained from UV/visible spectra. Two other sets of molecular parameters that may be calculated using the HMO method are the π-electron charge densities, Q, the measure of the π-electron charge localisation on each atom in the molecule, and the π-bond orders (P), the measure of the degree of π-overlap of atomic orbitals between each pair of atoms in the molecule. While the HMO method is simple and useful qualitatively, quantitative correlation with spectral data is reasonable only in a few special cases. The method has been shown to give an acceptable correlation between experimental and calculated electronic transition energies with series of aromatic hydrocarbons, and also with the set of linear polyenes, in the latter case provided that different bond β-values are used for the formal single and double bonds. In general, while the method gives reasonable predictions for hydrocarbons, it works much less well with molecules containing heteroatoms, such as O and N, and this means that it was never a useful tool for calculating the colour of dye molecules.

In view of the inadequacy of the HMO method, attention turned in time to rather more elaborate MO approaches that remained reliant on semi-empiricism. A particular molecular orbital approach that was extensively applied to the calculation of the colour properties of dye molecules in the later decades of the twentieth century is the Pariser–Pople–Parr (PPP) method.[38] Although the use of the PPP-MO method has now been essentially superseded by the wide range of more sophisticated computational chemistry methodologies that have emerged for calculation of the physical properties of organic molecules, as discussed later in this section, it is considered here in a level of detail that reflects its importance during that period, leading to a significant body of published data on organic colorants, and also because it provides a useful and relatively straightforward illustration of the principles involved in semi-empirical MO methodology. Like the HMO method, the PPP-MO method uses the LCAO approximation in combining atomic orbitals to form molecular orbitals as its basic mathematical premise. In addition, it retains the σ–π separation principle, essentially neglecting the influence of σ-electrons on the colour. The main way in which the PPP-MO method provides a refinement over the HMO method is by taking account of inter-electronic interaction energies, and in doing so specifically includes molecular geometry; both of these features are ignored by the HMO method.

The sequence of operations involved in a PPP-MO calculation is illustrated in the flow diagram shown in Figure 2.13. The method is illustrated for the case of 4-aminoazobenzene, **2.1c**. In this molecule, there are 15 relevant atoms and 16 π-electrons, each atom contributing a single electron to the π-system except for the amino nitrogen atom, which donates its lone pair. The molecular geometry may be

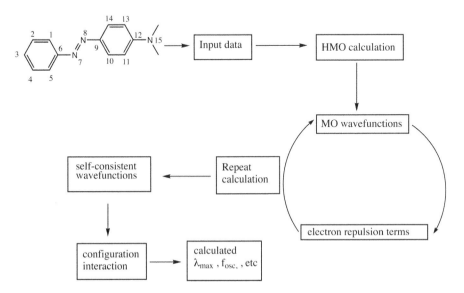

Figure 2.13 Sequence of stages in a PPP-MO calculation for compound **2.1c**.

specified in terms of all the relevant interatomic distances and bond angles. This stage of the process may be refined by including a molecular geometry optimisation in the calculation, for example using molecular mechanics (Section 2.9.3). Energy parameters, specifically valence state ionisation potentials (VSIP) and electron affinities (E_a) assigned to the relevant atoms and bond resonance integrals values (β) assigned to each pair of bonded atoms are treated as a set of semi-empirical parameters that are available from various literature sources and that may be modified to suit a particular type of molecule. Finally, to enable the calculations to be carried out, the π-electron densities (Q) for all relevant atoms and the π-bond orders (P) for all bonded atoms are required. This feature of the PPP-MO method raises the issue that the values of Q and P for specific molecules are not known initially, although they may be obtained after the calculations have been performed. The solution to this 'chicken and egg' issue is achieved by carrying out a preliminary HMO calculation, which gives an approximate set of Q and P values. These values are in turn used to set up and carry out the PPP-MO calculation and as a result a new, improved set of values is obtained. The process is then repeated until two successive calculations give a consistent result. This iterative process is referred to as a self-consistent field (SCF) method. The procedure leads to a set of molecular orbital energies from which, in principle, electronic transition energies may be calculated. However, since the PPP-MO method uses a set of data based on the ground state structure of the molecule, it predicts ground state energies rather better than excited state energies. To correct for the fact that molecular orbital energies may change after excitation by promotion of an electron from a lower energy occupied molecular orbital to a higher energy unoccupied molecular orbital, and hence to give improved excited state data, a procedure known as configuration interaction (CI) is carried out as the final stage of the calculation.

The PPP-MO method proved to be suitable for the treatment of large molecular structures and presented modest computing demands. As time progressed, commercial programs became available with user-friendly interfaces that were appropriate to the personal computing systems that were developing rapidly. The method was used successfully in calculating λ_{max} values for a wide range of dyes from virtually all of the chemical classes, the success probably owing much to the careful optimisation of the semi-empirical parameters carried out over several series of studies. As an example, Figure 2.14 illustrates the correlation of λ_{max} values measured experimentally in the same solvent and the values calculated by the PPP-MO approach

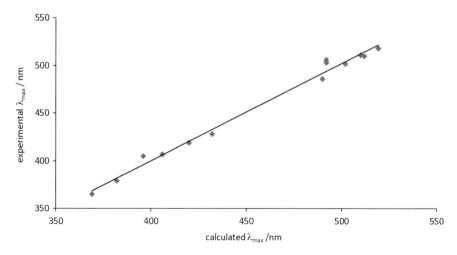

Figure 2.14 Correlation between calculated and experimental λ_{max} values for a series of nitroarylamine semi-permanent hair dyes.

Table 2.4 Experimental and PPP-MO calculated electronic spectral data for azobenzene and the isomeric amino derivatives.

Compound	λ_{max} (nm) (experimental.)	λ_{max} (nm) (PPP calcd)	$\varepsilon \times 10^{-3}$ ($l\ mol^{-1}\ cm^{-1}$)	f_{osc} (PPP calcd.)
Azobenzene, **2.1a**	320	344	21.0	1.21
2-Aminoazobenzene	414	424	6.5	0.67
3-Aminoazobenzene	412	415	1.3	0.17
4-Aminoazobenzene	387	392	24.5	1.31

for a series of nitroarylamine semi-permanent hair dyes (Chapter 10). As a further example, the method provides a reasonable account of substituent effects in the range of aminoazobenzene dyes, including compounds **2.1a–f** and **2.2a–f**, which have been discussed in terms of the valence-bond approach in Section 2.9.1.

Table 2.4 shows a comparison of experimental and PPP-MO calculated electronic spectral data for azobenzene and the three isomeric monoamino derivatives. Notably, the *ortho* isomer is observed to be most bathochromic, while the *para* isomer is least bathochromic. From a consideration of the principles of the application of the valence bond approach to the relationship between colour and molecular structure described in the previous section of this chapter, it might have been expected that the *ortho* and *para* isomers would be most bathochromic with the *meta* isomer least bathochromic. In contrast, the data contained in the table demonstrate that the PPP

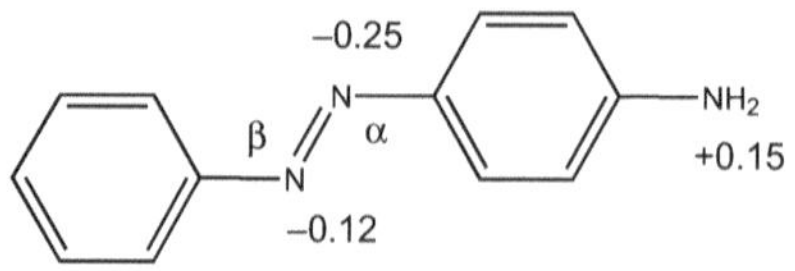

Figure 2.15   Some π-electron charge density differences between the ground and
first excited states calculated by the PPP-MO method for 4-aminoazo-
benzene, **2.1c**.

MO method is capable of correctly accounting for the relative bathochromicities of the amino isomers. An explanation for the failure
of the valence bond method to predict the order of bathochromicities
of the *o-*, *m-* and *p*-aminoazobenzenes emerges from consideration of
the changes in π-electron charge densities on excitation calculated
by the PPP-MO method as illustrated in Figure 2.15.

The valence bond representation of the ground and first excited
states of dye **2.1c**, illustrated in Figure 2.8, would suggest that a decrease in charge density on the amino nitrogen atom and an increase
in charge density on the β-nitrogen of the azo group would be observed on excitation. The changes in calculated charge densities
(Figure 2.15) indeed predict these effects, but suggest that there is
also a significant increase on the charge density of the α-nitrogen
atom of the azo group, an effect that may not be easily accounted for
using the valence bond approach. The results demonstrate that the
valence bond assumptions, particularly concerning the structure of
the first excited state, are not wholly accurate and as a result it is
perhaps not surprising that erroneous predictions sometimes arise.

Because molecular orbital methods are capable of calculating π-
electron charge densities both in the electronic ground states and
excited states of dye molecules, they are helpful in providing information on the nature of the electronic excitation process, by identifying the donor and acceptor groups and quantifying the extent of the
electron transfer. One application of this principle is that it allows the
dipole moments of the ground and first excited state to be calculated,
and this may be used to account for the influence of the nature of the
solvent on the λ_{max} value of a dye, an effect referred to as *solvatochromism*. For example, if the dipole moment of a dye molecule is
larger in the first excited state than in the ground state, then the effect
of a more polar solvent will be to stabilise the first excited state more
than the ground state. The consequence will be a bathochromic shift
of the absorption maximum as the solvent polarity is increased, in
accordance with the inverse relationship between the energy of the
electronic transition (ΔE) and the λ_{max} value, an effect known as

positive solvatochromism. As the numerous examples given in the previous section on the application of the valence bond approach to colour/structure relationships demonstrate, the first excited state in donor–acceptor chromogens is almost invariably more polar than the ground state, and thus such dyes as a rule exhibit positive solvatochromism. The reverse effect in which increasing solvent polarity causes a hypsochromic shift, referred to as negative solvatochromism, is less common.

Molecular orbital methods are capable of calculating not only the magnitude of the dipole moment change on excitation, but can also predict the direction of the electron transfer within the molecule. The vector quantity, which expresses the magnitude and direction of the electronic transition, is referred to as the *transition dipole moment*. For example, the calculated direction of the transition dipole moment of azo dye **2.1f** is illustrated in Figure 2.16.

The direction of the transition moment is of practical consequence in dyes developed for liquid crystal display systems. It is important for such applications that the direction of the transition moment is aligned with the molecular axis of the dye. Since this is the case with azo dye **2.1f**, the dye would appear to be a reasonable candidate as a liquid crystal display dye (see Chapter 11 for further discussion of this application of dyes).

The intensity of colour of a dye is dependent on the probability of the electronic transition. A familiar example of this principle is provided by the colours due to transition metal ions in solution, which are normally weak because the d–d transitions involved are 'forbidden', *i.e.*, of low probability. In contrast the π–π^* transitions due to organic dyes, which involve considerable charge transfer in donor/acceptor chromogens, are highly probable and thus give rise to much more intense colours. While most of the research published on the application of molecular orbital methods to dyes centres on their application to the calculation of λ_{max} values, the ability to predict the tinctorial strength of a dye is arguably of greater practical value since it is directly related to the economic viability of the dye. If, for example, a new dye has twice the colour strength of an existing dye, then

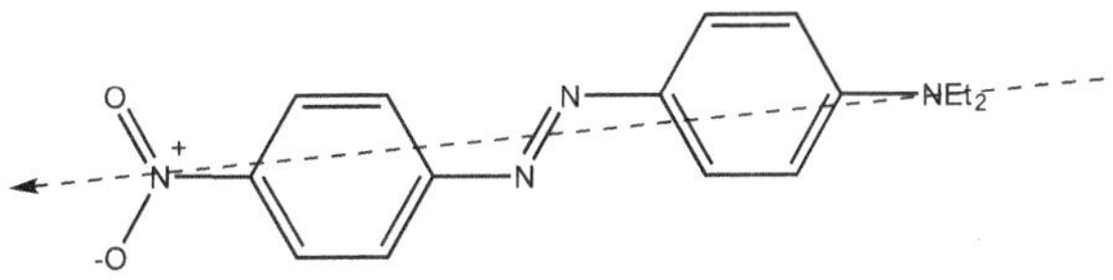

Figure 2.16 Direction of the transition moment in azo dye **2.1f**.

the dyer need only use half the quantity of that dye to obtain a given colour. Provided that the new dye costs less than double that of the existing dye, it will therefore be more cost-effective. The PPP-MO method is capable of providing a quantitative account of colour strength by calculating a quantity, f_{osc}, known as the oscillator strength. This parameter is given by the following equation:

$$f_{osc} = 4.703 \times 10^{29} \times M^2 / \lambda_M$$

where M is the transition dipole moment and λ_M is the mean wavelength of the absorption band.

In general, it has been observed that a reasonable correlation between f_{osc} and ε for broad classes of dyes may be achieved, although within specific classes the correlation is less good. It has been suggested that this may be because PPP parameterisation has been optimised for correlation with λ_{max} values rather than molar extinction coefficients. The correlation between f_{osc} and molar extinction coefficient (ε) values for a series of structurally related yellow azo pigments is illustrated in Figure 2.17.[39] The method correctly predicts the significantly higher tinctorial strength of the disazo compared with the monoazo pigments, although the correlation for individual species is less strong. Similarly, the data given in Table 2.4 demonstrate only a reasonable qualitative correlation between the calculated oscillator strengths and the experimental molar extinction

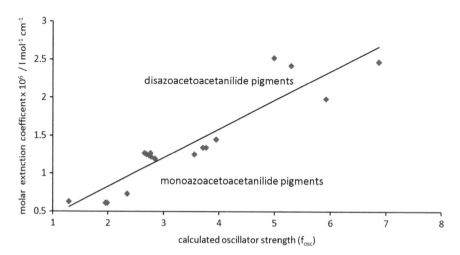

Figure 2.17 Correlation between calculated oscillator strength values and experimental molar extinction coefficient values for a series of monoazo and disazo pigments.

coefficients for the series of azo dyes in question. Nevertheless, the ability to predict broad trends in colour properties is a useful tool in the design of new coloured molecules. It may be argued that the oscillator strength gives a better measure of colour intensity than the molar extinction coefficient (ε) as it expresses the area under the absorption curve, whereas ε is profoundly dependent on the shape of the curve. It is thus only valid to relate the ε values to the intensity of colour for a series of dyes if the curves are of similar shape.

Brightness of colour is expressed by the width of the visible absorption band. This bandwidth is determined by the distribution of vibrational energy levels superimposed on the electronic ground and excited state energy levels. Broadening of the absorption bands may be caused in several ways. For example, an increase in the number and spread of energies of bond vibrations will generally lead to broader absorption bands. This argument may be used to provide an explanation as to why the relatively simple structure of heterocyclic azo dyes such as compounds **2.4** and **2.5** give brighter blue colours than the multi-substituted carbocyclic analogues, such as compound **2.3** with its increased number of vibrational levels. Some of the brightest colours are provided by the phthalocyanine chemical class (Chapter 5). The colour of these dyes and pigments owe their brightness in part to their rigid molecular structure both in the ground and excited states, with both states showing similar geometry and little vibrational fine structure. Because it is related to the vibrational characteristics of the molecules rather than their electronic structure, it is at first sight difficult to envisage how the PPP-MO might be useful in calculating bandwidth. An interesting approach to this problem has been provided by the application of an empirical extension of the method to the calculation of the Stokes shifts of fluorescent molecules.[40–42] Initially, a conventional calculation, using standard ground state molecular geometries and resonance integrals, is used to calculate an absorption λ_{max} value. From the calculated first excited state π-bond orders, excited state geometries and resonance integrals are derived. Subsequent calculations are carried out using these new parameters, iteratively, until convergence is achieved. The resulting λ_{max} value is equated to the fluorescence maximum and hence the Stokes shift is obtained as the difference between this value and the original result. Using the assumption that there is a simple relationship between absorption bandwidths and Stokes shifts – both parameters being dependent on vibrational energy levels in the two electronic states – the method has been adapted, with some success, to the calculation of bandwidths. While the method lacks rigorous theoretical justification, good correlations have been reported for a range of dye chemical classes.

The PPP-MO method clearly has deficiencies. The method assumes that the molecules are planar. While this is an important theoretical shortcoming, it is worth noting that the vast majority of commercial dyes and pigments have a molecular geometry that is more or less coplanar. Calculations are carried out based on π-electrons only and therefore cannot, except in a rather empirical way, account for some of the subtle effects of σ-electrons on colour. Among such effects that are commonly encountered are hydrogen bonding, steric interactions and the influence of substituents that do not contribute towards the π-system. As time progressed and more powerful computing facilities became accessible, attention inevitably turned towards the use of semi-empirical molecular orbital techniques that take into account all valence electrons, thus specifically including σ-electrons in the calculations, including CNDO, MNDO, MINDO, INDO and ZINDO, referring, respectively, to *c*omplete, *m*odified, *m*odified *i*ntermediate, *i*ntermediate and Zerner's *i*ntermediate *n*eglect of *d*ifferential Overlap, AM1 (Austin model 1) and PM3 (parameterised model 3). These methods differ from each other in the classes of integrals neglected in the calculations, and the type of parameterisation used. Programs providing these methods are now widely available, and they offer certain inherent advantages particularly in their applicability to a wider range of compounds, for example metal complexes, and also in cases where the dye molecules deviate significantly from planarity. All-valence electron methods also offer versatility in the range of physical properties that may be calculated, including thermodynamic properties such as heats of formation. Intuitively, it would seem reasonable to anticipate that such methods would lead to similar, conceivably better, correlations with experimental visible spectral data compared with those based on methods using π-electrons only. It is of interest, therefore, to note reports of comparisons of the different methods, some of which demonstrate that this is not necessarily the case.[43–45] Indeed, it has been reported that calculations based on methods such as CNDO, INDO and ZINDO may significantly underestimate the absorption maxima values for a range of typical organic dye structures.[43,44] The apparent superiority of the PPP-MO approach in terms of the correlation with experimental data that has been observed in these cases, especially in view of its lower level of sophistication, may well be due to the carefully optimised parameterisation that has been developed for the method, mainly on the basis of calibration against spectral data. Indeed, there is evidence that modification of the parameters within methods such as ZINDO can lead to improved correlations.[46]

Ab initio (meaning from the beginning) quantum mechanics calculations are the most rigorous and powerful theoretically, in that they attempt to solve Schrodinger's equation using minimal approximations. However, the methods have traditionally placed major demands on computing power, especially for the polyatomic systems that are typical of dye structures. An approach to *ab initio* calculations that has emerged as the most popular and versatile method for investigation of the electronic structure of atoms and molecules is referred to as *density functional theory* (DFT).[47,48] Using DFT methods, the properties of a many-electron system may be determined using *functionals*, a term that refers to functions of another function. In particular, DFT uses functionals of the spatially-dependent electron density in a system in order to simplify the problem of n electrons with $3n$ spatial coordinates. The fundamental principle utilised in the approach is that the ground state properties of a many-electron system are determined uniquely by an electron density that depends on only three spatial coordinates. This principle is refined further by extension into a time-dependent domain, referred to as *time-dependent density functional theory* (TD-DFT), which may be used to address the properties of excited states, a feature that is of obvious importance in the calculation of colour properties. Although DFT methods have been available since the 1970s, they were not considered sufficiently accurate for quantum chemistry calculations until into the 1990s when certain of the necessary approximations were refined to provide, for example, enhanced modelling of the exchange and correlation interactions. Consequently, DFT/TD-DFT methods for the calculation of electronic excitation energies and oscillator strengths, which do not present major computing demands, were introduced into computational chemistry packages. On the basis of their rapid growth in the early years of the twenty-first century, DFT/TD-DFT methods have been described as 'shooting stars' among quantum chemical techniques, with their excellent 'price/performance' ratio stimulating an ever-growing range of applications in almost all fields of chemistry.[47]

These methods, TD-DFT in particular, are reported to provide good correlations between calculated and experimental absorption maxima and oscillator strengths over a wide range of chemical types of dye, at least comparable to the predictions provided by the use of semi-empirical methods.[46,49–53] However, in view of their higher level of sophistication and enhanced theoretical rigour, the methods offer notable advantages especially in terms of their versatility, essentially allowing the determination of the wider range of energy parameters that are important in the interactions between molecules and light.[53] For example, the methods have been investigated for their ability to simulate vibrationally-resolved band shapes, thus providing the potential for more accurate correlation with experimental absorption spectra. Of particular note is the potential for calculation of the fluorescence properties of dyes, especially the emission properties on which technological applications of the dyes are commonly dependent. Indeed, the application of TD-DFT to the calculation of excited state profiles provides effectively the only reliable theoretical tool currently available for such studies. TD-DFT methods have been used to account for both absorption and emission wavelengths of the dyes, and thus the Stokes shift, which is a critical parameter for many functional fluorescent dye applications, for example in lasers and solar collectors (Chapter 11). In addition, progress has been made towards the calculation of fluorescence quantum yields, by estimating the extent of competing deactivation processes, notably vibrational relaxation. A further intrinsic advantage of TD-DFT methods is that they are capable of being coupled with models that take account of the often complex environment in which the dye molecule is located, for example the solvent or a polymeric or biological system. In this way, the method offers the potential to simulate the performance of a dye in a 'real' environment, which is of clear interest not only for traditional colour applications but also for specialist functional applications, especially for the use of dyes in biological systems. TD-DFT methods have also been used to probe the mechanisms underlying the electronic reorganisations that are triggered in photochromic colour change processes, which are growing in technological importance (see Chapter 11 for a discussion of photochromism).

2.9.3   Molecular Mechanics

Molecular modelling encompasses the range of theoretical methods and the associated computational techniques used to model or mimic molecular behaviour. These techniques, which include the quantum mechanical methods described in the previous section, are used widely in many areas of computational chemistry, materials science and biology to study molecular systems ranging in complexity from small molecules through to large biomolecules and material assemblies. The traditional methods used over many decades by chemists to assist their understanding of the structure of molecules by visual inspection commonly involved the construction of physical models of molecules based on sets of balls, representing atoms, and sticks or springs, representing bonds. Such methods have given way to the use of computational approaches that calculate energies and other physical properties of molecules and allow optimisation of molecular geometry, so that 3D projections of the molecular structure may be visualised on the computer monitor screen.[54]

Molecular mechanics uses classical Newtonian mechanics to model molecular systems. Essentially, the molecule is considered as if constructed from sets of balls and springs. Atoms are considered as spheres which are assigned a radius (usually the van der Waals' radius), polarisability and net charge, while bonds between atoms are simulated as springs with an equilibrium bond length. The equations of classical mechanics are used to calculate the interactions and corresponding energies. The set of equations and associated atomic and bond-related parameters used in the approach is referred to as a *force field*. There are many force fields available. There is no 'best' force field for all problems, and the selection must be determined based on the particular system of interest. However, MM2 is a widely used force field that appears to be suitable for application to dye molecules. The molecular mechanics energy, E_{MM}, is derived from several components that are combined additively. The essential components are terms representing the energy in bonds, in angles, from all van der Waals' interactions, in torsion angles and electrostatic interactions. Most force fields also include various miscellaneous energies, commonly referred to as *cross terms*. Molecular mechanics modelling, which does not present major computing demands, is most often used to calculate molecular energies and to assess the optimum molecular geometry using an energy minimisation process.

A deficiency of the molecular mechanics approach when used alone, however, is that it does not specifically include electrons in the calculations. This is especially relevant in the case of dye molecules, whose extensive delocalised π-systems mean that stereoelectronic effects may play an important role in determining molecular geometries. Thus, geometries may be refined further using molecular mechanics calculations in association with quantum mechanics calculations. The quantum mechanical approaches used for this purpose are generally parameterised all valence electron semi-empirical methods, commonly either AM1 (Austin model 1) or PM3 (parameterized model 3). These methods, in addition to calculating molecular geometries, bond lengths and bond angles, are capable of

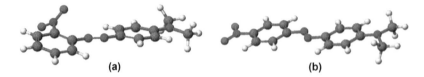

Figure 2.18 Molecular geometries of 2-nitrodimethylaminoazobenzene (a) and 4-nitrodimethylaminoazobenzene (b) calculated using MM2/AM1.

calculating a range of physical properties including heats of formation, ionisation potentials and dipole moments. The molecular geometries of two isomeric azo dyes calculated using MM2/AM1 by energy minimisation are shown in Figure 2.18. Clearly, the *p*-nitro isomer, as shown in Figure 2.18(b), adopts an essentially planar configuration, while stereoelectronic effects operating in the *o*-isomer, Figure 2.18(a), cause the molecule to deviate significantly from planarity, in agreement with the reasoning presented previously in this chapter for dyes **2.2e** and **2.2f**, the diethylamino homologues of these two dyes.

2.9.4   QSAR and QSPR

An alternative approach towards understanding the properties of molecules involves the development of QSAR (quantitative structure–activity relationships) and QSPR (quantitative structure–property relationships).[55] These are mathematical models that attempt to relate the structure-derived features of a molecule to its activity, common in biological systems, or physical properties. QSAR and QSPR approaches operate on the assumption that structurally similar compounds have similar activities or properties. They may be used in the analysis of structural characteristics that give rise to certain properties, thereby providing the potential to predict properties. As illustrated in the flow-chart in Figure 2.19, the development of a QSAR/QSPR model for a series of molecules starts with experimental data collection for a particular property. Subsequently, a representation of the molecules is developed using molecular descriptors which are judged to be useful for the particular modelling task. For validation of the model, the data set is divided into a training set and a testing set. In the learning phase, various modelling methods, such as multiple linear regression, logistic regression and neural networks, are used to construct models that describe the empirical relationship between the molecular structure and the property of interest. A final model based on optimised parameters is then subjected to a validation

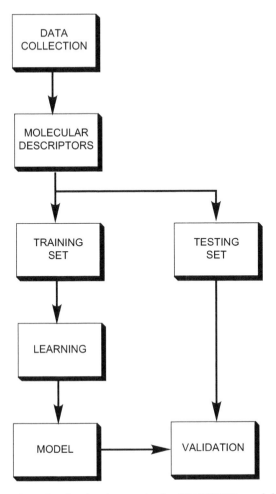

Figure 2.19 Flowchart for the development of a QSAR/QSPR model.

process with the testing set to ensure that the model is appropriate and useful. In recent years, growing numbers of reports have appeared on the use of these approaches applied to various features of the properties of dye molecules.[56] The approach has been used in some studies of dye–fibre interactions, for example to model the adsorption of anionic dyes, including direct dyes, on cellulosic fibres, in attempts to establish the relative importance of factors such as steric effects and electronic interactions in the binding between dye and fibre molecules.[57] As an example of the application in the use of the QSPR approach towards the development of relationships between molecular structure and colour properties of dyes, a series of nearly 200 azobenzene derivatives was investigated. A reasonable linear relationship was established between the N–N bond length and the λ_{max} values, while nonlinear relationships provided an even closer correlation.[58] Some studies have been reported on the development of QSAR models aimed at understanding the structure/toxicity relationships, an application for which the techniques appear to be well-suited. Some moderate success has been achieved in understanding the mutagenicity of series of azobenzene-based dyes and of aromatic amines.[59,60]

Computer-aided modelling approaches of various types, which were originally developed and refined in the search for new drug molecules, are thus now well established for the calculation of colour and other physical properties of individual dye and pigment molecules. However, the ability to predict the properties of pigments by calculation is significantly more complex because these colorants are applied as discrete crystalline particles, consisting of large numbers of molecules that are highly aggregated through strong intermolecular interactions. A modelling approach has been developed for pigments involving the principles of *crystal engineering*, as discussed further in Chapter 9.

2.10 COLOUR IN INORGANIC COMPOUNDS

The colours of inorganic compounds, as with organic molecules, are due to transitions between electronic energy levels as a result of the absorption of visible light. Several different mechanisms may be involved, as discussed in this section.

2.10.1 Colour in Metal Complexes (Coordination Compounds)

Complexes of the transition (d-block) metals with simple ligands are frequently coloured. These colours cannot be explained on the basis of the traditional valence bond (resonance) approach, in the way that has been discussed earlier in this chapter as applied to organic molecules. The explanation in this case is based on *crystal field theory*. This theory considers the effect of the surrounding ligands on the orbitals occupied by the d-electrons on the central transition metal ion. The three most common geometrical arrangements of the ligands in these metal complexes are octahedral (coordination number 6), tetrahedral and square planar (both coordination number 4). In each of these arrangements, the spherical field due to the five d-orbitals is distorted in such a way that they are no longer degenerate in terms of their energies. The effect of the spacial arrangement of the ligands is to split the orbitals into two (or more) levels with different energies. As a consequence, the transition metal complex may show colour if there is at least one electron in the lower level that may be promoted to the upper level, provided that there is a vacancy in the upper level. A simple example is the $Ti(H_2O)_6^{3+}$ ion. This complex cation contains a single d-electron (d^1 configuration), and as a result of a d–d (or crystal field) transition it appears red because of an absorption at 490 nm. In contrast, complexes of scandium, which has no d-electrons (d^0 configuration), and of zinc with its full complement of ten d-electrons (d^{10} configuration), are colourless, because no d–d electronic transitions are possible. The colours of many minerals and gemstones may be attributed to crystal field effects. Colours due to crystal field electronic transitions alone tend to be rather weak. Many chelated transition metal complexes, *i.e.*, coordination compounds involving ligands that possess more than one coordination site, are intensely coloured. This may be due to the π–π* transitions within coloured organic ligands, as exemplified by metal complex azo dyes (Chapter 3), phthalocyanines (Chapters 5 and 9) and the range of premetallised textile dyes (Chapter 7), or to ligand to metal charge transfer, as discussed in the next section.

2.10.2 Colour from Charge Transfer Transitions in Inorganic Materials

In contrast to the weak colours due to crystal field transitions, ligand to metal (L → M) charge transfer transitions give rise to extremely intense colours. These are transitions that take place in metal compounds whereby an electron in an orbital principally associated with the ligand is transferred, as a result of the energy provided by absorption of visible light, to an orbital that is mainly associated with the metal. Examples of inorganic species giving rise to charge transfer transitions include the deep purple permanganate ion (MnO_4^-) and the intense yellow chromate ion (CrO_4^{2-}), the latter being responsible for the colour of lead chromate pigments (Chapter 9). The deep red colour of complexes of iron with the thiocyanate ion (SCN^-) is also due ligand to metal charge transfer. Prussian blue is an example of mixed oxidation state transition metal compound, the colour of which is due to metal–metal electron transfer, in the case of this blue pigment from an Fe(II) atom to an adjacent Fe(III) atom (Chapter 9).

2.10.3 Colour in Inorganic Semiconductors

Cadmium sulfide is a yellow pigment of some commercial importance (Chapter 9). Since cadmium has a d^{10} electronic configuration, *i.e.*, a full shell of d electrons, the colour cannot be explained on the basis of crystal field effects. This is a prime example of an inorganic compound that owes its colour to its semiconductor behaviour.[61] In the ground state of these materials, an electron that undergoes the transition to a higher energy state, rather than occupying a discrete energy level, is part of a delocalised electron population that occupies a series of closely-spaced energy levels, referred to as the *valence band*. Absorption of light causes the electron to be promoted to an excited state, which consists of the electron population of the *conductance band*. In this case, the exciting photons of light need only the minimum energy that is required to overcome the gap between the valence and conductance bands. Consequently, photons of all energies above this minimum value will be absorbed. The electronic transition energy for cadmium sulfide is around 2.4 eV, corresponding to a wavelength of light of 517 nm, so that it absorbs all photons with energies of 2.4 eV or higher. Thus, it absorbs the entire blue–violet region of the spectrum, as well as green, and appears yellow. This type of electronic transition gives rise to steep reflectance curves, accounting for the bright colours of the pigments.

REFERENCES

1. M. P. Simunovic, *Eye*, 2010, **24**, 747.
2. K. Nassau (ed.), *Color in Science, Art and Technology*, Elsevier, Amsterdam, 1998.
3. D. A. Skoog, F. J. Holler and S. R. Crouch, *Principles of Instrumental Analysis*, Thomson Brooks/Cole, Pacific Grove, USA, 6th edn., 2007, pp. 169–173.
4. R. M. Christie, R. R. Mather and R. H. Wardman, *The Chemistry of Colour Application*, Blackwell Science, Oxford, 2000, ch. 8.
5. R. McDonald, *Colour Physics for Industry*, Society of Dyers and Colourists, Bradford, UK, 1997.
6. R. W. G. Hunt and M. R. Pointer, *Measuring Colour*, John Wiley & Sons Ltd., Chichester, 2011.
7. M. L. Gulrajani (ed.), *Colour Measurement: Principles, Advances and Industrial Applications*, Woodhead Publishing, Cambridge, 2010.
8. G. G. Guilbault, *Practical Fluorescence*, Marcell Dekker, New York, 1990.
9. J. R. Lakovicz, *Principles of Fluorescence Spectroscopy*, Springer, New York, 2006.
10. R. M. Christie, *Rev. Prog. Color.*, 1993, **23**, 1.

11. R. M. Christie, Fluorescent dyes, in *Handbook of Textile and Industrial Dyeing: Principles, Processes and Types of Dyes*, ed. M. Clark, Woodhead Publishing, Cambridge, 2011, vol 1, ch.17.

12. S. D. Higgins and A. D. Towns, *Chim. Oggi (Chem. Today)*, 2003, **21**, 13.

13. C. E. Moore, Luminescent organic pigments, in *The Pigment Handbook*, ed. P. A. Lewis, John Wiley & Sons, Inc., New York, 2nd edn, vol. 1, 1983, p. 859.

14. D. Barton and H. Davidson, *Rev. Prog. Color.*, 1974, **5**, 3.

15. A. E. Siegrist, H. Hefti, H. R. Meyer and E. Schmidt, *Rev. Prog. Color.*, 1987, **17**, 39.

16. M. Clark (ed.), *Handbook of Textile and Industrial Dyeing: Principles, Processes and Types of Dyes*, Woodhead Publishing, Cambridge, 2011, vols 1 and 2.

17. J. R. Aspland, *Textile Dyeing and Coloration*, American Association of Textile Chemists and Colourists, Raleigh, NC, 1997.

18. K. Hunger (ed.), *Industrial Dyes: Chemistry, Properties and Applications*, Wiley-VCH Verlag GmbH, Weinheim, 2nd edn, 2003.

19. W. Herbst and K. Hunger, *Industrial organic Pigments*, Wiley-VCH Verlag GmbH, Weinheim, 4th edn, 2006.

20. G. Buxbaum, *Industrial Inorganic Pigments*, Wiley-VCH Verlag GmbH, Weinheim, 2nd edn, 2008.

21. R. M. Christie, *The Organic and Inorganic Chemistry of Pigments*, Oil & Colour Chemsists Association, London, 2002.

22. *Colour Index International*, Society of Dyers and Colourists, Bradford, UK, 4th edn online. http://www.colour-index.com.

23. J. Griffiths, *Colour and Constitution of Organic Molecules*, Academic Press, London, 1976.

24. J. Griffiths, *Rev. Prog. Color.*, 1981, **11**, 7.

25. C. Graebe and C. Liebermann, *Ber.*, 1867, **1**, 106.

26. O. N. Witt, *Ber*, 1876, **9**, 522; 1888, **21**, 321.

27. J. T. Hewitt and H. V. Mitchell, *J. Chem. Soc. Trans.*, 1907, **91**, 1251.

28. W. Dilthey and R. Wizinger, *J. Prakt. Chem.*, 1928, **118**, 321.

29. C. R. Bury, *J. Am. Chem. Soc.*, 1935, **57**, 2115.

30. P. F. Gordon and P. Gregory, *Organic Chemistry in Colour*, Springer-Verlag, New York, 1983.

31. K. A. Bello and J. Griffiths, *J. Chem. Soc., Chem. Commun.*, 1986, 1639.

32. D. O. Hayward, *Quantum Mechanics for Chemists*, Royal Society of Chemistry, London, 2002.

33. J. J. Sakurai and J. J. Napolitano, *Modern Quantum Mechanics*, Addison-Wesley, London, 2010.

34. D. J. Griffiths, *Introduction to Quantum Mechanics*, Pearson Prentice Hall, London, 2005.

35. C. A. Coulson, B. O'Leary and R. B. Mallion, *Hückel Theory for Organic Chemists*, Academic Press, London, 1978.
36. K. Yates, *Hückel Molecular Orbital Theory*, Academic Press, New York, 1978.
37. M. Peric, I. Gutman and J. Radic-Peric, *J. Serb. Chem. Soc.*, 2006, **71**, 771.
38. J. Griffiths, *Dyes Pigments*, 1982, **3**, 211.
39. R. M. Christie and P. N. Standring, *Dyes Pigments*, 1989, **11**, 109.
40. W. Fabian, *Dyes Pigments*, 1985, **6**, 341.
41. Q. Xuhong, Z. Zhenghua and C. Kongchang, *Dyes and Pigments*, 1989, **11**, 13.
42. C. Lubai, C. Xing, H. Yufen and J. Griffiths, *Dyes and Pigments*, 1989, **10**, 123.
43. M. Adachi and S. Nakamura, *Dyes and Pigments*, 1991, **17**, 287.
44. A. El-Shafei, D. Hinks, H. S. Freeman and J. Lye, *AATCC Review*, 2001, **1**, 23.
45. G. Z. Li, J. Yang, H. F. Song, S. S. Yang, W. C. Lu and N. Y. Chen, *J. Chem. Information and Computer Sciences*, 2004, **44**, 2047.
46. J. Liu, Z. Chen and S. Yuan, *J. Zhejiang Univ., Sci. B*, 2005, **6**, 584.
47. W. Koch and M. C. Holthausan, *A Chemist's Guide to Density Functional Theory*, Wiley-VCH, Weinheim, 2000.
48. D. Stoll and J. A. Steckel, *Density Functional Theory: A Practical Introduction*, Wiley, 2009.
49. D. Guillaumont and S. Nakamura, *Dyes and Pigments*, 2000, **46**, 85.
50. D. Jacquemin, X. Assfeld, J. Preat and E. A. Perpete, *Mol. Phys.*, 2007, **105**, 325.
51. D. Jacquemin, E. Bremond, A. Planchat, I. Ciafini and C. Adamo, *J.Chem. Theory Comp.*, 2011, **42**, 1470.
52. A. El-Shafei, D. Hinks and H. S. Freeman, Molecular modelling and predicting dye properties, in *Handbook of Textile and Industrial Dyeing: Principles, Processes and Types of Dyes*, ed. M. Clark, Woodhead Publishing, Cambridge, 2011, vol. 1, ch. 7.
53. A. D. Laurent, C. Adamo and D. Jacquemin, *Phys. Chem. Chem. Phys.*, 2014, DOI: 10.1039/C3CP55336A.
54. P. Bladon, J. Gorton and R. B. Hammond, *Molecular Modelling: Computational Chemistry Demystified*, Royal Society of Chemistry, Cambridge, 2011.
55. F. Luan and M. N. Cordero, Overview of QSAR modelling in rational drug design', in *Recent Trends on QSAR in the Pharmaceutical Perceptions*, ed. M. T. H. Khan, Bentham Science Publishers, 2012, pp. 194–241.

56. F. Luan, X. Xu, H. Liu and M. N. Cordeiro, *Coloration Technol.*, 2013, **129**, 173

57. S. Timofei, W. Schmidt, L. Kurunkzi and Z. Simon, *Dyes Pigments*, 2000, **47**, 5.

58. B. Buttingsrud, B. K. Alsberg and P. O. Astrand, *Phys. Chem. Chem. Phys.*, 2007, **9**, 2226.

59. A. Garg, K. L. Bhat and C. W. Bock, *Dyes Pigments*, 2002, **55**, 35.

60. K. L. Bhat, S. Hayick, L. Sztandera and C. W. Bock, *QSAR Comb. Sci.*, 2005, **24**, 831.

61. J. Turley, *The Essential Guide to Semiconductors*, Prentice Hall, New York, 2002.

CHAPTER 3

Azo Dyes and Pigments

3.1 INTRODUCTION

Azo dyes and pigments constitute by far the most important chemical class of commercial organic colorant.[1–4] They account for around 60–70% of the dyes used in traditional textile applications (Chapters 7 and 8) and they occupy a similarly prominent position in the range of classical organic pigments (Chapter 9). Azo colorants, as the name implies, contain as their common structural feature the azo (–N=N–) linkage, which is attached at either side to two sp^2 carbon atoms. Usually, although not exclusively, the azo group links two aromatic ring systems. Most of the commercially important azo colorants contain a single azo group and are therefore referred to as monoazo dyes or pigments, but there are many that contain two (disazo), three (trisazo) or more such groups. In terms of their colour properties, azo colorants are capable of providing a virtually complete range of hues. There is no doubt though that they are significantly more important commercially in yellow, orange and red colours (*i.e.*, absorbing at shorter wavelengths), than in blues and greens. However, in more recent times the range of longer wavelength absorbing azo dyes has been extended, leading to the emergence of significant numbers of commercially important blue azo dyes as disperse dyes for application to polyester.[5] Indeed, there are even a few specifically-designed azo compounds that absorb beyond the visible in the near-infrared region of the spectrum.[6] The reader is directed to Chapter 2 for a discussion of colour/structure relationships in azo dyes. Azo dyes are capable of

Colour Chemistry, 2nd edition
By Robert M Christie
© R M Christie 2015
Published by the Royal Society of Chemistry, www.rsc.org

providing high intensity of colour, about twice that of the anthra-quinones for example (Chapter 4), and reasonably bright colours. They are capable of providing reasonable to very good technical properties, for example fastness to light, heat, water and other solvents, although in this respect they are often inferior to other chemical classes, for example carbonyl and phthalocyanine colorants, especially in terms of lightfastness.

Perhaps the prime reason for the commercial importance of azo colorants is that they are the most cost-effective of all the chemical classes of organic dyes and pigments. The reasons for this may be found in the nature of the processes used in their manufacture. The synthesis of azo colorants, which is discussed in some detail later in this chapter, brings together two organic components, a diazo component and a coupling component, in a two-stage sequence of reactions known as *diazotisation* and *azo coupling*. The versatility of the chemistry involved in this synthetic sequence means that an immense number of azo colorants may be prepared and this accounts for the fact that they have been adapted structurally to meet the requirements of most colour applications. On an industrial scale, the processes are straightforward, making use of simple multipurpose chemical plant. They are usually capable of production in high, often virtually quantitative, yields and the processes are carried out at or below ambient temperatures, thus presenting low energy requirements. The syntheses generally involve low cost, readily available commodity organic starting materials such as aromatic amines and phenols. The solvent in which the reactions are carried out is water, which offers obvious economic and environmental advantages over all other solvents. In fact, it is conceivable that azo dyes may assume even greater importance in the future as some of the other chemical types for which the synthetic routes do not offer such features, notably anthraquinones, become progressively less economic.[5] This chapter contains a discussion of the fundamental structural chemistry of azo colorants, including a description of the types of isomerism that they can exhibit, and the principles of their synthesis. In the final section, the ability of azo dyes to form metal complexes is discussed. Because of their prominence in most applications, numerous further examples of azo dyes and pigments will be encountered throughout this book.

3.2 ISOMERISM IN AZO DYES AND PIGMENTS

The structural chemistry of azo compounds is complicated by the possibilities of isomerism. Two types of isomerism are commonly

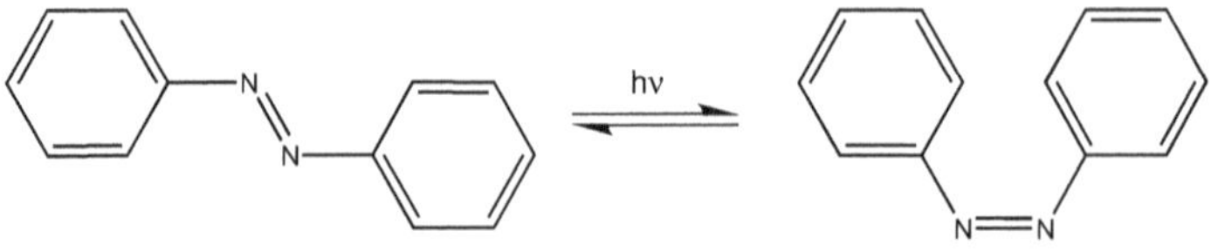

Figure 3.1   Photo-induced geometrical isomerism of azobenzene.

encountered with certain azo compounds: geometrical isomerism and tautomerism.

Some simple azo compounds, because of restricted rotation about the (–N=N–) double bond, are capable of exhibiting geometrical isomerism. The geometrical isomerism of azobenzene, the simplest aromatic azo compound, which may be considered as the parent system on which the structures of most azo colorants are based, is illustrated in Figure 3.1. The compound is only weakly coloured because it absorbs mainly in the UV region with a λ_{max} value of 320 nm in solution in ethanol, a feature that may be attributed to the absence of substituents that can act as auxochromes (Chapter 2). The compound exists normally as the trans, or (*E*), isomer, **3.1a**. This molecule is essentially planar in the solid state and probably also in solution and in the gas phase, a feature that is supported by calculations based on time-dependent density functional theory (TD-DFT).[7] When irradiated with UV light, the (*E*)-isomer undergoes conversion substantially into the cis or (*Z*)-isomer, **3.1b**, which may be isolated as a pure compound. In the dark, the (*Z*)-isomer reverts thermally to the (*E*)-isomer, which is thermodynamically more stable because of reduced steric congestion.[8] Some early disperse dyes, which were relatively simple azobenzene derivatives introduced commercially initially for application to cellulose acetate fibres, were found to be prone to *photochromism* (formerly referred to as *phototropy*), a reversible light-induced colour change. CI Disperse Red 1, **3.2**, is an example of a dye that has been observed under certain circumstances to give rise to this phenomenon.[9] When this dye is applied to cellulose acetate substrates, the colour weakens and changes in hue on exposure to direct sunlight. The original colour is restored when the material is kept in the dark.

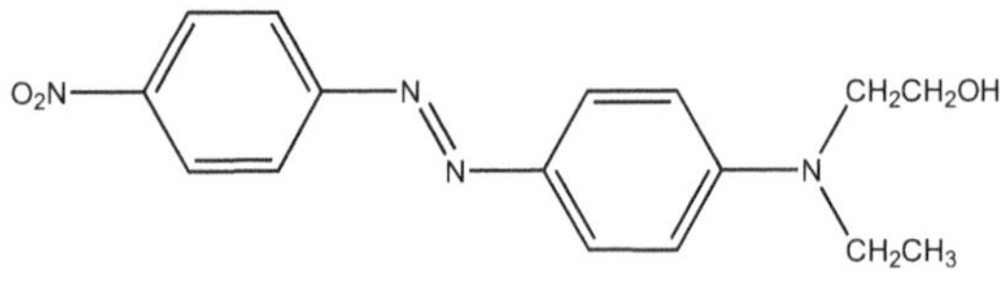

3.2

The phenomenon is explained by the colour changes associated with the (*E/Z*) isomerisation process, which may be initiated by exposure to light; the two geometrical isomers of the azo dyes have noticeably different colours. The effect, which is considered as a defect in traditional dyeing, is no longer encountered to any significant extent in the modern range of azo disperse dyes as the offending dyes have been progressively replaced by more stable products. Interestingly, however, there has been a significant revival of interest in dyes that can exhibit photochromism, especially when it is controllable and reversible, mainly for ophthalmic sun-screening applications such as the familiar spectacles that darken to become sunglasses on exposure to sunlight. The most important industrial photochromic dyes of this type are the spirooxazines and naphtho-pyrans (see Chapter 11 for further discussion on these and other chromic materials).

Many commercial azo colorants contain a hydroxyl group *ortho* to the azo group. As illustrated in Figure 3.2, this gives rise to intra-molecular hydrogen-bonding, which further stabilizes the (*E*)-isomer and effectively prevents its conversion into the (*Z*)-form.

Another important feature of azo compounds in which there is a hydroxyl group conjugated with (*i.e.*, *ortho* or *para* to) the azo group, and a large number of commercial azo dyes and pigments show this structural feature, is that they can exhibit tautomerism. The classical experimental evidence for hydroxyazo/ketohydrazone tautomerism in dyes of this type was provided by Zincke.[10] In 1884, he reported the reactions of benzenediazonium chloride with 1-naphthol and of phenylhydrazine with naphtho-1,4-quinone. It might have been expected that the former reaction would give hydroxyazo compound **3.3a**, and the latter would give the ketohydrazone tautomer **3.3b**. In the event, it was found that both reactions gave the same product, a tautomeric mixture of **3.3a** and **3.3b** (Scheme 3.1). Such tautomeric forms may nowadays be readily identified by their distinctive UV/visible, infrared and ^{1}H, ^{13}C and ^{15}N NMR spectral characteristics in solution,[11] and by X-ray crystallography in the solid state. In terms of

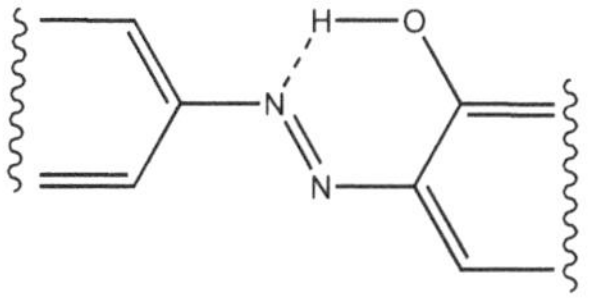

Figure 3.2 Intramolecular hydrogen bonding in *o*-hydroxyazo compounds.

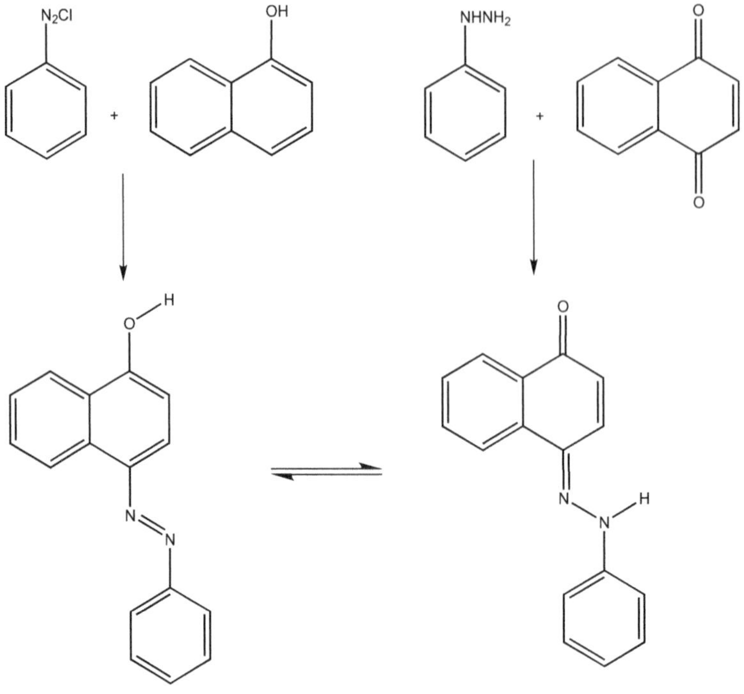

Scheme 3.1   Reaction scheme that provides evidence for tautomerism in some hydroxyazo dyes.

colour properties, the ketohydrazone isomers are usually bathochromic compared with the hydroxyazo forms, and give higher molar extinction coefficients.

Figure 3.3 shows the tautomeric forms of a further range of hydroxyazo dyes. In solution, the tautomers exist in rapid equilibrium although commonly one or other of the tautomers is found to predominate to an extent dependent on their relative thermodynamic stabilities.[12,13] In the case of 2-phenylazophenol, the azo isomer **3.4a** predominates over the hydrazone isomer **3.4b**. On the basis of a consideration solely of the summation of all of the theoretical bond energies in the molecule, it has been suggested that the hydrazone isomer would be expected to be more stable. However, this is outweighed by the reduced resonance stabilization energy in the hydrazone form, due to the loss of the aromatic character of one ring. In the case of hydroxyazonaphthalenes, 4-phenylazo-1-naphthol, **3.3**, and 1-phenylazo-2-naphthol, **3.5**, the reduction in the resonance stabilization energy in the ketohydrazone forms is due to the loss of

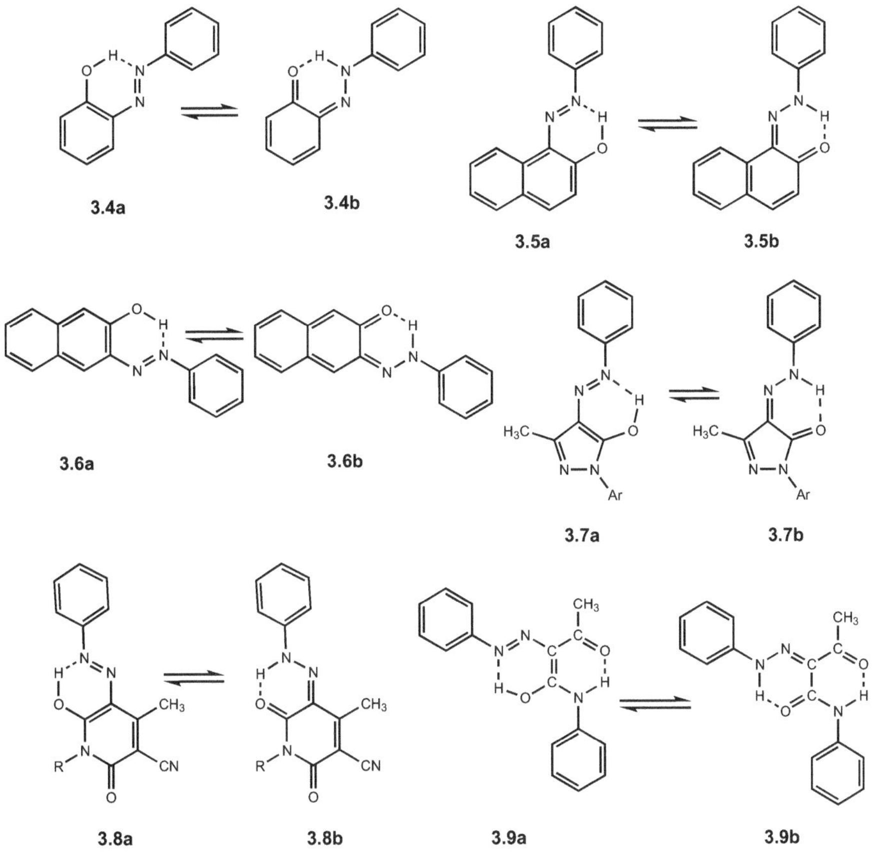

Figure 3.3 Tautomeric forms of azo dyes **3.4–3.9**.

aromaticity in only one of the naphthalene rings. In relative terms, the reduction in energy is less than in the benzene series, so that the two tautomers become much closer in stability. The position of the equilibrium, particularly in the case of compound **3.3**, then becomes heavily dependent on the environment in which the dye finds itself, most importantly on the nature of the solvent. Interestingly, 3-phenylazo-2-naphthol, **3.6**, exists exclusively in the azo form **3.6a**, because of the instability of hydrazone form **3.6b** in which there is complete loss of aromatic character of the naphthalene system. In the case of azo dyes derived from heterocyclic coupling components, such as the azopyrazolones, **3.7**, and the azopyridones, **3.8**, and from β-ketoacid-derived coupling components, such as the azoacetoacetanilides, **3.9**, the compounds exist exclusively in the hydrazone forms, **3.7b**, **3.8b** and **3.9b**, respectively. X-Ray crystallographic structure determinations have been carried out on a wide range of industrial azo pigments, and these compounds have been shown invariably to exist exclusively in the ketohydrazone form in the solid state (for a more detailed discussion of this feature see Chapter 9).

In *o*-hydroxyazo compounds, but not in *p*-hydroxyazo compounds, there is strong intramolecular hydrogen bonding both in the hydroxyazo and ketohydrazone forms, as illustrated for compounds **3.4**–**3.9** (Figure 3.3). A factor that contributes to the explanation for the predominance of the ketohydrazone isomer in many cases is that in this form the intramolecular hydrogen-bonding is significantly stronger than in the hydroxyazo form, because of the higher bond polarities in the ketohydrazone system. This may provide an explanation for the observations that in the case of 1-phenylazo-2-naphthol, **3.5**, the hydrazone form **3.5b** predominates while in the case of the isomeric 4-phenylazo-1-naphthol, **3.3**, in which there is no intramolecular hydrogen bonding, the two tautomers are of comparable stability. Intramolecular hydrogen-bonding is a commonly encountered stabilizing feature in a wide range of dyes and pigments of most chemical types, enhancing many useful technical properties. For example, lightfastness is usually improved. This has been explained by a reduction in the electron density at the azo group due to hydrogen bonding, which decreases its sensitivity towards photochemical oxidation. An additional positive feature of intramolecular hydrogen bonding is that it reduces the acidity of the hydroxyl group involved, and this consequently provides the dye with enhanced stability towards treatments with aqueous alkali such as used, for example, in the laundering of textiles.

3.3   SYNTHESIS OF AZO DYES AND PIGMENTS

Textbooks in general organic chemistry will illustrate that there are many ways of synthesizing azo compounds. However, almost without exception, azo dyes and pigments are made on an industrial scale by the same two-stage reaction sequence: diazotisation and azo coupling, as illustrated in Scheme 3.2.[1–4,14]

The first stage, diazotisation, involves the treatment of a primary aromatic amine, referred to as the *diazo component*, with sodium nitrite under conditions of controlled acidity and at relatively low

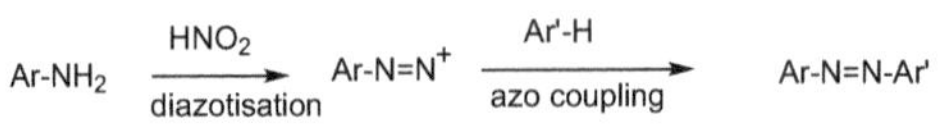

$$Ar\text{-}NH_2 \xrightarrow[\text{diazotisation}]{HNO_2} Ar\text{-}N\text{≡}N^+ \xrightarrow[\text{azo coupling}]{Ar'\text{-}H} Ar\text{-}N\text{=}N\text{-}Ar'$$

Scheme 3.2   Synthesis of azo colorants.

temperatures to form a diazonium salt. In the second stage of the sequence, azo coupling, the relatively unstable diazonium salt thus formed is reacted with a *coupling component*, which may be a phenol, an aromatic amine or a β-ketoacid derivative, to form the azo dye or pigment. The next two sections of this chapter deal separately with these two reactions, with emphasis on the practical conditions used for the reactions and on the reaction mechanisms. This sequence of reactions provides an interesting illustration for students of organic chemistry of the ways in which the selection of the optimum practical conditions for the reactions is heavily influenced by consideration of the reaction mechanisms that are operating.

3.3.1 Diazotisation

Diazotisation, the first stage of azo dye and pigment synthesis, involves the treatment of a primary aromatic amine $(ArNH_2)$, which may be carbocyclic or heterocyclic, with nitrous acid to form a diazonium salt $(ArN_2^+Cl^-)$.[14] Nitrous acid, HNO_2, is a rather unstable substance decomposing relatively easily by dissociation into oxides of nitrogen. It is therefore usually generated in the reaction mixture as required by treating sodium nitrite, a stable species, with a strong acid. The mineral acid of choice for many diazotisations is hydrochloric acid. This is because the presence of the chloride ion can exert a catalytic effect on the reaction under appropriate conditions, thus enhancing the reaction rate. Most primary aromatic amines undergo diazotisation with little interference from the presence of other substituents, although these may influence the reaction conditions required. When the reaction conditions are carefully controlled, diazotisation usually proceeds smoothly and in virtually quantitative yield. It is of considerable industrial importance that the reaction may be carried out in water, the reaction solvent of choice for obvious economic and environmental reasons.

Diazotisation is always carried out under strongly acidic conditions, but control of the degree of acidity is of particular importance in ensuring smooth reaction. The overall reaction equation for the diazotisation reaction using sodium nitrite and hydrochloric acid may be given as:

$$ArNH_2 + NaNO_2 + 2HCl \rightarrow ArN_2^+Cl^- + H_2O$$

Reaction stoichiometry therefore requires the use of two moles of acid per mole of amine. However, for several reasons, a somewhat greater excess of acid is generally used. One reason is that highly acidic conditions favour the generation from nitrous acid of the reactive nitrosating species that are responsible for the reaction, a feature which will emerge from the discussion of the reaction mechanism later in this chapter. A second reason is that acidic conditions suppress the formation of triazines as side-products, which may be formed as a result of N-coupling reactions between the diazonium salts and the aromatic amines from which they are formed. A practical reason for the use of acidic conditions is to convert the insoluble free amine ($ArNH_2$) into its water-soluble protonated form ($ArNH_3^+Cl^-$). However, too strongly acidic conditions are avoided so that the position of the equilibrium is not too far in favour of the protonated amine and allows a reasonable equilibrium concentration of the free amine, which under most conditions is the reactive species, as discussion of the reaction mechanism will demonstrate. There is therefore an optimum level of acidity for the diazotisation of a particular aromatic amine, which depends on the basicity of the amine in question. In the case of aniline derivatives, electron-withdrawing groups, such as the nitro group, reduce the basicity of the amino group. As a consequence, for example, the diazotisation of 4-nitroaniline requires much more acidic conditions than aniline itself. Very weakly basic amines, such as 2,4-dinitroaniline, require extremely acidic conditions. They are usually diazotised using a solution of sodium nitrite in concentrated sulfuric acid, which forms nitrosyl sulfuric acid ($NO^+HSO_4^-$). Heterocyclic aromatic amines, such as aminothiophenes, aminothiazoles and aminobenzothiazoles, are of importance as diazo components particularly in the synthesis of azo disperse dyes. Diazotisation of these amines can prove problematic. Generally, the use of concentrated acids is required due to the reduction in basicity of the amine by the heterocyclic system and as a result of protonation of heterocyclic nitrogen atoms, and also because the diazonium salts are sensitive to hydrolysis in dilute acids.

It is especially important to control the acidity when aromatic diamines are treated with nitrous acid to form either the mono- or bis-diazonium salts, a process of some importance in the synthesis of disazo dyes and pigments (see later). *p*-Phenylenediamine is an example of a diamine in which either one or both of the amino groups may be diazotised by careful selection of reaction conditions. The use of dilute hydrochloric acid can result in smooth formation of the monodiazonium salt. The use of nitrosyl sulfuric acid is required to diazotise the second amino group, since the strong electron-withdrawing effect of the diazo group in the monodiazonium salt reduces the basicity of the amino group which remains.

It is critically important in diazotisation reactions to maintain careful control of the temperature of the reaction medium. The reactions are generally carried out in the temperature range 0–5 °C, necessitating the use of ice-cooling. In certain cases, for example with some heterocyclic amines, even lower temperatures are desirable, although temperatures that are too low can cause the reactions to become impracticably slow. Efficient cooling is therefore essential, not least because the reactions are invariably strongly exothermic. One reason for the need for low temperatures is that higher temperatures promote the decomposition of nitrous acid, giving rise to the formation of oxides of nitrogen. In the industrial manufacture of azo dyes and pigments, the presence of these strongly oxidising species in the exhaust gases is highly undesirable as they can cause combustible materials with which they may come into contact to catch fire, with potentially disastrous consequences! The main reason for maintaining low temperatures, however, is the instability of diazonium salts. The diazonium cation, although stabilised by resonance, decomposes readily with the evolution of nitrogen, the principal decomposition product being the phenol, from reaction with water (Scheme 3.3).

Normally, amines are diazotised using a *direct* method that involves the addition of sodium nitrite solution to an acidic aqueous solution of the amine. Aromatic amines that also contain sulfonic acid groups, for example, sulfanilic acid (4-aminobenzene-1-sulfonic acid), are widely used in the synthesis of water-soluble azo dyes and of metal salt azo pigments. Because these amines often dissolve with difficulty in aqueous acid, they are commonly diazotised using an *indirect* method, which involves dissolving the compound in aqueous alkali as the sodium salt of the sulfonic acid, introducing the appropriate quantity of sodium nitrite and then adding this combined solution with stirring to the dilute acid, ensuring that a low temperature is maintained by ice cooling.

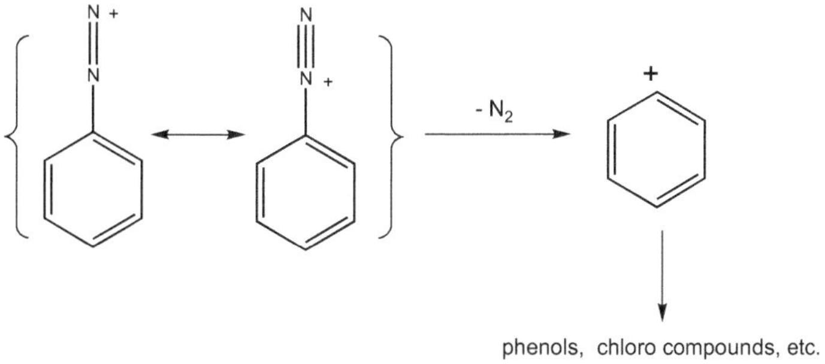

phenols, chloro compounds, etc.

Scheme 3.3 Thermal decomposition of diazonium salts.

The quantity of sodium nitrite used in diazotisation is usually the equimolar amount required by reaction stoichiometry or a very slight excess. A large excess of nitrite is avoided because of the instability of nitrous acid and since high concentrations can promote diazonium salt decomposition. When direct diazotisation is used, the sodium nitrite is usually added at a controlled rate such that a slight excess is maintained throughout the course of the reaction. In practice, this is monitored easily by the characteristic blue colour nitrous acid gives with starch/potassium iodide paper. When diazotisation is judged to be complete, any remaining nitrous acid excess is usually destroyed prior to azo coupling to avoid side-products due to C-nitrosation of the coupling components. This is commonly achieved by addition of sulfamic acid, which reacts as follows:[15]

$$NH_2SO_3H + HNO_2 \rightarrow N_2 + H_2SO_4 + H_2O$$

Because diazonium salts are relatively unstable species, they are almost always prepared in solution as required and used immediately to synthesise an azo dye or pigment. It is generally inadvisable to attempt to isolate diazonium chlorides as they may decompose explosively in the solid state. It is, however, possible to prepare stabilised diazonium salts, which may be handled reasonably safely in the solid state. This is achieved by the use of alternative counteranions, which are much larger in size and less nucleophilic than the chloride anion. The most commonly used stabilised diazonium salts are derived from tetrafluoroborates (BF_4^-), tetrachlorozincates $(ZnCl_4^{2-})$ and salts obtained from the di-anion of naphthalene-1,5-disulfonic acid. One use of stabilised diazonium salts is in the *azoic dyeing* of cotton. This process involves the impregnation of cotton fibres with a solution of a coupling component and subsequent treatment with a solution of the stabilised diazonium salt to form an azo pigment, which is trapped mechanically within the cotton fibres. Azoic dyeing, which many years ago was an important method for producing washfast dyeings on cotton, is of relatively limited importance today having largely been superseded by processes such as vat dyeing and reactive dyeing (Chapter 7).

An outline general mechanism for the diazotisation of an aromatic amine is given in Scheme 3.4.[16] The first stage in the reaction is *N*-nitrosation of the amine, the nitrosating species being represented in the scheme as Y–N=O. It has been shown that various species may be responsible for nitrosation, depending on the nature of the aromatic amine in question and on the conditions employed for the reaction. The nitrosating species may be the nitroso-acidium ion $(H_2O^+\text{-NO})$,

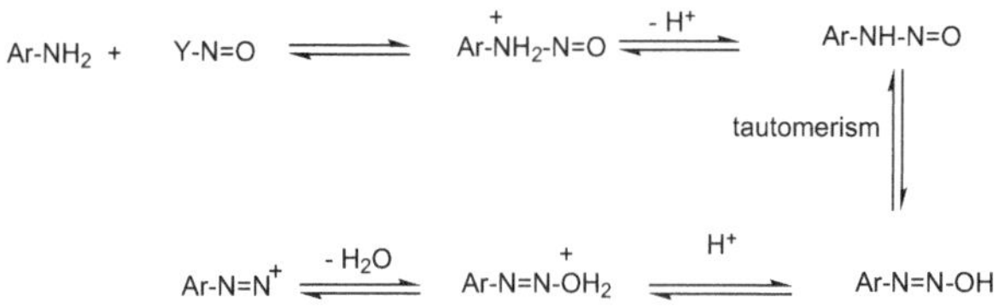

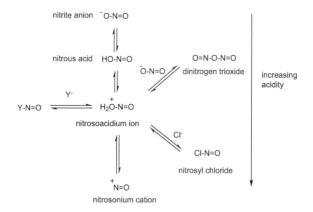

Scheme 3.4 Mechanism of diazotisation.

Scheme 3.5 Equilibria involved in the formation of nitrosating species from nitrous
acid.

nitrosyl chloride (NOCl), dinitrogen trioxide (N_2O_3) or the nitrosonium
cation (NO^+). The formation of each of these species from nitrous acid
is illustrated in the series of equilibria shown in Scheme 3.5.

The diazotisation reaction provides a classical example of the
application of physical chemistry in the elucidation of the detail of
organic reaction mechanisms. In particular, the results of studies of
the kinetics of diazotisation have proved especially informative
in establishing the nature of the nitrosating species and the rate-
determining step for particular cases. For diazotisation in dilute acids
such as sulfuric and perchloric, where the anion is relatively weakly
nucleophilic, the nitrosating species has been shown to be dinitrogen
trioxide under conditions of low acidity, and the nitroso-acidium ion
at higher acidities. In hydrochloric acid, the rate of diazotisation
shows a marked increase as a result of catalysis by the chloride anion.
The kinetics of the reaction in this last case is consistent with a
mechanism involving nitrosation of the free amine by reaction with
nitrosyl chloride.

In most of the cases discussed so far, the rate-determining step of
the reaction is nitrosation of the free amine. When diazotisation is

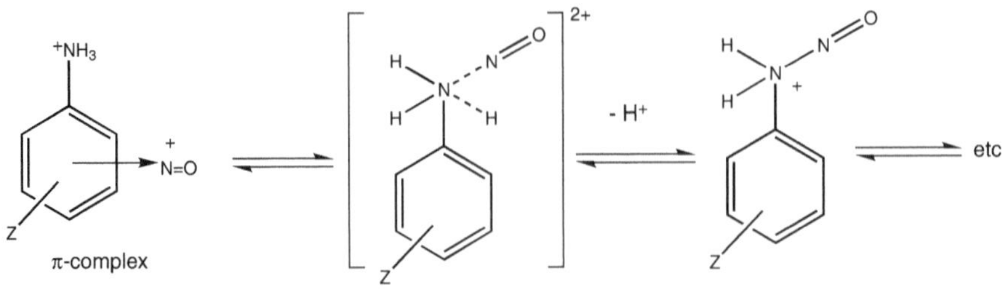

Scheme 3.6 Mechanism of nitrosation in concentrated acids.

carried out in concentrated acids, the nitrosonium cation, NO^+, is the nitrosating species. In this case, an exchange mechanism has been proposed, as illustrated in Scheme 3.6, in which the initial step is reaction of the nitrosonium cation with the protonated amine to form a π-complex. The deprotonation step that follows the exchange becomes rate-determining.

3.3.2   Azo Coupling

Azo coupling is an example of aromatic electrophilic substitution in which the electrophile is the diazonium cation, ArN_2^+.[10,17–19] Electrophilic substitution reactions, of which nitration, sulfonation and halogenation are arguably the best-known examples, are the most commonly encountered group of reactions of aromatic systems. However, the diazonium cation is a relatively weak electrophile and will therefore only react with aromatic systems that are highly activated to electrophilic attack by the presence of strongly electron-releasing groups. The most commonly encountered strongly electron-releasing groups are the hydroxyl and amino groups, and this in turn means that the most common compounds that are capable of undergoing azo coupling, referred to as *coupling components*, are either phenols or aromatic amines (primary, secondary or tertiary). There is a third type of coupling component, commonly a β-ketoacid derivative, in which coupling takes place at a reactive methylene group. The azo coupling reaction between benzenediazonium chloride and phenol is illustrated in Scheme 3.7.

   To ensure that the azo dyes and pigments are obtained in high yield and purity, careful control of experimental conditions is essential to minimise the formation of side products. It is a useful feature of both diazotisation and azo coupling reactions that they may be carried out in water as the reaction solvent. Temperature control, which is so critical in diazotisation reactions, is generally less important in the case of azo coupling. The reactions are normally carried out at or just below ambient temperatures. There is usually little advantage in

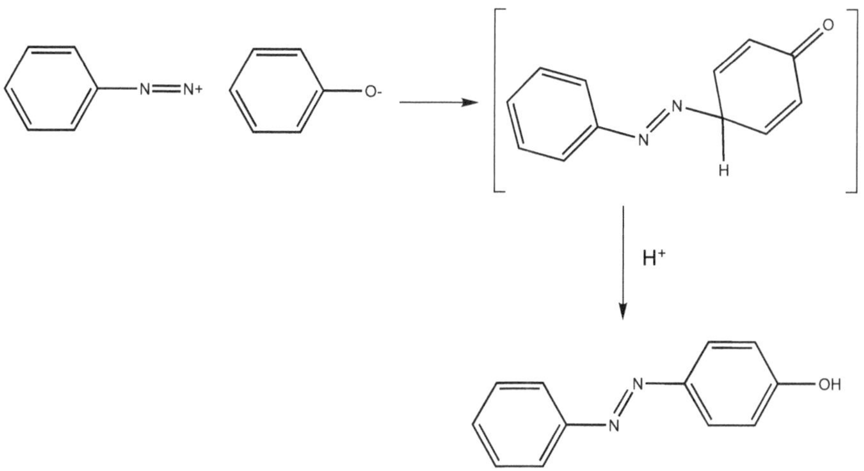

Scheme 3.7 Azo coupling reaction between benzenediazonium chloride and phenol.

raising the temperature, other than in a few special cases, since this tends to increase the rate of diazonium salt decomposition more than the rate of coupling.

The experimental factor that requires most careful control in azo coupling is pH. There is usually an optimum pH range for a specific azo coupling reaction, which is principally dependent on the particular coupling component used. Phenols are usually coupled under alkaline conditions, in which case the phenol (ArOH) is converted predominantly into the phenolate anion (Ar–O⁻). There are two reasons why this facilitates the reaction. This first is a practical reason in that the anionic species is more water-soluble than the phenol itself. A second and arguably more important reason is that the –O⁻ group is more powerfully electron-releasing than the –OH group itself and hence much more strongly activates the system towards electrophilic substitution. Strongly alkaline conditions are generally avoided, however, as they promote diazonium salt decomposition. In addition, these conditions can cause conversion of the diazonium cation (ArN_2^+) into the diazotate anion (Ar–N=N–O⁻), a species that is significantly less reactive than the diazonium cation in azo coupling. Generally, it is desirable to carry out the reaction at the lowest pH at which coupling takes place at a reasonable rate. In the case of aromatic amines as coupling components, weakly acidic to neutral conditions are commonly used. The pH is selected such that the amine is converted substantially into the more water-soluble protonated form $(ArNH_3^+)$, but at which there is a significant equilibrium concentration of the free amine $(ArNH_2)$, which is more reactive to towards azo coupling. Reactive methylene-based coupling components undergo azo coupling *via* the enolate anion, the concentration of which increases with increasing pH. These compounds are also frequently coupled at weakly acidic to neutral pH values, under which conditions a sufficiently high concentration of enolate anion exists for the reaction to proceed at a reasonable rate and side-reactions due to diazonium salt decomposition are minimised. Commonly, the rate of addition of the diazonium salt solution to the coupling component is controlled carefully to ensure that an excess of diazonium salt is never allowed to build up in the coupling medium, in order to minimise side-reactions due to diazonium salt decomposition, especially when higher pH conditions are required. This is especially important in the synthesis of azo pigments, insoluble compounds from which the removal of impurities is difficult.

Figure 3.4 illustrates the structures of a selected range of coupling components commonly used in the synthesis of azo dyes and pigments. In the figure, the position(s) at which azo coupling normally takes place is also indicated. The coupling position is governed by the normal substituent directing effects, both electronic and steric, that

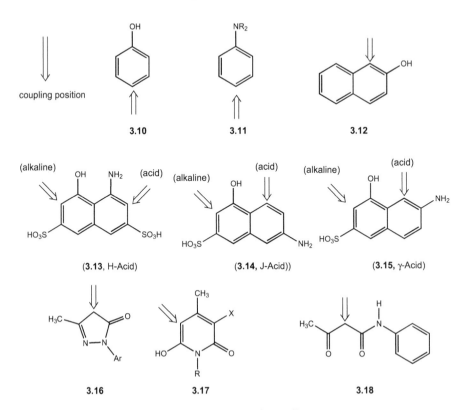

Figure 3.4   Structures of some commonly-used coupling components.

are encountered for aromatic electrophilic substitution. These effects, together with other aspects of the reaction mechanisms involved in electrophilic aromatic substitution, are dealt with at length in most organic chemistry textbooks, and so are not considered further here. The coupling components include the relatively simple benzene derivatives, phenol, **3.10**, and aniline, **3.11**, naphthalene derivatives **3.12–3.15**, some heterocyclic compounds such as the pyrazolones **3.16** and pyridones **3.17**, while the β-keto acid derivatives are exemplified by acetoacetanilide **3.18**. Many coupling components, such as compounds **3.10–3.12** and **3.16–3.18,** are capable of a single azo coupling reaction to give a monoazo colorant. Several coupling components, for example naphthalene derivatives **3.13–3.15**, contain both an amino and a hydroxyl group in separate rings. These compounds are useful because they are capable of reacting twice with diazonium salts, thus providing a route to disazo colorants. As an example, 1-amino-8-hydroxynaphthalene-3,6-disulfonic acid, **3.13**, referred to trivially as H-Acid, is used in the synthesis of a number of important azo dyes. The position of azo coupling with this coupling component may be controlled by careful choice of pH. Under alkaline conditions, the hydroxyl group is converted into the phenolate anion $(-O^-)$, which is more electron releasing than the amino group. In contrast, under weakly acidic conditions it exists un-ionised as the –OH group, which is less electron-releasing than the amino group.[20] A feature of naphthalene chemistry is that a substituent exerts its maximum electronic effect in the ring to which it is attached, so that under alkaline conditions azo coupling is directed into the ring containing the hydroxyl group, while under weakly acidic conditions reaction takes place in the ring containing the amino group. This selectivity of the coupling position, also shown by coupling components such as J-acid, **3.14**, and γ-acid, **3.15**, allows the preparation of a range of unsymmetrical disazo dyes by a controlled sequential procedure involving two separate azo coupling reactions.

3.4 STRATEGIES FOR AZO DYE AND PIGMENT SYNTHESIS

In the case of monoazo dyes and pigments, the strategy for synthesis is straightforward, involving appropriate selection of diazo and coupling components, and choice of reaction conditions in accordance with the chemical principles presented in the previous two sections of this chapter. For azo colorants containing more than one azo group, the situation is more complex and it becomes even more critical that the synthetic strategy and reaction conditions are selected carefully to ensure that a pure product is obtained in high yield.

A system, in common use as a systematic treatment of the possible synthetic strategies leading to polyazo compounds, has been proposed by the Society of Dyers and Colourists (SDC).[21] The system has undoubted merit as a method of classification, although rigorous justification of some of the symbolism used may be questioned. To describe the strategies, it is appropriate at this stage to define the nature of the various reacting species as follows:

A: primary aromatic **A**mine: *i.e.*, a normal diazo component;
D: primary aromatic **D**iamine: *i.e.*, a tetrazo component;
E: coupling component capable of reaction with one diazonium ion: an **E**nd component;
Z: coupling component capable of reaction with more than one diazonium ion;
M: coupling component containing a primary aromatic amino group that, after an azo coupling reaction, may be diazotised and used in a second azo coupling: a **M**iddle component.

3.4.1 Synthesis of Monoazo Dyes and Pigments

The synthesis of monoazo dyes and pigments is represented by the symbolism:

$$A \rightarrow E$$

In using this terminology, it should be emphasised that the arrow has the meaning 'diazotised and then coupled with' rather than its usual meaning in organic reaction sequences. Thus, for example, the first stage in the synthesis of CI Disperse Orange 25, **3.19**, is the diazotisation of 4-nitroaniline, using sodium nitrite and aqueous hydrochloric acid at temperatures less than 5 °C. The diazonium salt thus formed is reacted with *N*-ethyl-*N*-β-cyanoethylaniline, under weakly acidic conditions since the coupling component is a tertiary aromatic amine.

3.19

3.4.2 Synthesis of Disazo Dyes and Pigments

The situation becomes more complex when two separate diazotisations and azo couplings are required. Four separate strategies leading to disazo colorants may be identified, using the SDC terminology and symbolism, exemplified as follows as strategies (a)–(d).

(a) $A^1 \rightarrow Z \leftarrow A^2$

In this the first strategy, two primary aromatic amines (A^1 and A^2) are diazotised and reacted separately under appropriate pH conditions and in an appropriate sequence with a coupling component that has two available coupling positions. As an example, the synthesis of CI Acid Black 1, **3.20**, a bluish-black dye commonly used to in the dyeing of wool, is as follows. 4-Nitroaniline is diazotised under the usual conditions and the diazonium salt reacted in the first coupling reaction under weakly acidic conditions with H-Acid, **3.13**, under which conditions azo coupling is directed into the ring containing the amino group. Secondly, aniline is diazotised and the resulting diazonium salt reacted with the monoazo intermediate to form the disazo dye. For the second coupling, alkaline conditions are used since a phenolic coupling component is involved. In general, reactions of this type are carried out in this sequence because of the good solubility of the monoazo intermediate in alkali and the improved selectivity of the process when carried out in this way.

3.20

(b) $E^1 \leftarrow D \rightarrow E^2$

In this strategy, a primary aromatic diamine is diazotised twice (tetrazotised) and coupled separately with two coupling components (E^1 and E^2). In the case of CI Pigment Yellow 12, **3.21**, which is an important bright yellow pigment used extensively in printing inks, the product is symmetrical (a bis-ketohydrazone). This greatly simplifies the synthetic procedure since the two coupling reactions may be carried out simultaneously. 3,3′-Dichlorobenzidine (1 mole) is tetrazotised (bis-diazotised) and coupled under weakly acidic to neutral conditions with acetoacetanilide, **3.18**, (2 moles) to give the product directly. When an unsymmetrical product is required, a much more careful approach to the synthesis is essential to ensure that the unsymmetrical product is not contaminated by quantities of the two possible symmetrical products. One approach that may be used involves the diazotisation, by careful choice of conditions, of one amino group of the diamine D followed by coupling with the first component E^1. The monoazo intermediate formed contains the second amino group, which may be diazotised followed by coupling with component E^2. Alternatively, the synthesis may be achieved by tetrazotisation of the diamine followed by two sequential azo coupling reactions with careful selection of pH conditions to control the outcome of the reactions. In the case of CI Direct Blue 2, **3.22**, for example, the synthesis could start, in principle, with the tetrazotisation of benzidine (4,4'-diaminobiphenyl). The resulting tetrazonium salt is first coupled with γ-Acid, **3.15**, under acidic conditions and then the monoazo intermediate is reacted with H-Acid, **3.13**, under alkaline conditions. This particular example, by way of illustration, uses benzidine as the tetrazo component. Formerly, benzidine was an important tetrazo component, particularly for the manufacture of direct dyes for the dyeing of cotton. However, this compound has for many years now been recognised as a potent human carcinogen and its use in colour manufacture has long since ceased in the developed world (Chapter 12).

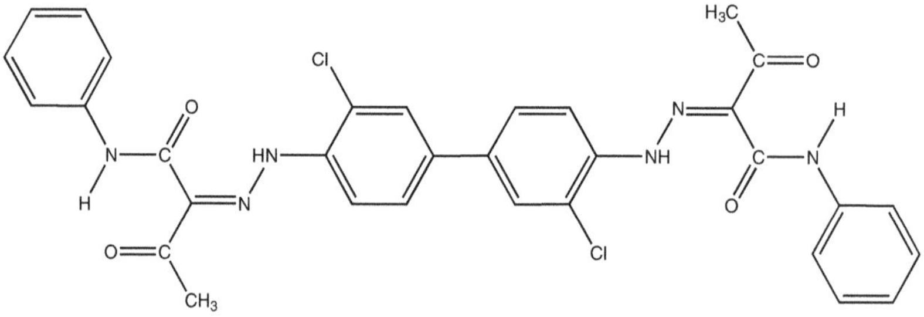

3.21

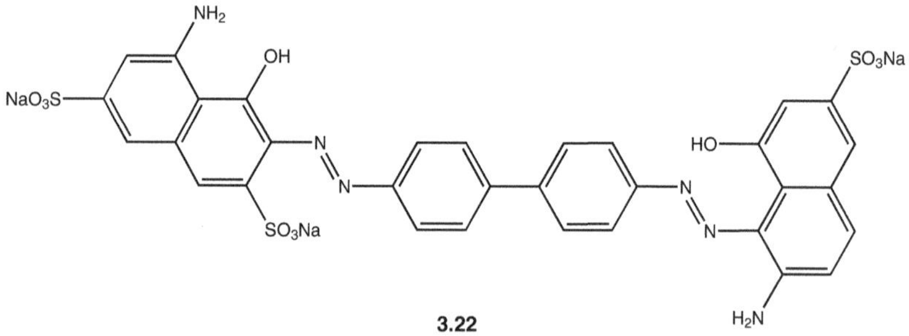

3.22

(c) A→M→E

This third strategy in azo colorant synthesis makes use of the feature that a primary aromatic amine has the potential to be used both as a coupling component and as a diazo component. As an example, in the synthesis of CI Disperse Yellow 23, **3.23**, aniline, as the diazo component, is first diazotised and the resulting diazonium salt reacted under weakly acidic conditions with aniline, as the coupling component, to give 4-aminoazobenzene. A practical complication with this stage is the inefficiency of the coupling reaction with aniline, and the formation of side-products due to an N-coupling reaction. In an improved method, aniline is first reacted with formaldehyde and sodium bisulfite to form the methyl-ω-sulfonate derivative (ArNHCH$_2$SO$_3$Na), which couples readily. After coupling, the labile methyl-ω-sulfonate group may be removed easily by acid hydrolysis to give 4-aminoazobenzene. This amine is then diazotised and the resulting diazonium salt is reacted with phenol under alkaline conditions to give the disazo dye **3.23**.

3.23

(d) A^1→Z–X–Z←A^2

The products from the fourth strategy are structurally not dissimilar from those obtained by the strategy given in strategy (b). However, in this strategy the disazo colorant is synthesised by linking together two molecules of a monoazo derivative by some chemical means. As an example, the synthesis of CI Pigment Red 166, **3.24**, a disazo condensation pigment that exists structurally as a bis-ketohydrazone, is shown in Scheme 3.8. The monoazo compound **3.26** containing a carboxylic acid group is prepared by an azo coupling reaction with 3-hydroxy-2-naphthoic acid, **3.25**, as the coupling component. The acid **3.26** is then converted into the acid chloride, **3.27**, followed by a condensation reaction between the acid chloride (2 moles) and *p*-phenylenediamine (1 mole), a reaction that requires elevated temperatures in a high boiling organic solvent, to form pigment **3.24**. In principle, a simpler and more cost-effective route using strategy (b), involving azo coupling of the diazonium salt (2 moles) with the appropriate bifunctional coupling component

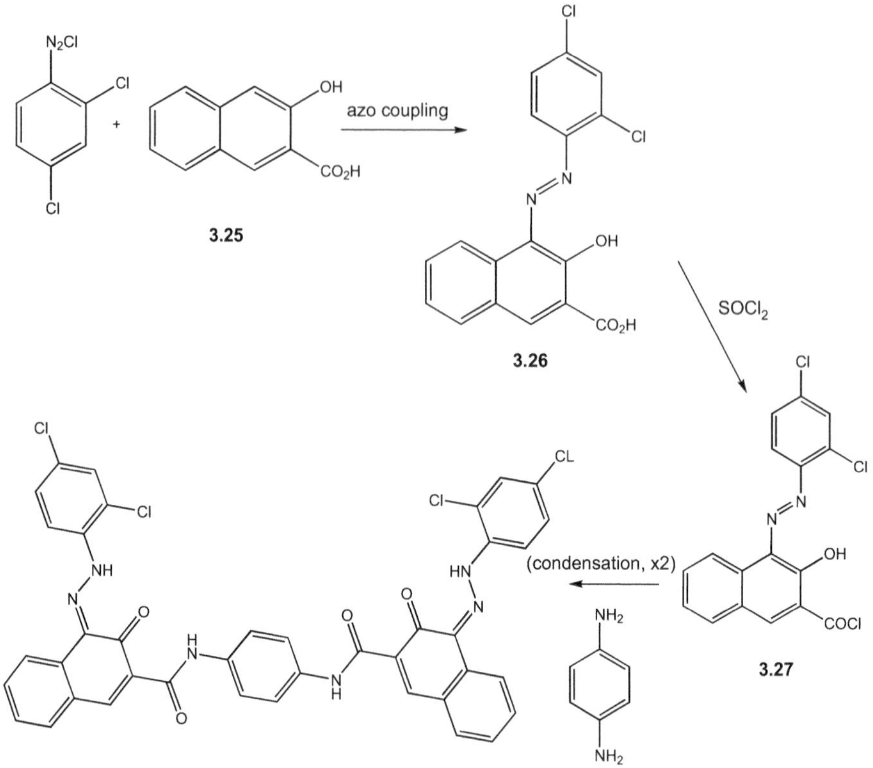

Scheme 3.8   Synthesis of a disazo condensation pigment.

(1 mole), might be proposed. In practice, this route fails because the monoazo derivative formed from the first azo coupling reaction is so insoluble that the second coupling reaction cannot take place.

3.4.3   Synthesis of Dyes and Pigments containing more than two Azo Groups

The strategies leading to dyes with several azo groups are in essence extensions of the methods leading to disazo colorants. As the number of separate diazotisations and azo coupling reactions required for the synthesis of a polyazo dye increases, so does the number of potential strategies. There are a number of commercial trisazo dyes, most being brown or black direct dyes for application to cotton. There are five strategies leading to trisazo colorants, which may be illustrated using the accepted symbolism as:

(a) $E \leftarrow D \rightarrow Z \leftarrow A$

(b) $E^1 \leftarrow D \rightarrow M \rightarrow E^2$

(c) $A \rightarrow M^1 \rightarrow M^2 \rightarrow E$

(d) $A^1 \rightarrow M \rightarrow Z \leftarrow A^2$

(e) $A^1 \rightarrow \underset{\underset{A^3}{\uparrow}}{Z} \leftarrow A^2$

As an example of strategy (a), as listed above, the synthesis of CI Direct Black 38, **3.28**, may be represented as follows:

3.28

m-phenylenediamine $\leftarrow$ benzidine $\rightarrow$ H-Acid (**3.13**) $\leftarrow$ aniline

A few azo dyes containing four, five and six azo groups, referred to respectively as tetrakisazo, pentakisazo and hexakisazo dyes, are known although they are progressively less important commercially and, naturally, the synthetic strategies are more complex. In addition, because of the increased likelihood of side-reactions, the products are increasingly less likely to be pure species.

3.5 METAL COMPLEX AZO DYES AND PIGMENTS

Metal complex formation has been a prominent feature of textile dyeing from very early times, since it was recognised that the technical performance, including fastness to washing and light, of many natural dyes could be enhanced by treatment with certain metal ions, a process known as *mordanting* (Chapter 1). Mordant dyeing is only used to a certain extent today, although it is restricted mainly to the complexing of certain azo dyes on wool with chromium(III) (Chapter 7). However, because of increasing sensitivity towards the toxicological and environmental issues associated with the use of heavy metals in particular, processes of this type continue to decline in importance (Chapter 12). There are basically two types of metal complex azo dyes: those in which the azo group is coordinated to the metal (medially metallised) and those in which it is not (terminally metallised). The discussion in this section is restricted to the former type, which is by far the most important commercially.[22]

The most important metal complex azo dyes are formed from the reaction of transition metal ions with ligands in which the *ortho* positions adjacent to the azo group contain groups capable of coordinating with the metal ion. The most important group in this respect is the hydroxyl (–OH) group, although carboxy (–CO$_2$H) and amino (–NH$_2$) groups can also be used. The most important transition metals used commercially to form metal complex azo dyes are copper(II), cobalt(III) and, especially, chromium(III). The reaction of an *o,o'*-dihydroxyazo compound, which acts as a tridentate ligand, with chromium(III) is illustrated in Figure 3.5. The azo group is capable of coordination to the metal ion through only one of the two azo nitrogen atoms, utilising its lone pair in bonding. While transition metal complex azo chemistry has proved to be of considerable industrial importance in textile dyes, curiously, and perhaps rather surprisingly, it has provided very limited success in commercial organic pigments, although other types of metal complexes, notably phthalocyanines, are of immense significance in this respect (Chapters 5 and 9).

1:1 Copper complexes of azo dyes are used widely in both reactive dyes (Chapter 8) and direct dyes (Chapter 7) for cellulosic fibres. In these dyes, the copper complexes adopt four-coordinate square planar geometry, with the three coordinating sites of the dye occupying corners of the square and the fourth occupied by a monodentate ligand, commonly water (Figure 3.6). The most important cobalt and chromium complexes of azo dyes adopt six-coordinate octahedral geometry, with the six positions occupied by coordination with two

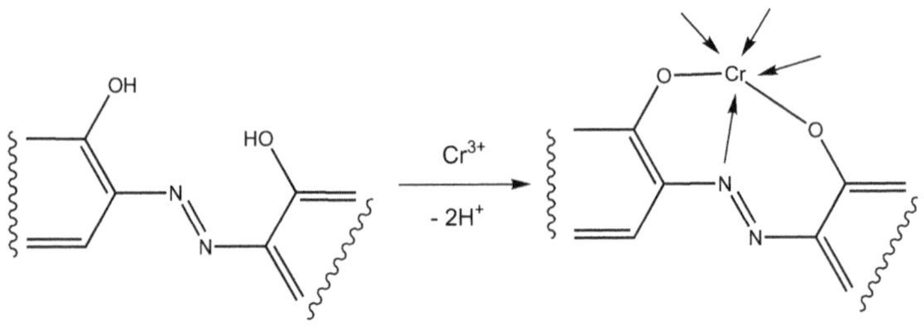

Figure 3.5 Complex formation between an *o,o'*-dihydroxyazo compound and chromium(III).

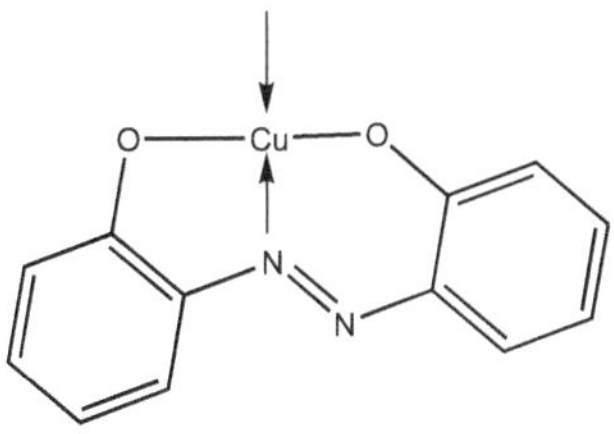

Figure 3.6 Square-planar azo copper complex.

tridentate azo dye ligands. These products are referred to as 2:1 premetallised dyes and they are of considerable importance in the dyeing of protein fibres such as wool (Chapter 7). The importance of the octahedral complexes of these particular transition metals may be attributed to the high kinetic stability of the complexes, which results from the d^3 electronic configuration of Cr(III) and the low-spin d^6 electronic configuration of Co(III).

The 2:1 octahedral complexes present many opportunities for isomerism. The isomerism in a number of metal complex azo dyes has been characterised using various techniques, including ^{1}H and ^{15}N NMR spectroscopy[23,24] and X-ray crystallography,[25] and in some cases isomers have been separated chromatographically.[26] The geometrical isomerism of some metal complex azo dyes is illustrated schematically in Figure 3.7. Several geometrical isomers are possible: one meridional (*mer*) isomer (I), in which the two dye molecules are mutually perpendicular, and five facial (*fac*) (II–VI) isomers in which the molecules are parallel. The meridional arrangement is favoured by complexes of *o,o'*-dihydroxyazo systems that contain a 5:6 chelate ring system, while the facial arrangement is favoured in *o*-carboxy-*o'*-hydroxyazo systems in which there is a 6:6 chelate ring system (Figure 3.8). Because of their asymmetry, the geometrical isomers, apart from (VI) which is centrosymmetric, also give rise to pairs of optically-active enantiomers. Other types of isomerism in azo metal complexes that have been identified include positional isomerism as a result of the possibility of coordination to either (but not both) of the azo nitrogen atoms, and isomerism as a result of different states of hybridisation (sp^2 or sp^3) of the azo nitrogen atoms.

There are several particular technical advantages associated with the formation of coloured metal complexes. Commonly, the transition metal complexes of a coloured organic ligand exhibit lightfastness that is significantly better than that of the free ligand. An explanation that has been offered for this effect is that coordination

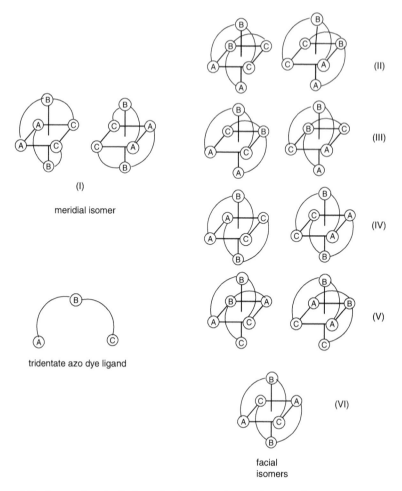

(I)

meridial isomer

tridentate azo dye ligand

(II)

(III)

(IV)

(V)

(VI)

facial
isomers

Figure 3.7   Isomerism in 2:1 octahedral metal complex azo dyes.

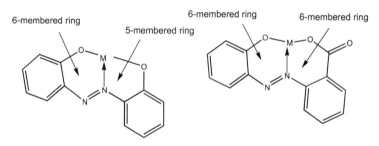

Figure 3.8   Chelate ring formation in metal complex azo dyes.

with a transition metal ion reduces the electron density at the chromophore, which in turn leads to improved resistance to photo-chemical oxidation. Other effects that may have a part to play in the enhanced lightfastness of transition metal complexes are steric protection of the chromophore towards degrading influences and the ability of transition metal ions to quench excited states that otherwise might undergo photochemical decomposition. In addition, the larger size of the metal complex molecules compared with the free ligand generally gives rise to improved washfastness properties in textile dyes as a result of stronger interactions with the fibre. On the other hand, the colours of the metal complexes are almost invariably duller than those of the azo dye ligand, a feature that limits their usefulness. The reduction in the brightness of the colour which accompanies metal complex formation is due to a broadening of the visible absorption band. There are several possible reasons for this effect. Broadening may due to the presence of a number of isomers, each with a slightly different absorption band. Alternatively, it may be due to overlap of the absorption band associated with the $\pi-\pi^*$ transitions of the ligand with those arising from metal ion d–d transitions or from ligand–metal charge transfer transitions.

REFERENCES

1. C. V. Stead, *Rev. Prog. Color.*, 1967–69, **1**, 23.
2. P. F. Gordon and P. Gregory, *Organic Chemistry in Colour*, Springer-Verlag, New York, 1983, ch. 3.
3. C. V. Stead, Chemistry of azo colorants, in *Colorants and Auxiliaries: Organic Chemistry and Application Properties*, ed. J. Shore, Society of Dyers and Colourists, Bradford, 1990, vol. 1, ch. 4.
4. H. Zollinger, *Color Chemistry: Syntheses, Properties and Applications of Organic Dyes and Pigments*, Wiley-VCH Verlag GmbH, Weinheim, 3rd edn, 2003, ch. 7.
5. O. Annen, R. Egli, R. Hasler, B. Henzi, H. Jacob and P. Matzinger, *Rev. Prog. Color.*, 1987, **17**, 72.
6. K. A. Bello and J. Griffiths, *J. Chem. Soc., Chem. Commun.*, 1986, 1639.
7. L. Briquet, D. P. Vercauteren, E. A. Perpete and D. Jacquemin, *Chem. Phys. Lett.*, 2006, **417**, 190.
8. C. Ciminelli., G. Granucci and M. Persico, *Chem.–Eur. J.*, 2004, **10**, 2327.
9. W. R. Brode, J. H. Gould and G. M. Wyman, *J. Am. Chem. Soc*, 1953, **75**, 1856.

10. T. Zincke and H. Bindewald, *Ber. Dtsch. Chem. Ges.*, 1884, **17**, 3026.
11. V. Machacek, A. Lycka, P. Simunek and T. Weidlich, *Magn. Reson. Chem.*, 2000, **38**, 293.
12. S. Stoyanov and L. Antonov, *Dyes Pigments*, 1988, **10**, 33.
13. S. Stoyanov and L. Antonov, *Dyes Pigments*, 1995, **28**, 31.
14. H. Zollinger, *Diazo Chemistry I: Aromatic and Heteroaromatic Compounds*, VCH, Weinheim, 1994.
15. J. Fitzpatrick, T. A. Meyer, M. E. O'Neill and D. L. H. Williams, *J. Chem. Soc., Perkin Trans.*, 1984, **2**, 927.
16. J. H. Ridd, *J. Soc. Dyers Colourists*, 1965, **81**, 355.
17. H. R. Schwander, *Dyes Pigments*, 1982, **3**, 133.
18. H. Zollinger, *Helv. Chim. Acta*, 1955, **38**, 1597, 1617 and 1623.
19. I. Szele and H. Zollinger, *Top. Curr. Chem.*, 1983, **112**, 1.
20. R. Kaminski, U. Lauk, P. Skrabal and H. Zollinger, *Helv. Chim. Acta*, 1983, **66**, 2002.
21. E. N. Abrahart, *Dyes and their Intermediates*, Edward Arnold, London, 1977.
22. F. Beffa and G. Back, *Rev. Prog. Color.*, 1984, **14**, 33.
23. G. Schetty and W. Kuster, *Helv. Chim. Acta*, 1974, **57**, 2149.
24. A. Lycka, J. Jirman and A. Cee, *Magn. Reson. Chem.*, 1990, **28**, 408.
25. R. Grieb and A. Niggli, *Helv. Chim. Acta*, 1965, **48**, 317.
26. G. Schetty and W. Kuster, *Helv. Chim. Acta*, 1961, **44**, 219.

CHAPTER 4

Carbonyl Dyes and Pigments

4.1 INTRODUCTION

The chemical class of colorants that is second in importance to azo dyes and pigments is characterized by the presence of a carbonyl (C=O) group, which may be regarded as the essential chromophoric unit. The vast majority of carbonyl dyes and pigments contain two or more carbonyl groups that, as illustrated in Figure 4.1, are linked to one another through a conjugated system, frequently an aromatic ring system.[1,2]

Carbonyl colorants are found in a much wider diversity of structural arrangements than is the case with azo dyes and pigments. The most important group of carbonyl dyes and pigments are the anthraquinones, to which a substantial part of this chapter is devoted. Other types that are of commercial importance in particular application classes include indigoids, benzodifuranones, coumarins, naphthalimides, quinacridones, perylenes, perinones and diketopyrrolopyrroles. In contrast to azo dyes, carbonyl dyes are found in nature and, for example, dyes such as indigo and alizarin, an anthraquinone derivative, are amongst the most important natural dyes for textiles (Chapter 1). Synthetic carbonyl dyes and pigments are capable of providing a wide range of colours, essentially covering the entire visible spectrum. However, a major reason underlying the industrial importance of carbonyl colorants is that they are capable of giving long wavelength absorption with relatively short conjugated systems. This feature applies especially to some anthraquinone and indigoid derivatives, which are thus of particular importance in the blue shade

Colour Chemistry, 2nd edition
By Robert M Christie
© R M Christie 2015
Published by the Royal Society of Chemistry, www.rsc.org

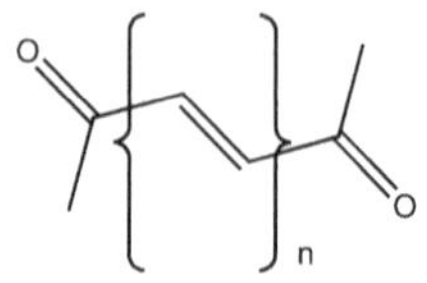

Figure 4.1   General structural arrangement in most carbonyl colorants.

area. In terms of fastness properties, carbonyl dyes and pigments are often superior to their azo counterparts. They are thus frequently the colorants of choice when high technical performance is demanded by a particular application. In many textile dye application classes, carbonyl dyes (especially anthraquinones) rank second in significance to azo dyes. A particular textile application class dominated by carbonyl dyes is the vat dye class, a group of dyes applied to cellulosic fibres, such as cotton (Chapter 7). In the vat dyeing of cotton, the ability of the carbonyl groups to undergo reduction to a water-soluble form that is capable of transferring to the fibre and subsequently re-oxidised to its original system is utilized. Azo dyes are inappropriate as vat dyes because reduction of the azo group is irreversible. A range of carbonyl pigments, including the quinacridones, perylenes, perinones and diketopyrrolopyrroles, impart extremely high performance in their application. These products owe their high fastness to light, heat and solvents principally to the ability of the carbonyl group to participate in intra- and intermolecular hydrogen bonding (Chapter 9).

   The chemistry involved in the manufacture of carbonyl colorants is generally more elaborate and much less versatile than is the case with azo dyes. Often the synthetic sequence involves multiple stages and the use of specialist intermediates. Consequently, the number of commercial products is more restricted and they tend to be rather more expensive. Indeed, certain carbonyl dyes, notably some anthraquinones, are becoming progressively less important commercially, particularly for traditional textile applications, as a wider range of azo dyes absorbing at longer wavelengths has emerged and as the cost differential between azo dyes and carbonyl dyes increases. In the following sections of this chapter, the characteristic structural features of the most important types of carbonyl colorants are reviewed and an overview of some of the more important synthetic routes is presented.

4.2   ANTHRAQUINONES

A common arrangement of the carbonyl groups in coloured molecules gives rise to a group of compounds known as *quinones*. These may be defined as cyclohexadienediones, *i.e.*, compounds containing two ketone carbonyl groups and two double bonds in a six-membered ring. The simplest quinones are *o*- and *p*-benzoquinones, **4.1** and **4.2**, respectively. Derivatives of the benzoquinones and of the naphtho-1,4-quinone system **4.3** are of relatively limited interest as colouring materials. This is probably due, at least in part, to the instability resulting from the presence of an alkene-type double bond. By far the most important quinone colorants are the anthraquinones, or more correctly anthrax-9,10-quinones. Anthraquinones, as demonstrated by the parent compound **4.4**, contain a characteristic system of three linear fused six-membered rings in which the carbonyl groups are located in the central ring and the two outer rings are fully aromatic. Anthraquinone colorants can give rise to the complete range of hues. However, within any particular application class, they are often more important as violets, blues and greens, thus complementing the azo chemical class, which generally provides the most important yellows, oranges and reds. They frequently give brighter colours than azo dyes, but the colours are often weaker, one reason why they are, in general, less cost-effective than azo dyes. Anthraquinone dyes are capable of providing excellent lightfastness properties, generally superior to azo dyes.

4.1   **4.2**   **4.3**   **4.4**

There is a wide diversity of chemical structures of anthraquinone colorants.[1-3] Many anthraquinone dyes are found in nature, perhaps the best known being alizarin, 1,2-dihydroxyanthraquinone, the principal constituent of madder (Chapter 1).[4] Naturally-based anthraquinone dyes are of limited current commercial importance, although synthetic alizarin, as CI Mordant Red 11, is used to an extent in the dyeing and printing of natural fibres. Many of the current commercial range of synthetic anthraquinone dyes are simply substituted derivatives of the anthraquinone system. For example, a number of the most important red and blue disperse dyes for application to polyester fibres are simple non-ionic anthraquinone molecules, containing substituents such as amino, hydroxyl and methoxy, and several sulfonated derivatives are commonly used as acid dyes for wool.

There are a large number of anthraquinones that are structurally more complex and polycyclic in nature. In this book, provided that the anthraquinone nucleus is recognisable somewhere in the structure, the colorant is classed as a member of the anthraquinone class, although some texts consider these annellated derivatives separately. These polycyclic anthraquinones belong largely to the vat dye textile application class where their large extended planar structure is an important feature for their application (Chapter 7). There are around 200 different anthraquinonoid vat dyes in use commercially, of widely different structural types, and covering the entire spectrum from yellow through blue to black. Figure 4.2 shows a selection of these compounds, including both carbocyclic and heterocyclic types. The examples presented include indanthrone, **4.5a** (CI Vat Blue 4), violanthrone, **4.5b** (CI Vat Blue 20), pyranthrone, **4.6a** (CI Vat Orange 9),

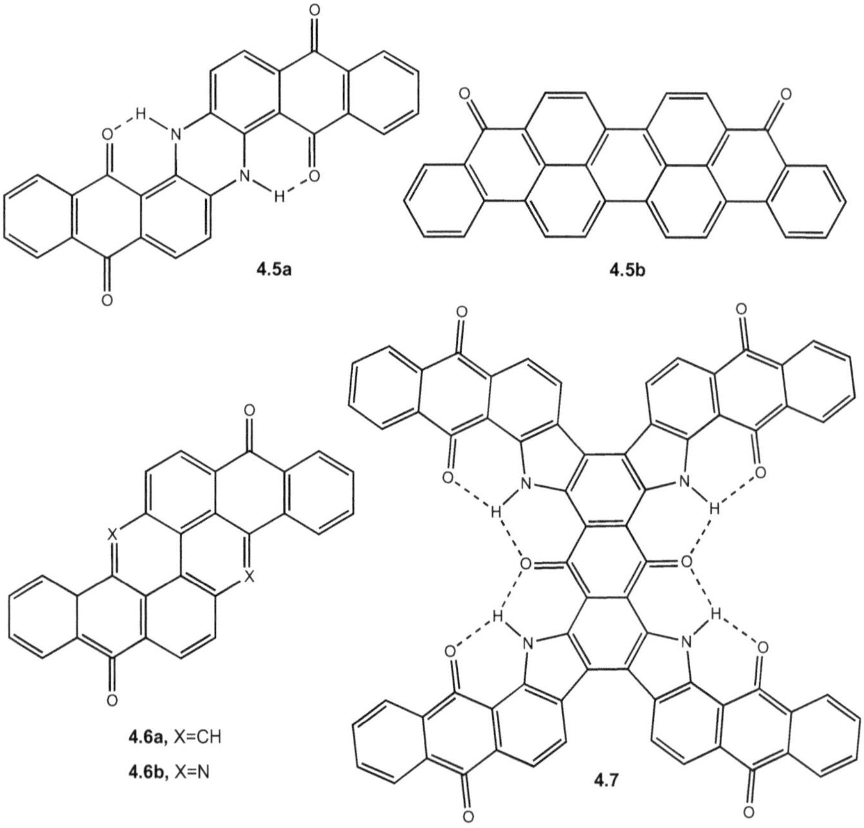

4.5a

4.5b

4.6a, X=CH
4.6b, X=N

4.7

Figure 4.2 Some polycyclic anthraquinone vat dyes and pigments.

flavanthrone, **4.6b** (CI Vat Yellow 1) and CI Vat Green 8, **4.7**, a derivative containing 19 fused rings. While the anthraquinone chromophoric grouping is second only in importance to the azo chromophore in the chemistry of textile dyes, it is of considerably less importance in pigments. This is probably because the traditional role of anthraquinones in many dye application classes, which is to provide lightfast blues and greens, is more successfully adopted by the phthalocyanines (Chapter 5) in the case of pigments. However, the insolubility and generally good fastness properties of vat dyes stimulated considerable effort into the selection of suitable examples of the colorants for use, after conversion into the appropriate physical form, as pigments (in fact, commonly referred to as vat pigments) for paint, printing ink and plastics applications. Of the known vat dyes, only about 25 have been fully converted into pigment use, and less than this number are of real commercial significance. The range of anthraquinone pigments includes some of the longest-established vat dyes, notably indanthrone, **4.5a** (CI Pigment Blue 60), together with some of its halogenated derivatives, and flavanthrone, **4.6b** (CI Pigment Yellow 24).

The parent compound, anthraquinone, **4.4**, is only weakly coloured, its strongest absorption being in the UV region (λ_{max} 325 nm). The UV/visible spectral data for a series of substituted anthraquinones **4.4a–h** are given in Table 4.1, and these illustrate the effect of the substituent pattern on the colour. The introduction of simple electron-releasing groups, commonly amino or hydroxyl, into the anthraquinone nucleus gives rise to a bathochromic shift that is dependent on the number and position of the electron-releasing group(s) and their relative strengths (OH < NH$_2$ < NHAr). They are thus typical donor–acceptor systems, with the carbonyl groups as the acceptors and the electron-releasing auxochromes as the donors.[3] By choice of appropriate substitution patterns, dyes are obtained which may absorb in

Table 4.1 Absorption maxima for some substituted anthraquinones.

Compound	Substituent	λ_{max} (nm) in MeOH
4.4a	1-OH	402
4.4b	2-OH	368
4.4c	1-NH$_2$	475
4.4d	2-NH$_2$	440
4.4e	1-NHPh	500
4.4f	1,4-diNH$_2$	590
4.4g	1,4,5,8-TetraNH$_2$	610
4.4h	1,4-DiNHPh	620

any desired region of the visible spectrum. From the data, it is clear that the electron-releasing groups exert their maximum bathochromic effect in the α-positions (1-, 4-, 5-, 8-) rather than the β-positions (2-, 3-, 6-, 7-). This is one reason why substitution in α-positions is preferred in anthraquinone dyes. In addition, α-substituents give dyes with higher molar extinction coefficients and, very importantly, they enhance technical performance, especially lightfastness, by virtue of their participation in intramolecular hydrogen bonding with the carbonyl groups. The 1,4-substitution pattern is particularly significant in providing blue dyes of commercial importance in various textile dye application classes (Chapter 7).

The effect of substituents on colour in substituted anthraquinones may be rationalised using the valence-bond (resonance) approach, in a similar way as has been presented previously for a series of azo dyes (see Chapter 2 for details). For the purpose of explaining the colour of the dyes, it is assumed that the ground electronic state of the dye most closely resembles the most stable resonance forms, the normal Kekulé-type structures, while that the first excited state of the dye more closely resembles the less stable, charge-separated forms. Some relevant resonance forms for anthraquinones **4.4**, **4.4c**, **4.4d** and **4.4f** are illustrated in Figure 4.3. The ground state of the parent compound, **4.4**, is assumed to resemble closely structures such as I, while charge-separated forms, such as structure II are assumed to make a major contribution to the first excited state. Structure II is clearly unstable due to the carbocationic centre. In the case of aminoanthraquinones **4.4c** and **4.4d**, donation of the lone pair from the amino nitrogen atom markedly stabilises the first excited states, represented respectively by structures III and IV, lowering their energy and leading, as a consequence of the inverse relationship between the difference in energy and the absorption wavelength, to a pronounced bathochromic shift. Two reasons may be proposed for the enhanced bathochromicity of the 1-isomer, **4.4c**, compared with the 2-isomer, **4.4d**. Firstly the proximity of positive and negative charges in space in the first excited state structure III gives rise to a degree of electrostatic stabilisation. Secondly, the intramolecular hydrogen bonding in structure III, which is not present in structure IV, serves to increase the electron-releasing power of the lone-pair on the nitrogen atom and to increase the electron-withdrawing effect of the carbonyl group; both effects stabilise the excited state relative to the ground state and lead to a more pronounced bathochromic effect. The 1,4-diamino compound, **4.4f**, is particularly bathochromic because of an extensively resonance-stabilised first excited state, involving structures V and VI.

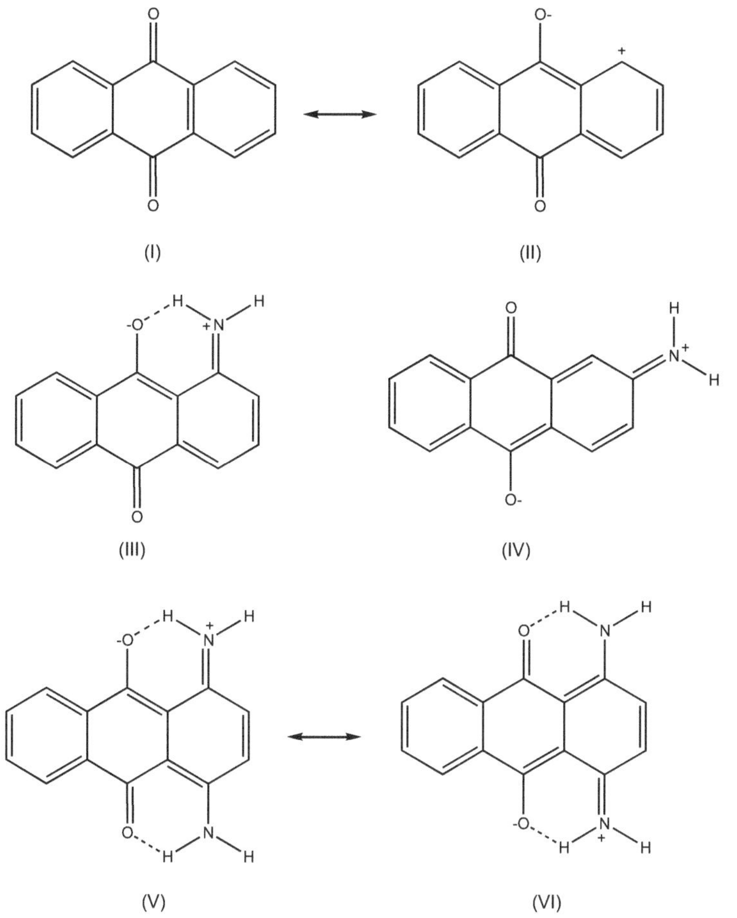

Figure 4.3 Some relevant resonance forms for anthraquinones 4.4 (I, II), 4.4c (III), 4.4d (IV) and 4.4f (V, VI) (see also Table 4.1).

4.3 INDIGOID DYES AND PIGMENTS

Indigo, **4.8**, the parent system of this group of colorants, is one of the oldest known natural dyes (Chapter 1) obtained from plant sources such as *Indigofera tinctoria* in Asia and from dyers' woad (*Isatis tinctoria*) in Europe. Natural indigo dyeing is still practised quite widely as a traditional craft process in Asia for textiles and clothing, often with symbolic status, and generally starts with the fermentation of extracts of the leaves harvested from the plants.[5] The naturally-occurring material that generates indigo is usually indican, the colourless glucoside of indoxyl. During fermentation, enzymatic processes generate indoxyl, **4.9**, or other related precursors, which undergo oxidation in air leading to indigo within the fibre. Nowadays, most indigo for industrial use is produced synthetically, and the dye is referred to as CI Vat Blue 1. The synthetic routes are outlined later in this chapter.

4.8

4.9

The structure of indigo was first proposed by von Baeyer, although he initially suggested that it had the (*Z*)- (or *cis*) configuration **4.10**.[6] X-Ray crystal structure determination carried out rather later confirmed that the molecule in fact exists as the (*E*)- (or *trans*) isomer **4.8**.[7] Although hundreds of indigoid derivatives have been synthesised and evaluated as dyes over the years, relatively few of these have achieved commercial importance and indeed the parent compound remains by far the most important member of the class. Indigo may be applied to cellulosic fabrics as a vat dye imparting an attractive blue colour and providing reasonably good fastness to washing and light. However, because its technical properties were inferior to those provided by higher performing blue vat dyes, such as indanthrone, **4.5a**, indigo came close to being phased out as a commercial dye in the first half of the twentieth century. The revival in its popularity came about in the 'flower power' era of the 1960s as denim, especially blue jeans, became fashionable. This particular application makes positive use of 'defects' of indigo as a textile dye. The typical ring dyeing of cross-sections of the fibres leads to an unevenly dyed appearance, while low abrasion resistance provides the fashionable 'washed-out' look. The 'stonewashed' appearance is produced by a physical abrasion process in which the fabric is rotated with pumice stones, or alternatively using enzymes to degrade the indigo. Indeed, the lasting success of indigo as a commercial dye is heavily dependent on the continuing popularity of blue denim as a fashion trend.

4.10

A question which has intrigued colour chemists for years is why indigo, as a relatively small molecule, absorbs at such long wavelengths. The colour of indigo depends crucially on its environment.[2,8] In the gas phase, the only situation in which a dye will effectively owe its colour to single molecules, indigo is red ($\lambda_{max} = 540$ nm). In solution, indigo exhibits pronounced positive solvatochromism; in nonpolar solvents, it is violet (*e.g.*, in CCl_4, $\lambda_{max} = 588$ nm), while in polar solvents it is blue (*e.g.*, in DMSO, $\lambda_{max} = 620$ nm). In the solid state and when applied to fabric as a vat dye, it is blue. Amorphous and crystalline forms of indigo in the solid state differ significantly in their absorption maxima ($\lambda_{max} = 650$ *versus* 675 nm). X-Ray structure analysis of the molecule has demonstrated that, in the solid state, the molecules are highly aggregated by intermolecular hydrogen bonding, each indigo molecule being attached to four others. It is clear that the intermolecular hydrogen bonding is a major factor in providing the bathochromic shift of colour compared to the monomolecular state.[9] It also explains its low solubility and relatively high melting point (390–392 °C).

The indigo molecule has been the subject of numerous studies using quantum mechanical calculations of varying complexities and sophistication over many years. A theoretical study using an *ab initio* complete active space-self-consistent field (CAS-SCF) calculation provided an excellent fit with the experimental spectra,[10] even correctly predicting shoulders, while time dependent density functional theory (TD-DFT) calculations have even been applied to account for solvent effects and to provide a simple model to describe the solid state properties.[8] Various concepts have been advanced, supported by analysis of the quantum mechanics calculations, to explain the long wavelength absorption of indigo in terms of its molecular structure.[11] These approaches have essentially aimed at identifying the particular feature of the structure of the molecule that is primarily responsible for the visible absorption. Since the outer benzene rings play a secondary role in determining the colour of indigo, the concepts generally focus on the central core of the molecule. The most commonly-invoked explanation is that the basic structural unit responsible for the colour of indigo is an arrangement consisting of two electron donor groups (NH) and two electron acceptor groups (C=O) 'cross-conjugated' through an ethene bridge as illustrated in Figure 4.4.[12–14] This so-called *H-chromophore* arrangement gives rise to a particularly bathochromic absorption.

UV/visible spectral data for some indigo derivatives, which illustrate the nature of the substituent effects, are given in Table 4.2. Indigo is a dye of the donor/acceptor type. The valence-bond (resonance) approach to colour/structure relationships, as described in Chapter 2 applied to azo dyes, may be used to explain the particular stability of the first excited state of the indigo molecule, and hence its bathochromicity, invoking important resonance contributions from charge separated structures VII–X, in each of which charge is accommodated in a stable location, *i.e.*, negative charge on oxygen, positive charge on quaternary nitrogen, as illustrated in Figure 4.5. Resonance structures VII and VIII are especially stable due to the retention of aromatic character of both benzene rings. The molecular symmetry means that these two forms are identical in energy. Forms IX and X are also equal in energy, but are less stable due to the loss of aromatic character in one ring. Form XI, referred to as a quadrupole structure,

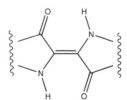

Figure 4.4   H-chromophore unit of indigo.

Table 4.2   UV/visible spectral data (λ_{max} values (nm)) for solutions of some symmetrical di-substituted indigo derivatives **4.8** and **4.8a–d** in 1,2-dichloroethane. The positions of substituents may be identified with reference to the numbering system given for structure 4.8.

Substituent	5,5'-Isomer	6,6'-Isomer
H	**4.8**, 620	**4.8**, 620
NO_2	**4.8a**, 580	**4.8b**, 635
OCH_3	**4.8c**, 645	**4.8d**, 570

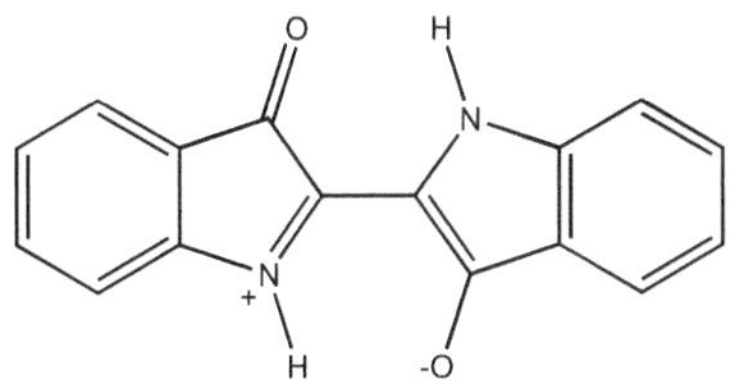

Figure 4.5 Valence bond (resonance) approach to indigo.

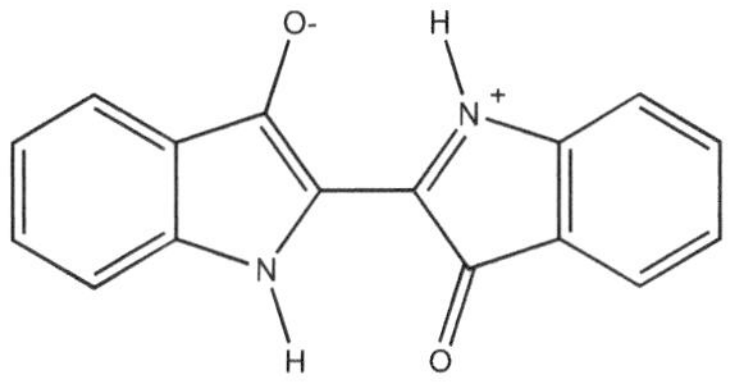

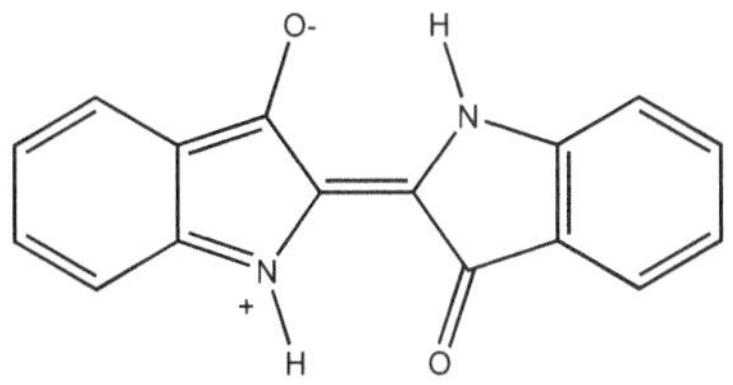

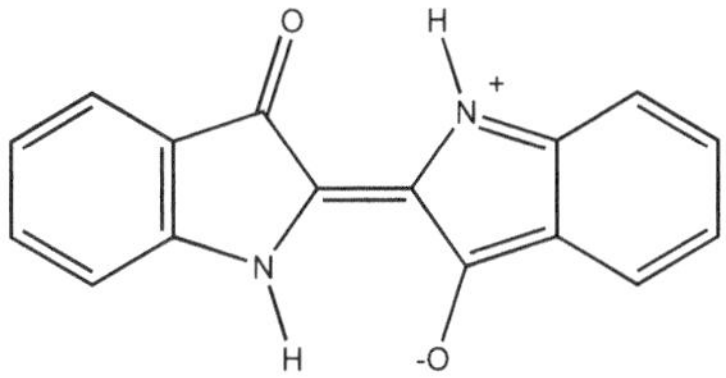

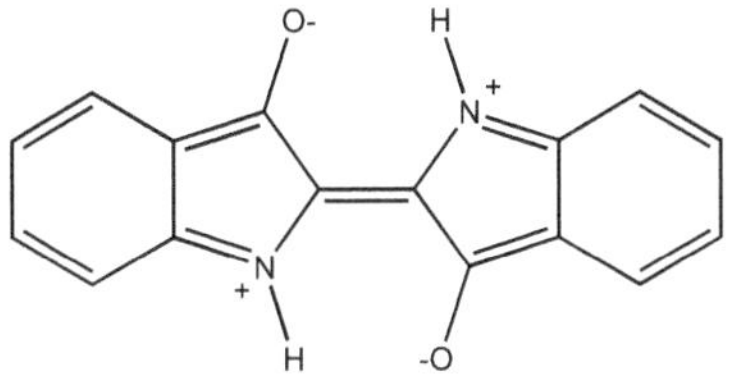

has a negative charge on both oxygens and a positive charge on both nitrogens. Compared to many other dyes, this constitutes a very stable excited state, so that the energy difference (ΔE) compared to the ground state is small, and consequently indigo absorbs at long wavelengths. The effect of substituents on the colour of indigo may also be explained using this approach. For example, an electron-releasing group in the 5- (or 7)-positions, which are *ortho* or *para* respectively to the NH group stabilises the first excited state by increasing the electron density on the nitrogen atom and a bathochromic shift results. In contrast, an electron-withdrawing group at these positions destabilises the first excited state by increasing the positive charge at an already electron-deficient nitrogen atom, causing a hypsochromic shift. For substituents at the 4- and 6-positions the situation is reversed.

A range of other indigoid systems, represented by the general structure **4.11** are known, in which one or both of the NH groups are replaced by other heteroatoms capable of donating a lone pair of electrons into the π-system. These include the yellow oxindigo, **4.11a**, X=Y=O, the red thioindigo, **4.11b**, X=Y=S, and the violet selenoindigo, **4.11c**, X=Y=Se, together with mixed derivatives such as compound **4.11d** (X=NH, Y=S). In this group, thioindigo is a useful red vat dye and some of its halogenated derivatives are used as pigments, but the others are of limited commercial significance. Indirubin, **4.12**, is an isomer of indigo that is encountered as a minor red constituent in samples of natural indigo. While indirubin has no value as a textile dye, some interest in the dye has developed in recent years because of its biological activity and that of some of its derivatives. Quantum mechanical calculations based on a modified PPP-MO approach predict the bathochromicity of indigo compared with indirubin, giving calculated λ_{max} values of 605 and 543 nm, respectively, in reasonable qualitative agreement with experimental values (603 and 552 nm). The calculations also demonstrate that, in the case of indigo, excitation gives rise to symmetrical charge transfer from the heterocyclic nitrogen donor groups to the carbonyl acceptor groups, which is consistent with the concept of cross conjugation in the H-chromophore (Figure 4.4), and with the valence-bond approach illustrated in Figure 4.5. However, with indirubin, **4.12**, charge transfer is demonstrated to be essentially localised in one half of the molecule. The absence of cross-conjugation in this case thus explains its absorption at shorter wavelengths.[15]

4.11

4.12

4.4 BENZODIFURANONES

The benzodifuranones constitute one of the most recently introduced groups of carbonyl colorants. They were launched commercially in the late 1980s by ICI as disperse dyes for application to polyester.[16,17] This group of dyes, of which compound **4.13**, CI

Disperse Red 356, is a representative example, is capable of providing a range of colours although the most important products are
red. They provide bright, intense red shades on polyester with good
fastness to light, sublimation and washing. For good build-up of
colour on polyester, it has been found that asymmetric products,
typified by dye **4.13**, perform better than when the molecules are
symmetrically substituted.

4.13

4.5 FLUORESCENT CARBONYL DYES

Several carbonyl dyes are characterised by their strong fluorescence.[18]
Derivatives of coumarin provide the most important industrial
fluorescent dyes. Fluorescent coumarin dyes are known that absorb
and emit in most parts of the visible spectrum, although most commercial products are yellow with a green fluorescence. Fluorescent
brightening agents (FBAs) based on coumarins that absorb in the UV
region and emit in the visible region are also used commercially. The
coumarin fluorescent dyes invariably contain an electron-releasing
substituent, most commonly the diethylamino group, in the 7-position,
and electron-withdrawing substituents in the 3-position. The coumarins are thus typical donor–acceptor dyes.[19] The most widely-
used dyes contain a benzimidazolyl (**4.14a**), benzoxazolyl (**4.14b**,
4.14d, **4.14e**) or benzothiazolyl (**4.14c**) group as the acceptor in the 3-
position. These coumarins provide a series of important industrial
disperse dyes that allow synthetic fibres, especially polyester, to be
dyed in fluorescent yellow shades. CI Disperse Yellow 232, **4.14d**, for
example, dyes polyester to give a brilliant fluorescent greenish-yellow
hue with good fastness to light, sublimation and washing. Other
important types of fluorescent carbonyl dyes include the aminonaphthalimides, such as CI Disperse Yellow 11, **4.15**, and some
sterically hindered perylenes, which are considered in the next

section of this chapter. Fluorescent carbonyl dyes and pigments find a
wide range of uses in textiles, plastics, paint and printing inks, especially where high visual impact is desirable, such as in advertising
and safety applications. They are also used in the detection of flaws in
engineered articles, solar collector systems, dye lasers and a host of
analytical and biological applications.

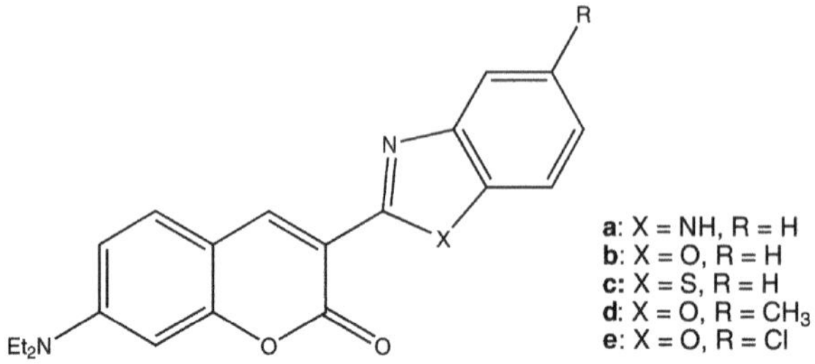

a: X = NH, R = H
b: X = O, R = H
c: X = S, R = H
d: X = O, R = CH₃
e: X = O, R = Cl

4.14

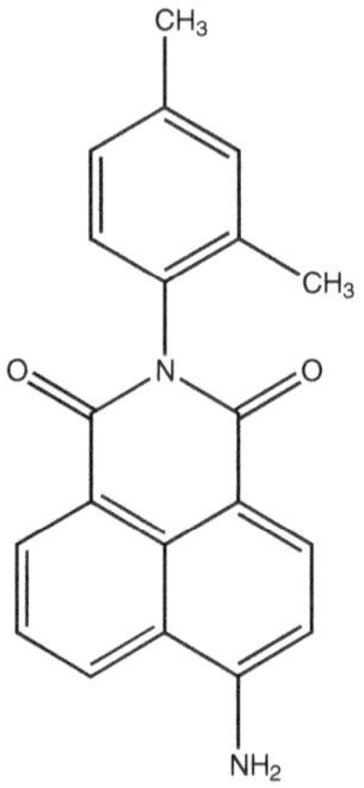

4.15

4.6   CARBONYL PIGMENTS

Carbonyl pigments exist in a wide diversity of structural arrangements
and provide a range of colours essentially covering the entire visible
spectrum.[20,21] This group of pigments includes quinacridones,
anthraquinones, thioindigos, perylenes, perinones, quinophthalones,
isoindolines and diketopyrrolopyrroles. Carbonyl pigments generally
impart extremely high technical performance so that they are often
suitable for applications, such as automotive paints, which place severe demands on performance in terms of colouristics and fastness to

light, heat and solvents. An important factor in providing this level of performance is the ability of the carbonyl group to participate in intra- and intermolecular hydrogen bonding. The range of carbonyl pigment chemical types is discussed in this section, with the exception of those anthraquinones that are used as vat pigments and thioindigo derivatives, both of which have been discussed in previous sections of this chapter.

The quinacridones constitute one of the most important chromophoric systems developed for pigment applications following the phthalocyanines (Chapter 5). Linear trans quinacridones were first discovered in 1935, but their potential as pigments was not realised until the late 1950s, when they were introduced commercially by Du Pont. The pigments offer outstanding fastness properties, similar to those of copper phthalocyanine, in the orange, red and violet shade areas.[22] Structurally, quinacridones consist of a system of five fused alternate benzene and 4-pyridone rings. Several geometrical arrangements of such a system are possible, *e.g.* structures **4.16–4.18**, but the outstanding technical properties are only given by compounds that possess the linear trans arrangement, **4.16**. Compounds **4.16a–c** are the products of most significant commercial importance.

4.16a: R^1 = H; R^2 = H
4.16b: R^1 = H; R^2 = CH$_3$
4.16c: R^1 = H; R^2 = Cl
4.16d: R^1 =Cl; R^2 = H

4.17

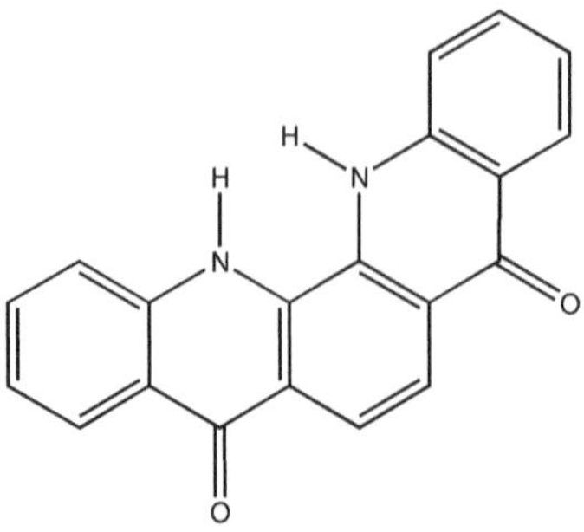

4.18

At first sight, it may seem somewhat surprising that such small molecules should provide such a high degree of heat, light and chemical stability, and also insolubility. These properties have been explained by strong two-dimensional molecular association due to hydrogen bonding in the crystal structure between N–H and C=O groups, as illustrated in Figure 4.6.[23,24] There is considerable evidence from related

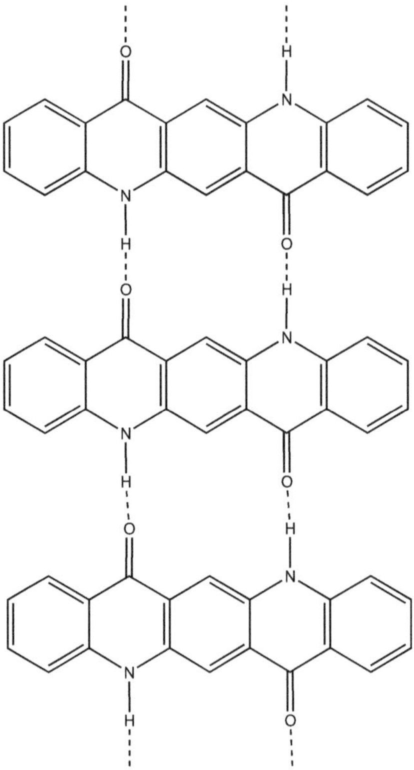

Figure 4.6   Intermolecular association by hydrogen bonding in the crystal lattice of linear trans quinacridone, **4.16a**.

derivatives for the importance of intermolecular H-bonding in determining the properties of the quinacridones. For example, the inferior properties of the angular quinacridone **4.18** and the 4,11-dichloro derivative **4.16d** may be attributed to steric and geometric constraints, which reduce the efficiency of the hydrogen bonding. In addition, the *N,N*-dimethyl derivatives, in which no H-bonding is possible, are reported to be soluble in organic solvents. A further factor that plays a part in the technical performance of quinacridone pigments is the strong dipolar nature of the pyridone rings arising from a major contribution from resonance forms such as **4.19**, leading to strong intermolecular dipolar association throughout the crystal structure.

4.19

The origin of the colour of quinacridone pigments provides a useful example of the influence of crystal structure effects, which have not yet been explained fully in fundamental terms. Interestingly, in solution the quinacridones exhibit only weak yellow to orange colours. The intense red to violet colours of the pigments in the solid state are therefore presumably determined largely by interactions between molecules in the crystal lattice structure. The quinacridones show *polymorphism*, the ability of a compound to exist in different crystal structural arrangements, and this has a profound effect on the colour of the pigments. The parent compound, **4.16a**, exists in three distinct polymorphic modifications each one with its own characteristic X-ray powder diffraction pattern. Two of these forms, the α- and β- modifications are red, while the γ-form is violet. The α-form is the least stable to polymorphic change and has not been commercialised.

There is little doubt that one of the most significant developments in organic pigments in the late twentieth century was the discovery and commercialisation of pigments based on the 1,4-diketopyrrolo[3,4-*c*]pyrrole (DPP) system.[25] The essential structural feature of this group of pigments, of which CI Pigment Red 254 (**4.20**) is a representative commercial example, are the two fused five-membered ketopyrrole rings. DPP pigments, by appropriate substituent variation, are capable of providing orange through red to bluish-violet shades. However, the DPP pigments which have gained most commercial prominence provide brilliant saturated red shades of outstanding durability for automotive paint and other high performance applications. In addition, their excellent thermal stability means that they are of considerable interest for the pigmentation of plastics. The pigments are, like the quinacridones, remarkable in producing such an excellent range of fastness properties from such small molecules. X-Ray structural analysis has demonstrated that this is due to the strong intermolecular forces, due hydrogen bonding and dipolar interactions, which exist throughout the crystal structure, similar to those involved with the quinacridones.[26]

4.20

Several high-grade perylene pigments, mostly reds but also including blacks, are of importance. The pigments may be represented by the general structure **4.21**, in which the imide nitrogen substituents, R, may be alkyl or aryl groups. An interesting observation in the perylene series is that small structural changes in the side-chain can lead to quite profound colour differences. The N,N′-dimethyl compound, for example, is red while the corresponding diethyl derivative is black. X-Ray diffraction studies have now been applied to an extensive range of perylenes in an attempt to characterise the effect of differences in the crystal lattice structure on the light absorption properties of the pigments, a phenomenon that has been described as *crystallochromy*.[27–30] As an example, the N,N′-dimethyl compound consists of a parallel arrangement of molecules in stacks, whereas in the N,N′-diethyl compound the molecules are in stacks twisted with respect to one another with considerably more overlapping of the perylene ring systems in neighbouring molecules. Some perylenes with bulky substituents, such as the *t*-butyl group, on the imide nitrogens are useful highly efficient and stable fluorescent dyes. Perylene fluorescent dyes are capable of providing the level of stability, especially to light, required for use in highly demanding applications, such as in solar energy collectors, and for such applications they are superior to many of the other chemical classes of fluorescent dyes, including coumarins, aminonaphthalimides and Rhodamines.[18] The strong fluorescence of these dyes has been attributed to the structural rigidity of the molecules. In a rigid molecule, loss of energy from excited states by intramolecular thermal motion is minimised, favouring fluorescent emission over non-radiative energy loss. A group of carbonyl pigments, structurally related to the perylenes, are the perinones. Two isomeric perinone pigments are manufactured, CI Pigment Orange 43, **4.22**, the trans isomer, and CI Pigment Red 194, **4.23**, the cis isomer, the former being an especially important high performance product, particularly for plastics applications.

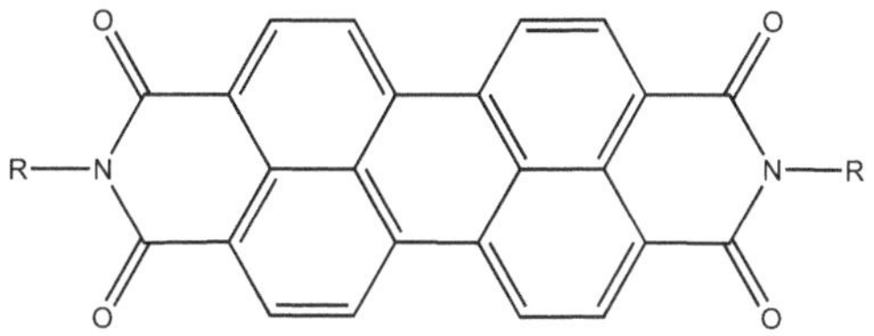

4.21

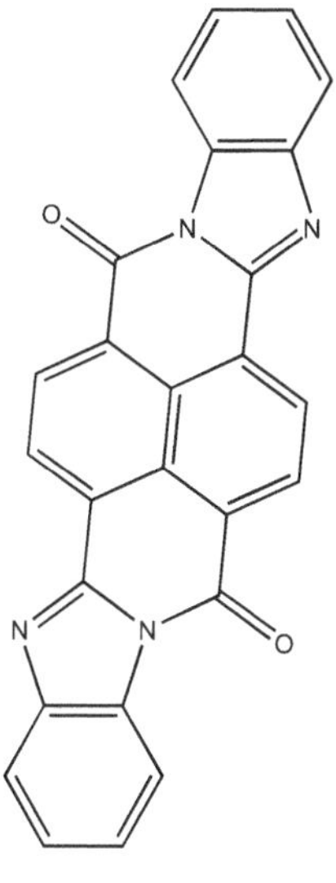

4.22

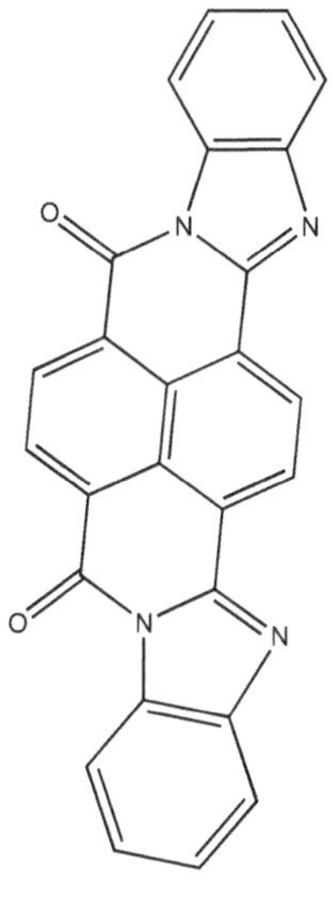

4.23

Examples of other types of carbonyl pigment of industrial importance include the quinophthalone **4.24**, CI Pigment Yellow 138, a high performance greenish-yellow product, and the reddish-yellow iso-indoline, CI Pigment Yellow 139, **4.25**.

4.24

4.25

4.7   THE QUINONE–HYDROQUINONE REDOX SYSTEM

The quinone–hydroquinone system represents a classical example of a fast, reversible redox system.[31] This type of reversible redox reaction is characteristic of many inorganic systems, such as the interchange between oxidation states in transition metal ions, but it is relatively uncommon in organic chemistry. The reduction of benzoquinone to hydroquinone formally involves the transfer of two hydrogen atoms. In practice, the reversible reaction involves the stepwise transfer of two electrons from the reducing agent as shown in Scheme 4.1.

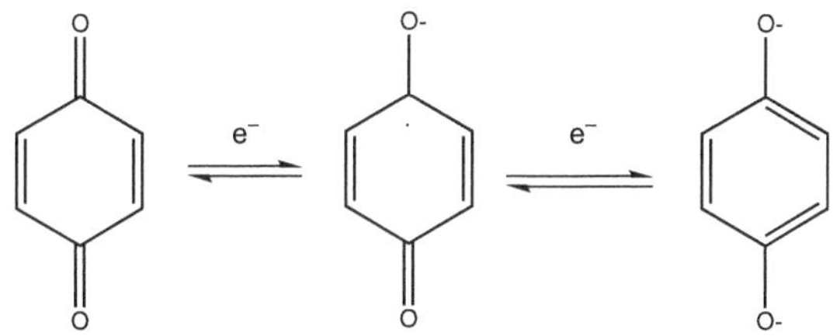

Scheme 4.1 Benzoquinone/hydroquinone redox system.

Commonly, these redox reactions are carried out in alkaline media, in which case the unprotonated semi-quinone is first formed and the dianion is the final reduction product as illustrated in the scheme. In media where protons are available a series of acid/base equilibria are also involved. A wide range of dicarbonyl compounds undergo this type of reaction, including the anthraquinones, indigoid derivatives and perinones. The reduction/oxidation of carbonyl dyes is used practically in the chemistry of vat dyeing of cellulosic fibres (Chapter 7). In this process, the insoluble colorant is treated with a reducing agent in an alkaline medium to give the reduced or *leuco* form. After application of the leuco form of the dye to the fibre, the process is reversed and the coloured dye is generated by oxidation. The reducing agent most commonly used is sodium dithionite ($Na_2S_2O_4$), while appropriate oxidising agents include atmospheric oxygen and hydrogen peroxide.

4.8 SYNTHESIS OF CARBONYL COLORANTS

The synthesis of carbonyl colorants uses a wide diversity of chemical methods, in which each individual product essentially requires its own characteristic route. This is in complete contrast to the synthesis of azo dyes and pigments (Chapter 3) where a common reaction sequence is universally used. The subject is too vast to attempt to be comprehensive in a text of this nature. The following section, therefore, presents an overview of some of the fundamental synthetic strategies that may be used to prepare some of the more important types of carbonyl colorant.

4.8.1 Synthesis of Anthraquinones

The synthesis of anthraquinone colorants may effectively be envisaged as involving two general stages. The first stage involves the construction of the anthraquinone ring system and in the second phase the anthraquinone nucleus is elaborated to produce the desired structure. Frequently, the latter involves substitution reactions, but group interconversion and further cyclisation reactions may also be employed. Although the chemistry of the synthesis of most anthraquinone dyes and pigments is long established, some of the mechanistic details of the individual reactions remain unexplained, at least in detail.

There are two principal ways of constructing the anthraquinone system that are successful industrially. The first of these is the oxidation of anthracenes and the second involves a Friedel–Crafts acylation route. Anthracene, **4.26**, a readily available raw material, may be oxidised to give anthraquinone, **4.4**, in high yield as illustrated in Scheme 4.2. The most important oxidising agents used in this process are sodium dichromate and sulfuric acid (chromic acid) or nitric acid. This route is of considerable importance for the synthesis of the parent anthraquinone, but it is of much less importance for the direct synthesis of substituted anthraquinones. The main reasons for this are that there are few substituted anthracenes readily available as starting materials and also because many substituents would be susceptible to the strongly oxidising conditions used.

An alternative route to anthraquinone, which involves Friedel–Crafts acylation, is illustrated in Scheme 4.3. This route uses benzene and phthalic anhydride as starting materials. In the presence of aluminium(III) chloride, a Lewis acid catalyst, these compounds react to form 2-benzoylbenzene-1-carboxylic acid, **4.27**. Intermediate **4.27** is then heated with concentrated sulfuric acid, under which conditions cyclisation to anthraquinone (**4.4**) takes place. Both stages of this reaction sequence involve Friedel–Crafts acylation reactions. In the first stage the reaction is intermolecular, while the second step, in which cyclisation takes place, involves an intramolecular reaction. In contrast to the oxidation route, the Friedel–Crafts route offers considerable versatility. Various substituted benzene derivatives and phthalic anhydrides may be brought together as starting materials leading to a wide range of substituted anthraquinones. For example, the use of toluene rather than benzene leads directly to

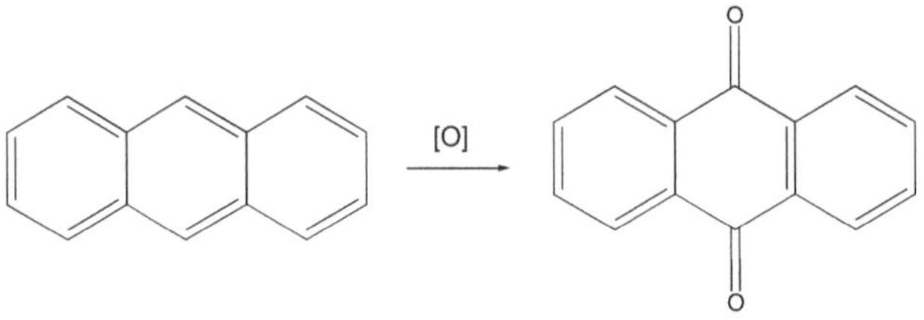

4.26   4.4

Scheme 4.2 Oxidation of anthracene.

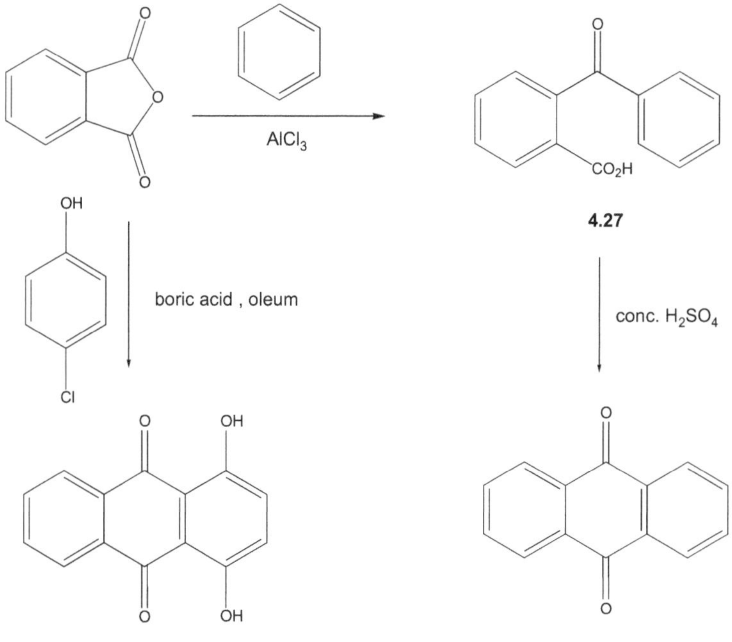

Scheme 4.3 Friedel–Crafts route to anthraquinones.

2-methylanthraquinone. A further industrially important example is also illustrated in Scheme 4.3. Starting from 4-chlorophenol and phthalic anhydride, two inexpensive and readily available starting materials, 1,4-dihydroxyanthraquinone, quinizarin, **4.28**, an important intermediate in the synthesis of several anthraquinone dyes may be synthesised efficiently in a 'one-pot' reaction. Boric acid in oleum is used as the catalyst and solvent system in this case. During the course of the reaction in which the anthraquinone nucleus is formed, a hydroxyl group replaces the chlorine atom. The detailed mechanism by which this takes place has not been fully established.

The outer rings of the anthraquinone molecule are aromatic in nature and as such are capable of undergoing substitution reactions. The reactivity of the rings towards substitution is determined by the fact that they are attached to two electron-withdrawing carbonyl groups. The presence of these groups deactivates the aromatic rings toward electrophilic substitution. Nevertheless, using reasonably vigorous conditions, anthraquinone may be induced to undergo electrophilic substitution reactions, notably nitration and sulfonation. Scheme 4.4 illustrates a series of electrophilic substitution reactions of anthraquinone. Nitration of anthraquinone with mixed

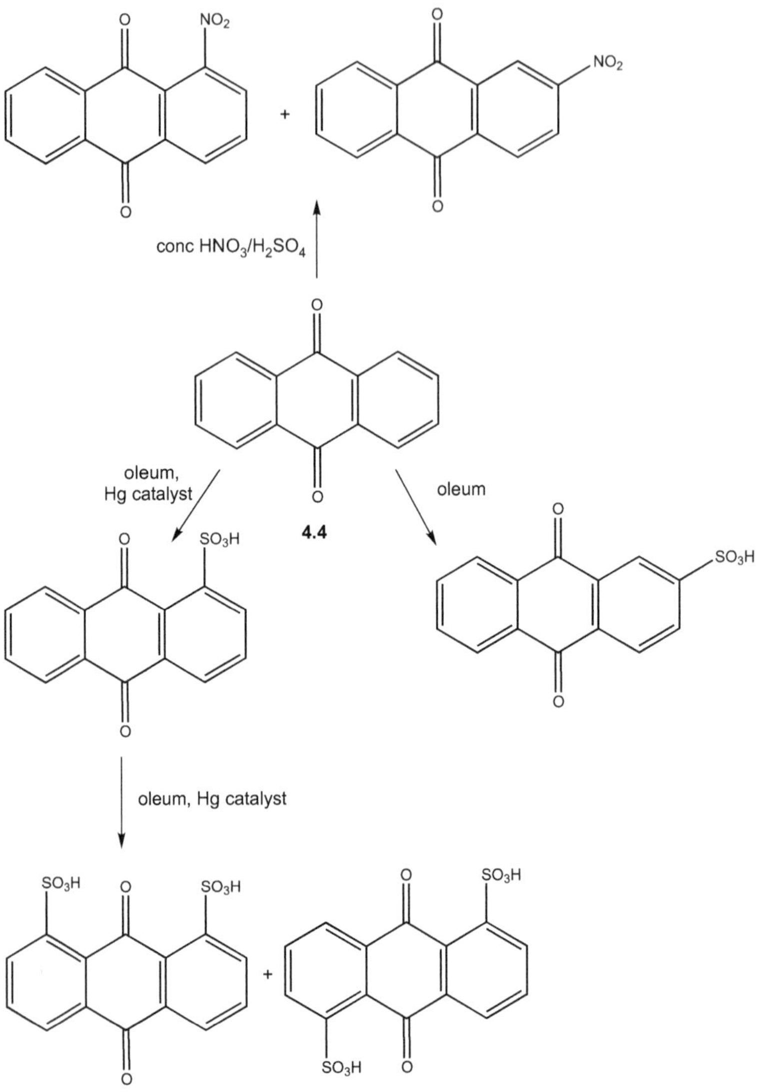

Scheme 4.4 Some useful electrophilic substitution reactions of anthraquinone, **4.4**.

concentrated nitric and sulfuric acids leads to a mixture of the 1- and 2-nitroanthraquinones, which is difficult to separate and so this process is of limited use. However, nitration of some substituted anthraquinones, particularly some hydroxy derivatives, can give more useful products. Sulfonation of anthraquinone requires the use of oleum at elevated temperatures, and under these conditions reaction leads mainly to the 2-sulfonic acid. However, in the presence of a mercury(II) salt as catalyst, the 1-isomer, a much more useful dye intermediate, becomes the principal product. A probable explanation for these observations is that in the absence of a catalyst the 2-position is preferred sterically. In the presence of the catalyst, it has been proposed that mercuration takes place first preferentially at the 1-position, and the mercury is displaced subsequently by the electrophile responsible for sulfonation. Disulfonation of anthraquinone in the presence of mercury salts leads to a mixture of the 1,5- and 1,8-disulfonic acids, which are easily separated and these are also useful dye intermediates.

Nucleophilic substitution reactions, in which the aromatic rings are activated by the presence of the carbonyl groups, are commonly used in the elaboration of the anthraquinone nucleus, particularly for the introduction of hydroxyl and amino groups. These substitution reactions are often catalysed either by boric acid or transition metal ions. As an example, amino and hydroxyl groups may be introduced into the anthraquinone system by nucleophilic displacement of sulfonic acid groups. Another example of an industrially useful nucleophilic substitution is the reaction of 1-amino-4-bromoanthraquinone-2-sulfonic acid (bromamine acid, **4.29**) with aromatic amines, as shown in Scheme 4.5, to give a series of useful water-soluble blue dyes. The displacement of bromine in these reactions is catalysed by the presence of copper(II) ions.

An important route to 1,4-diaminoanthraquinones, represented by structure **4.31**, is illustrated in Scheme 4.6. Quinizarin, **4.28**, is first reduced to leucoquinizarin, which has been shown to exist as the diketo structure **4.30**. Condensation of compound **4.30** with two moles of an amine, followed by oxidation leads to the diaminoanthraquinone **4.31**. Boric acid is a useful catalyst for this reaction, particularly when less basic amines are used.

The syntheses of three polycyclic anthraquinones, indanthrone, **4.5a**, pyranthrone, **4.6a**, and flavanthrone, **4.6b**, are illustrated in Scheme 4.7. Despite the structural complexity of the products, the

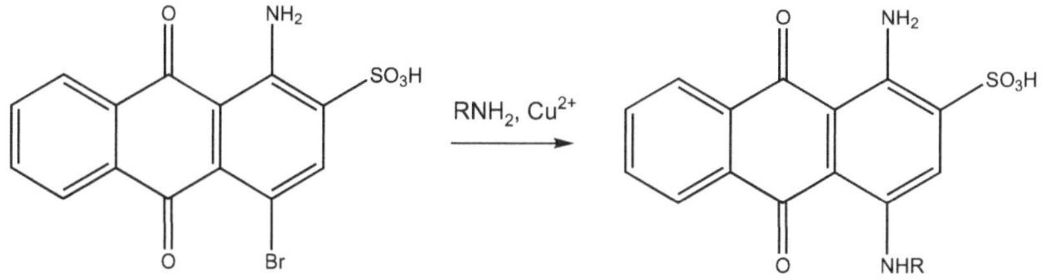

Scheme 4.5 Reaction of bromamine acid with aromatic amines.

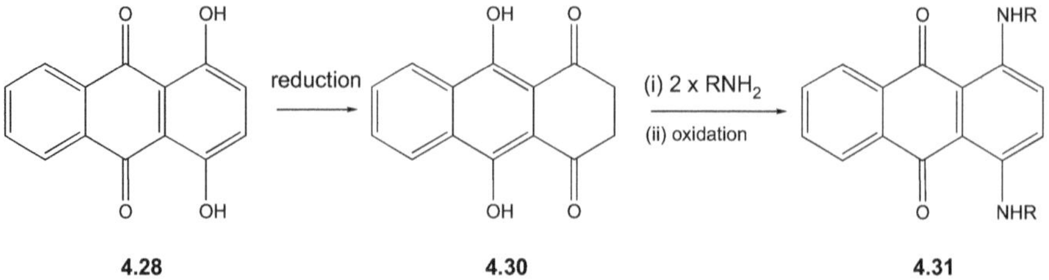

Scheme 4.6 Formation of 1,4-diaminoanthraquinones from quinizarin.

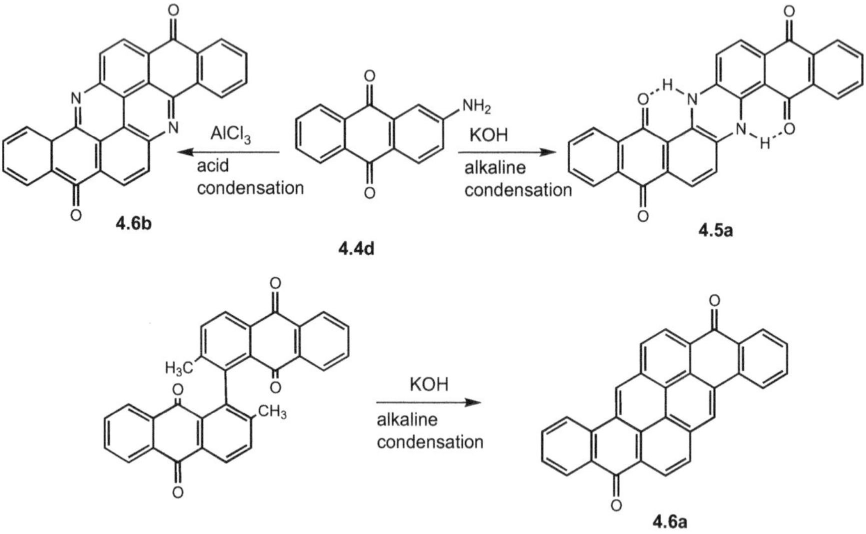

Scheme 4.7 Syntheses of the polycyclic anthraquinones indanthrone, **4.5a**, pyran-throne, **4.6a**, and flavanthrone, **4.6b**.

syntheses of these types of compound are often quite straightforward involving, for example, condensation or oxidative cyclisation reactions. For example, the blue vat dye indanthrone, **4.5a**, is prepared by fusion of 2-aminoanthraquinone, **4.4d**, with either sodium or potassium hydroxide at around 220 °C. Curiously, the same dye can be prepared from 1-aminoanthraquinone, **4.4c**, in a similar way, although the 2-isomer is the usual industrial starting material. Pyranthrone, **4.6a**, is also prepared by an alkaline fusion process, starting in this case from 2,2′-dimethyl-1,1′-dianthraquinonyl. The methyl groups in this molecule are sufficiently acidic as a result of activation by the carbonyl groups to be ionised and subsequently give rise to cyclisation by a reaction analogous to an aldol condensation. Flavanthrone, **4.6b**, may also be prepared by a condensation reaction starting from 2-aminoanthraquinone, **4.4d**. The yellow vat dye is obtained from fusion of compound **4.4d** with alkali at temperatures of around 300 °C or by acid-catalysed condensation, for example using aluminium(III) chloride. However, its industrial manufacture usually involves a similar condensation process, either acid or base catalysed, starting from 1-chloro-2-acetylaminoanthraquinone.

4.8.2 Synthesis of Indigoid Colorants

Indigo, **4.8**, was for many centuries obtained from natural sources. The chemical structure of indigo was first proposed by von Baeyer in 1869, and eleven years later he reported the first successful synthesis, a multistage route starting from *o*-nitrocinnamic acid. The first successful commercial synthesis of indigo, attributed to Heumann in 1897, is shown in Scheme 4.8. In this classical synthesis, phenylglycine-*o*-carboxylic acid, **4.32**, is converted by fusion with sodium hydroxide at around 200 °C, in the absence of air, into indoxyl-2-carboxylic acid, **4.33**. This material readily decarboxylates and oxidises in air to indigo. A much more efficient synthesis, which forms the basis of the manufacturing method in use today, is due originally

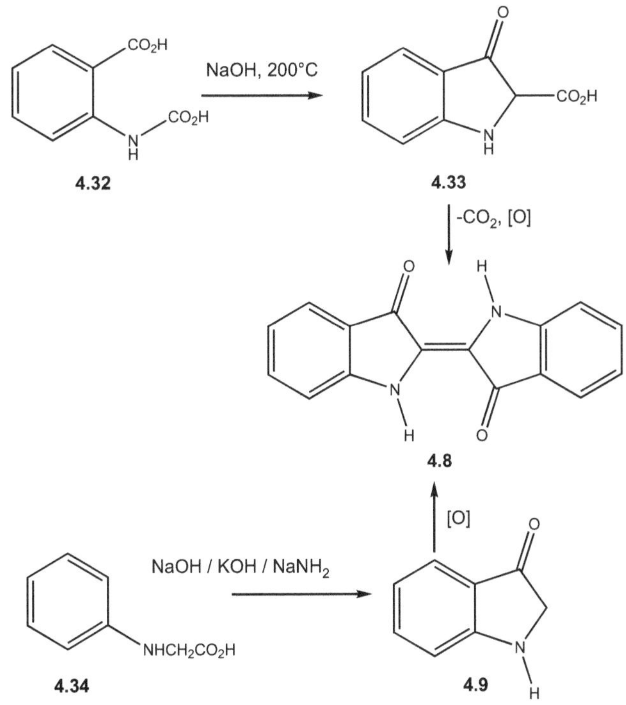

Scheme 4.8 Synthetic routes to indigo, **4.8**.

to Pfleger (1901). In this route, also illustrated in Scheme 4.8, the more readily available starting material phenylglycine, **4.34**, is treated in an alkaline melt of sodium and potassium hydroxides containing soda-mide. This process leads directly to indoxyl, **4.9**, which undergoes spontaneous oxidative dimerisation in air to indigo. Thioindigo, **4.11b**, is best prepared from *o*-carboxybenzene-thioglycolic acid, **4.35**, by a route analogous to Heumann's indigo synthesis. The final oxidation step in this case uses sulfur, rather than oxygen, as the oxidising agent.

4.35

4.8.3   Synthesis of Benzodifuranones and Coumarin Dyes

As has been the case with most new chromogenic systems, the ben-zodifuranones were discovered in the laboratory by chance. A con-siderable amount of industrial development was needed subsequently to establish commercially-viable synthetic routes to the dyes, and the difficulties experienced explain, at least in part, the significant time gap between the initial discovery and the appearance of the first commercial dyes. In particular, processes were required to provide the unsymmetrical molecules, which proved to have superior dyeing properties as disperse dyes for polyester. A successful approach, as illustrated in Scheme 4.9, involves the reaction of the *p*-hydroquinone, **4.36**, with mandelic acid, **4.37a**, to give intermediate **4.38**, which may subsequently be reacted with the substituted man-delic acid, **4.37b**, to give the benzodifuranone, **4.13**.[17]

The greenish yellow fluorescent coumarin dyes represented by structure **4.14** may be synthesised in several ways. One important general route involves the base-catalysed condensation of aldehyde **4.39** with the appropriate cyanomethyl compound, **4.40**, followed by acid hydrolysis of the resulting imine, **4.41**, as shown in Scheme 4.10.

4.8.4   Synthesis of Carbonyl Pigments

Several methods of synthesis of the quinacridones are reported. Each of these involves several stages, accounting at least in part for the somewhat higher cost of these pigments. The two most important methods to the parent compound are outlined in Scheme 4.11. In both routes, the starting material, diethyl succinylsuccinate, **4.42**,

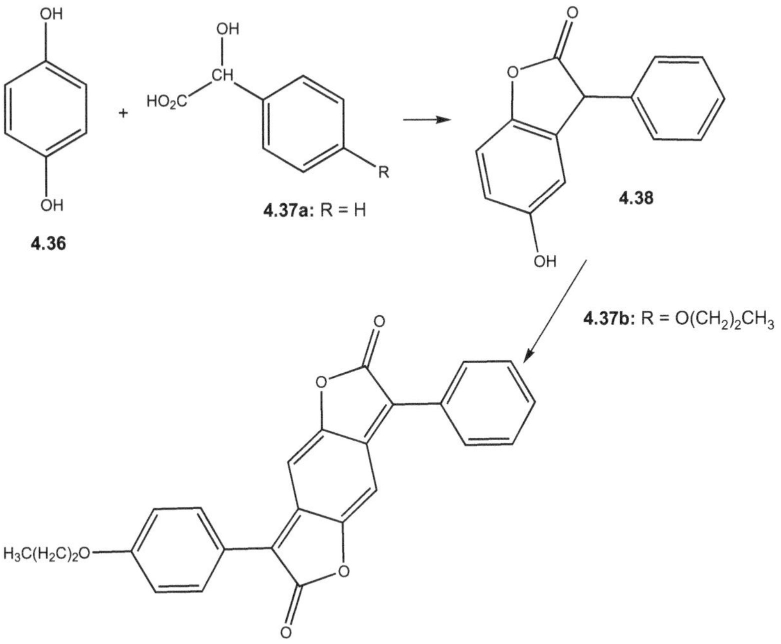

Scheme 4.9 Synthesis of benzodifuranone dye **4.13**.

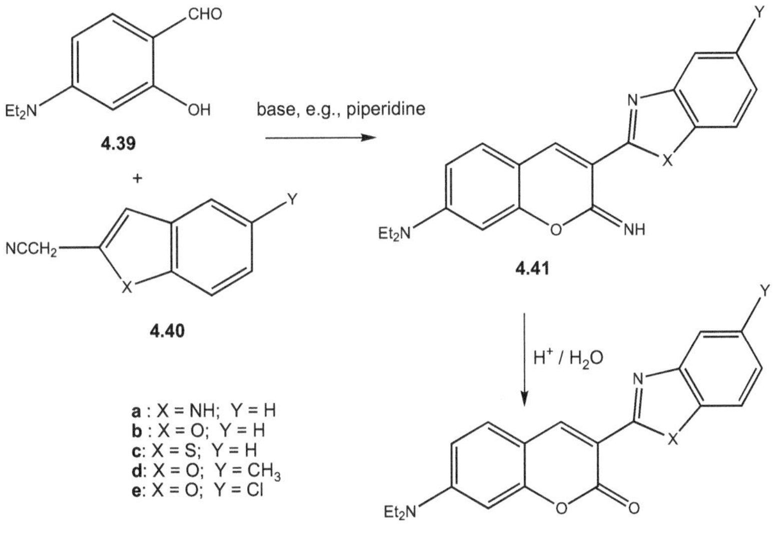

Scheme 4.10 Synthesis of fluorescent coumarin dyes **4.14a–e**.

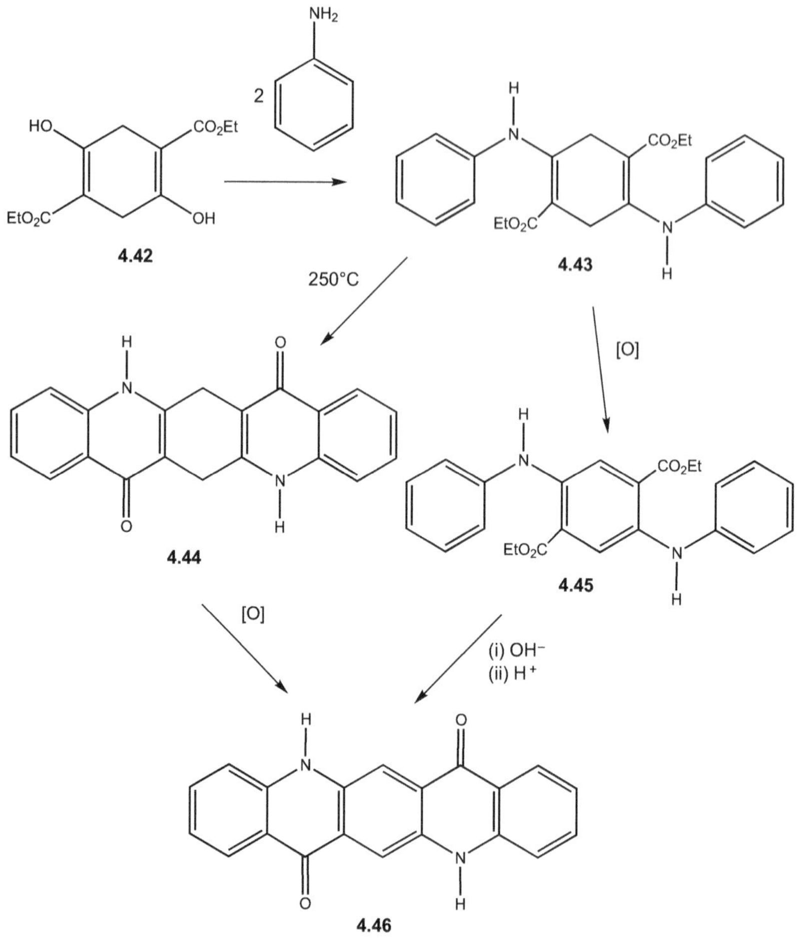

Scheme 4.11 Synthetic routes to linear trans quinacridone **4.46**.

(1 mole), which may be prepared by a base-catalysed self-condensation of a succinic acid diester, is condensed with aniline (2 moles) to form the 2,5-diphenylamino-3,6-dihydroterephthalic acid diester, **4.43**. Diester **4.43** undergoes ring closure at an elevated temperature in a high boiling solvent to give the dihydroquinacridone, **4.44**, which is relatively easily oxidised (*e.g.*, with sodium 3-nitrobenzene-1-sulfonate) to the quinacridone, **4.46**. Alternatively, compound **4.43** may be oxidised to the 2,5-diarylaminoterephthalate diester, **4.45**. Base hydrolysis of the diester, followed by ring closure by treatment, for example, with polyphosphoric acid gives the quinacridone.[32]

The formation of a DPP molecule was first reported in 1974 as a minor product in low yield from the reaction of benzonitrile with ethyl

bromoacetate and zinc.[33] A study by research chemists at Ciba Geigy
into the mechanistic pathways involved in the formation of the molecules led to the development of an efficient 'one-pot' synthetic procedure to DPP pigments from readily available starting materials
(Scheme 4.12). The reaction involves the treatment of diethyl succinate
(1 mole) with an aromatic cyanide (2 moles) in the presence of a strong
base. The reaction proceeds through the intermediate **4.47**, which may
be isolated and used to synthesise unsymmetrical derivatives.[25,26]

Perylenes **4.21** are diimides of perylene-3,4,9,10-tetracarboxylic acid
and may be prepared by reaction of the bis-anhydride of this acid (**4.48**)
(1 mole) with the appropriate amine (2 moles) in a high-boiling solvent
as illustrated in Scheme 4.13. The synthesis of perinones **4.22** and **4.23**
involves condensation of naphthalene-1,4,5,8-tetracarboxylic acid with

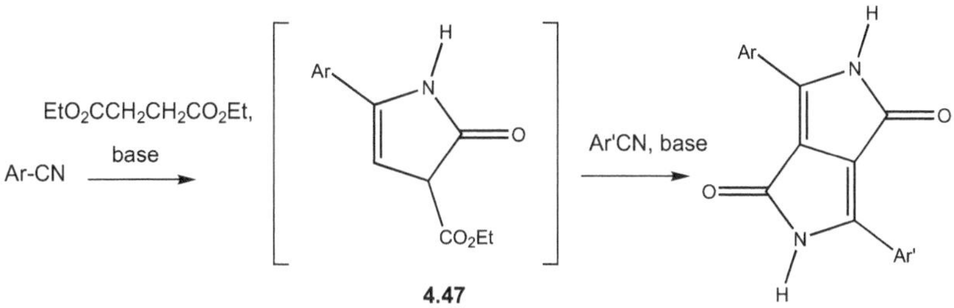

Scheme 4.12 Synthesis of DPP (1,4-diketopyrrolo[3,4-*c*]pyrrole) pigments.

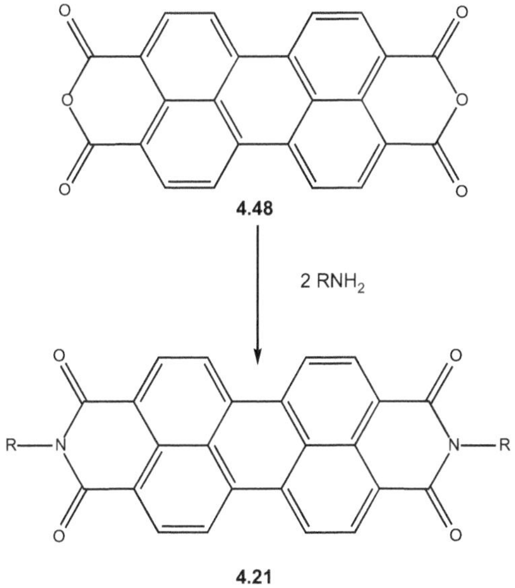

Scheme 4.13 Synthesis of perylenes **4.21**.

o-phenylenediamine in refluxing acetic acid. This affords a mixture of the two isomers, which may be separated by a variety of methods, generally involving their differential solubility in acids and alkalis. Quinophthalone pigment **4.24** may be prepared by the reaction of 8-amino-2-methylquinoline, **4.49**, with tetrachlorophthalic anhydride (**4.50**) (2 moles) (Scheme 4.14). Isoindoline pigment **4.25** is prepared from the condensation of 1-amino-3-iminoisoindoline (**4.51**) (1 mole) with barbituric acid (**4.52**) (2 moles) as illustrated in Scheme 4.15.

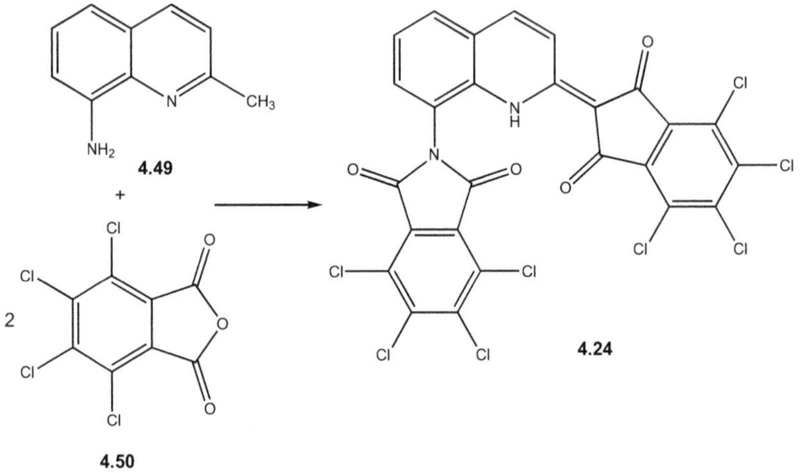

Scheme 4.14 Synthesis of quinophthalone pigment **4.24**.

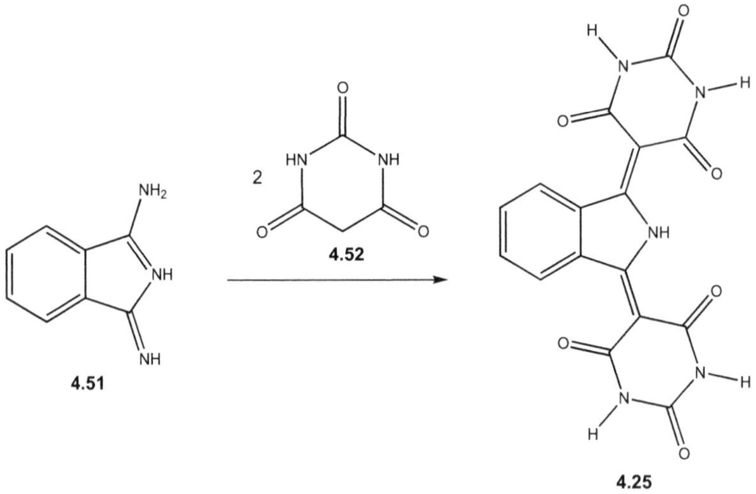

Scheme 4.15 Synthesis of isoindoline pigment **4.25**.

REFERENCES

1. G. Hallas, in *Colorants and Auxiliaries: Organic Chemistry and Application Properties*, ed. J. Shore, Society of Dyers and Colourists, Bradford, 1990, vol. 1, ch. 6.
2. H. Zollinger, *Color Chemistry: Syntheses, Properties and Applications of Organic Dyes and Pigments*, Wiley-VCH Verlag GmbH, Weinheim, 3rd edn, 2003, ch 8.
3. P. F. Gordon and P. Gregory, *Organic Chemistry in Colour*, Springer-Verlag, New York, 1983, ch. 4.
4. R. Chenciner, *Madder Red: A History of Luxury and Trade, Plant Dyes and Pigments*, Curzon, Richmond, UK, 2000.
5. J. Balfour-Paul, *Indigo*, British Museum Press, London, 1998.
6. A. Baeyer and V. Drewson, *Chem. Ber.*, 1883, **16**, 2205.
7. A. Reis and W. Schneider, *Z. Kristallogr.*, 1828, **184**, 269.
8. D. Jacquemin, J. Preat, V. Wathelet and E. A. Perpete, *J. Chem. Phys.*, 2006, **124**, 074014.
9. H. von Eller, *Bull. Soc. Chim. Fr.*, 1955, 1438.
10. L. Serrano-Andres and B. O. Roos, *Chem.–Eur. J.*, 1997, **3**, 717.
11. R. M. Christie, *Biotechnic Histochem.*, 2007, **82**, 51.
12. M. Klessinger and W. Luttke, *Tetrahedron*, 1963, **19**, 315.
13. M. Klessinger, *Tetrahedron*, 1966, **22**, 3355.
14. M. Klessinger, *Dyes Pigments*, 1982, **3**, 235.
15. R. M. Christie, in *Indirubin the Red Shade of Indigo*, ed. L. Meijer, N. Guyard, L. A. Skaltsounis and G. Eisenbrand, Life in Progress Editions, Roscoff, France, 2006, ch 10.
16. C. W. Greenhalgh, J. L. Carey and D. F. Newton, *Dyes Pigments*, 1980, **1**, 103.
17. C. W. Greenhalgh, J. L. Carey, N. Hall and D. F. Newton, *J. Soc. Dyers Colourists*, 1994, **110**, 178.
18. R. M. Christie, *Rev. Prog. Color.*, 1993, **23**, 1.
19. R. M. Christie and C. H. Lui, *Dyes Pigments*, 2000, **47**, 79.
20. W. Herbst and K. Hunger, *Industrial Organic Pigments*, Wiley-VCH Verlag GmbH, Weinheim, 4th edn, 2006.
21. R. M. Christie, *The Organic and Inorganic Chemistry of Pigments*, Oil & Colour Chemists Association, London, 2002.
22. E. E. Jaffe, *J. Oil Colour Chem. Assoc.*, 1992, **75**, 24.
23. G. Lincke, *Dyes Pigments*, 2000, **44**, 101.
24. G. D. Potts, W. Jones, J. F. Bullock, S. J. Andrews and S. J. Maginn, *J. Chem. Soc., Chem. Commun.*, 1994, 2565.
25. A. Iqbal, L. Cassar, A. C. Rochat, J. Pfenninger and O. Wallquist, *J. Coatings Technol.*, 1988, **60**, 1.

26. A. Iqbal, M. Jost, R. Kirchmayr, A. C. Rochat, J. Pfenninger and O. Wallquist, *Bull. Soc. Chim. Belg.*, 1988, **97**, 615.
27. F. Graser and E. Hadicke, *Liebigs Ann. Chem.*, 1980, 1994.
28. F. Graser and E. Hadicke, *Liebigs Ann. Chem.*, 1984, 483.
29. F. Graser and E. Hadicke, *Acta Crystallogr., Sect. C*, 1986, **42**, 189.
30. P. Zugenmaier, J. Duff and T. L. Bluhm, *Cryst. Res. Technol.*, 2000, **9**, 1095.
31. L. Michaelis and M. Schubert, *Chem. Rev.*, 1938, **22**, 437.
32. S. S. Labana and L. L Labana, *Chem. Rev.*, 1967, **67**, 1.
33. D. G. Farnum, G. Mehta, G. G. I. Moore and F. P. Siegal, *Tetrahedron Lett.*, 1974, **29**, 2549.

CHAPTER 5

Phthalocyanines

5.1 INTRODUCTION

The phthalocyanines represent the most important chromophoric system developed during the twentieth century.[1–3] Historically, the most important event was probably their accidental discovery around 1928 by a dye manufacturing company in Grangemouth, Scotland. However, there is little doubt that researchers prior to this had observed the formation of phthalocyanines, although the significance of their observations was not fully recognised. In 1907, von Braun and Tscherniak were engaged in a study of the chemistry of *o*-cyano-benzamide, **5.1**, and discovered that when this compound was heated a trace amount of a blue substance was obtained.[4] This compound was almost certainly metal-free phthalocyanine, **5.2**. In 1927, de Diesbach and von der Weid reported that when 1,2-dibromobenzene was treated with copper(ı) cyanide in boiling quinoline for eight hours, a blue product was obtained in reasonable yield.[5] This was almost certainly the first preparation of copper phthalocyanine (CuPc), **5.3**. They obtained the molecular formula of the compound from elemental analysis and noted its remarkable stability to alkali, concentrated acids and heat but were unable to propose a structure. In 1928, in the manufacture of phthalimide by Scottish Dyes (later to become part of ICI) from the reaction of phthalic anhydride with ammonia in a glass-lined reactor, the formation of a blue impurity was observed in certain production batches. This contaminant was isolated as a dark blue, insoluble crystalline substance. Ultimately, the

Colour Chemistry, 2nd edition
By Robert M Christie
© R M Christie 2015
Published by the Royal Society of Chemistry, www.rsc.org

compound proved to be iron phthalocyanine (FePc), the source of the
iron being the wall of the reactor, which became exposed due to a flaw
in the glass lining. An independent synthesis involving passing ammonia gas through molten phthalic anhydride in the presence of iron
filings confirmed the findings. Following this discovery, the colour
manufacturing industry was quick to recognise the unique properties of
the compounds and to exploit their commercial potential. The phthalocyanines have subsequently emerged as one of the most extensively
studied classes of compounds, because of their intense, bright colours,
their high stability and their unique molecular structure.[6–9]

5.1

5.2   5.3

5.2   STRUCTURE AND PROPERTIES OF PHTHALOCYANINES

Elucidation of the structure of the phthalocyanines followed some pioneering research into the chemistry of the system by Linstead of
Imperial College, University of London.[10–14] The structure that we now
recognise was first proposed from the results of analysis of a number of
metal phthalocyanines, which provided the molecular formulae, and
from an investigation of the products of degradation studies. Finally,
Robertson confirmed the structure as a result of one of the classical
applications of single-crystal X-ray crystallography.[15–17]

   The phthalocyanine system, which may be considered as the tet-
raaza derivative of tetrabenzoporphin, is planar, consisting of four
isoindole units connected by four nitrogen atoms forming an internal
16-membered ring of alternate carbon and nitrogen atoms. Most

phthalocyanines contain a central complexed metal atom, derivatives having been prepared from most of the metals in the periodic table. The central metal atom is in a square-planar environment. The phthalocyanines are structurally related to the natural pigments chlorophyll, **5.4**, and haemin, **5.5**, which are porphyrin derivatives. However, unlike these natural colorants, which have limited stability, the phthalocyanines exhibit exceptional stability and they are in fact probably the most stable of all synthetic organic colorants. Copper phthalocyanine, used here as an example, is usually illustrated as structure **5.3**, which contains three benzenoid and one *o*-quinonoid outer rings. However, it has been established that the molecule is centrosymmetric and this means that structure **5.3** should be regarded as only one of a large number of resonance forms contributing to the overall molecular structure. The extensive resonance stabilisation of the phthalocyanines may well account for their high stability. The phthalocyanines are aromatic molecules, a feature that has been attributed to the 18 π-electrons in the perimeter of the molecules. Phthalocyanines, together with porphyrins, are referred to in general as aza[18]annulenes, the term *annulene* denoting a conjugated cyclic system of methine groups.

5.4
R = CH₃: chlorophyll a
R = CHO: chlorophyll b

5.5

The metal phthalocyanines in general show brilliant, intense colours. The UV/visible absorption spectrum of metal-free phthalocyanine (**5.2**) in 1-chloronaphthalene shows two absorption bands of similar intensity at 699 and 664 nm. The corresponding spectra of metal phthalocyanines, however, show a single narrow major absorption band, a feature that has been explained by their higher symmetry compared with the metal-free compound and it is the nature of this absorption that gives rise to the brilliance and intensity of their colour.[18] The colours of traditional phthalocyanine dyes and pigments are restricted to blues and greens, although recent years have seen the development of several derivatives whose absorption is extended into the near-infrared region of the spectrum. Phthalo-cyanines provide an example of the polyene rather than the donor/ acceptor chromogenic type, and in this sense there is a contrast with azo and carbonyl dyes, which are described in the previous two chapters. As a consequence, the valence bond (resonance) approach may not be applied so readily to provide an explanation of their colour (Chapter 2). The phthalocyanines and structurally related systems have been extensively investigated theoretically by molecular orbital methods and these approaches, which range from semi-empirical methods through to *ab initio* calculations, have provided a successful account of the molecular structure and light absorption properties of the system.[18–21]

The UV/visible spectra of phthalocyanines in solution show two strong absorption bands. The longer wavelength band in the visible region is referred to as the *Q* band, while the *B* band (or *Soret* band) is in the UV region. The position of the visible absorption band of metal phthalocyanines is dependent on the nature of the central metal ion, the substituent pattern on the outer rings and the degree of ring annelation.[22,23] Among the most extensively investigated phthalo-cyanines are the complexes of the first transition series metals, iron, cobalt, nickel, copper and zinc. Within this series, the colour is affected little by the nature of the central metal ion, in which case the λ_{max} values are to be found in the range 670–685 nm. The most hypsochromic of the series of unsubstituted metal phthalocyanines is PtPc (λ_{max} 652 nm) while the most bathochromic is PbPc (λ_{max} 714 nm). Neither of these is of particular interest commercially due either to economic or toxicity considerations, but the bath-ochromic effect of the vanadyl derivative, VOPc (λ_{max} 701 nm), is of interest from the point of view of extending the absorption range of phthalocyanines. Substituents on the outer aromatic rings almost invariably shift the absorption band to longer wavelengths. Copper hexadecachlorophthalocyanine, for example, absorbs at 720 nm in 1-chloronaphthalene, giving rise to its green colour. There is some considerable interest in phthalocyanines in which the absorption band is extended into the near-infrared region for applications such as optical data storage and security printing (Chapter 11). This may be achieved in a number of ways. For example, the arylthio group causes a much more pronounced bathochromic shift than the halogens and several copper poly(arylthiophthalocyanine) derivatives have been patented for applications that make use of their intense absorption in the near-infrared region of the spectrum. Alternatively, extending the outer ring system by annelation shifts the absorption band bathochromically. Copper naphthalocyanine, **5.6**, for example, absorbs at 784 nm, while the corresponding vanadyl derivative gives a λ_{max} of 817 nm. A bathochromic shift of the absorption may also be observed when the compound is in the solid state, compared with the solution phase. For example, titanyl phthalocyanines may be produced in a number of crystal phases absorbing in the range 780–830 nm.

5.6

While textile dyes based on phthalocyanines are of rather limited importance, the phthalocyanines provide by far the most important blue and green organic pigments. In particular, copper phthalo-cyanine, **5.3**, CI Pigment Blue 15, is by far the most important blue pigment, finding almost universal use as a colorant in a wide range of paint, printing ink and plastics applications. In fact, there is a con-vincing argument that it is the most important of all organic pig-ments. It owes this dominant position to its intense brilliant blue colour and excellent technical performance. The pigment exhibits exceptional stability to light, heat, solvents, alkalis, acids and other chemicals. Among the features which demonstrate this high stability are the ability of the material to sublime unchanged at temperatures above 500 °C, and the observation that it dissolves without decomposition in concentrated sulfuric acid, from which solutions it may be recovered. In addition, copper phthalocyanine is a relatively low cost product since, despite its structural complexity, its manufacture (see next section) is straightforward, giving high yields from inexpensive, commodity starting materials.

Copper phthalocyanine exhibits *polymorphism*, which refers to the ability of a material to adopt different crystal structural arrangements or phases. The most important crystal phases are the α- and β-forms, and several other forms have been reported. Both the α- and β-forms are of commercial importance. The two forms exhibit different hues, the α-form being reddish-blue while the β-form is greenish-blue. The β-form is the more stable form, particularly towards organic solvents. The α-form has a tendency to convert in the presence of certain solvents into the β-form with a corresponding change in shade, unless it is stabilised, for example by the incorporation of a single chlorine substituent. β-CuPc is of particular importance as the cyan pigment used most commonly in printing inks while the α-form is more important in surface coatings and plastics applications. A simplified structural comparison between the α- and β-phases is shown in Figure 5.1. In both cases, the molecules are arranged in stacks with the intermolecular interactions defined mainly by π–π interactions within the stacks. It has been suggested that in the 'herring-bone' arrangement of CuPc molecules in the crystal structure of the β-form the copper atom at the centre of each molecule is coordinated to nitrogen atoms in adjacent molecules forming a distorted octahedron, a coordination geometry that is particularly favoured in complexes of copper. No such octahedral

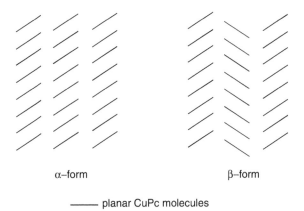

α–form   β–form

——— planar CuPc molecules

Figure 5.1   Polymorphism of copper phthalocyanine (CuPc).

coordination is possible in the parallel arrangement of molecules in the crystal structure of α-CuPc, a factor which may contribute to the lower stability of this polymorphic form.[24]

Several other phthalocyanines are used commercially as pigments. The most important green organic pigments are the halogenated copper phthalocyanines CI Pigment Green 7, in which the 16 ring hydrogen atoms of the CuPc molecule are replaced virtually completely by chlorine, and CI Pigment Green 36, a designation that incorporates a range of bromo- and bromochloro-copper phthalocyanines. The hue of these pigments becomes progressively yellower with increasing bromine substitution. The green pigments are complex statistical mixtures of polyhalogenated products, in contrast to the unsubstituted blue pigment, which is a pure species, and consequently differences in crystal form are not observed. The phthalocyanine greens exhibit the same outstanding colouristic and technical performance as the blue pigments from which they are derived and find equally widespread use in the coloration of paints, printing inks and plastics. Although phthalocyanine complexes have been prepared from virtually every metallic element in the periodic table, only the copper derivatives are of any significant commercial importance as pigments, simply because the copper compounds give the best combination of colour and technical properties. However, metal-free phthalocyanine, **5.2**, finds some use as a greenish-blue pigment of high stability.

In view of the immense commercial importance of phthalocyanines as pigments, it is perhaps surprising that only a few are important as textile dyes. This may be primarily due to the size of the molecules, which is too large to allow penetration into many fibres, especially the synthetic fibres polyester and polyacrylonitrile. However, there are a few polysulfonated copper phthalocyanines on which turquoise/cyan direct dyes for cotton and paper are based. These dyes provide bright

5.7

shades and have good lightfastness. An example is CI Direct Blue 86, **5.7**, which is a disulfonated copper phthalocyanine, although it is likely to be a mixture of positional isomers. CI Direct Blue 199 is a similar dye of complex composition containing sulfonate and sulfonamide groups. In addition, turquoise reactive dyes for cotton incorporate the copper phthalocyanine system as the chromophoric unit (Chapter 8).

While the industrial importance of phthalocyanines is dominated by traditional applications as pigments and, to a lesser extent, textile dyes, they have been extensively investigated for a wide range of other applications because of their unique light absorption, electronic and chemical properties and their high stability. As functional colorants, they are of interest in electrochromic systems, electrophotography, optical data storage, organic solar cells and as photosensitisers for photodynamic therapy of cancer[25-27] (Chapter 11). They are also important in various chemical sensing systems and as reaction catalysts.[28-30]

5.3 SYNTHESIS OF PHTHALOCYANINES

The synthesis of metal phthalocyanines requires essentially the presence of three components: a phthalic acid derivative, such as phthalic anhydride, phthalimide, phthalonitrile or *o*-cyanobenzamide, a source of nitrogen (in cases where the phthalic acid derivative does not itself contain sufficient nitrogen) and an appropriate metal derivative. Commonly the reaction requires high temperatures and may be carried out in a high boiling solvent or as a 'dry bake' process. In this way, using appropriate starting materials and reaction conditions, virtually the entire range of metal phthalocyanines may be prepared. Substituted phthalocyanines are prepared either by using an appropriately substituted phthalic acid derivative as a starting material, or by substitution reactions carried out on the unsubstituted derivatives. Metal-free phthalocyanines are conveniently prepared by subjecting certain labile metal derivatives, such as those of sodium or lithium, to acidic conditions. The method of synthesis is discussed further in this section for the case of copper phthalocyanine, because of the particular importance of this product. Although the structure of copper phthalocyanine is rather complex, its synthesis is remarkably straightforward. It may be prepared in virtually quantitative yield from readily available, low cost starting materials. Two chemically related methods, the phthalic anhydride and phthalonitrile routes, are commonly used for its manufacture. Both involve simultaneous

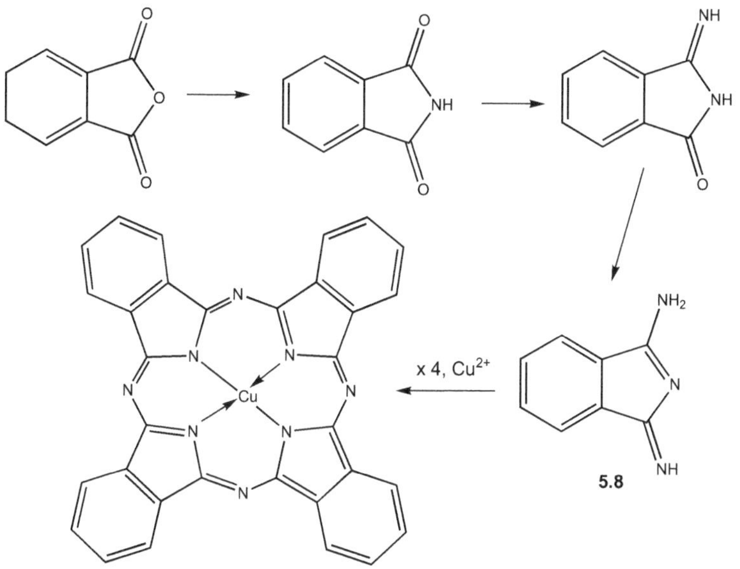

Scheme 5.1 Phthalic anhydride route to CuPc.

synthesis of the ligand and metal complex formation in a template procedure.

(a) The phthalic anhydride route
In the most commonly-encountered version of this method, phthalic anhydride is heated with urea, copper(I) chloride and a catalytic amount of ammonium molybdate in a high boiling solvent. An outline of the process is given in Scheme 5.1. Urea acts as the source of nitrogen in the process, the carbonyl group of the urea molecule being displaced as carbon dioxide. Mechanistic schemes have been proposed to explain the course of this synthesis but much of the detail remains to be established unequivocally. In essence, phthalic anhydride reacts with urea or products of its decomposition or polymerisation, resulting in progressive replacement of the oxygen atoms by nitrogen and, ultimately, the formation of the key intermediate 1-amino-3-imino-isoindoline, **5.8**. The presence of ammonium molybdate is essential to catalyse this part of the sequence. Subsequently, this intermediate undergoes a tetramerisation with cyclisation aided by the presence of the copper ion to form copper phthalocyanine.

(b) The phthalonitrile route
In this process, phthalonitrile, **5.9**, is heated to around 200 °C with copper metal or a copper salt, with or without a solvent. A mechanism

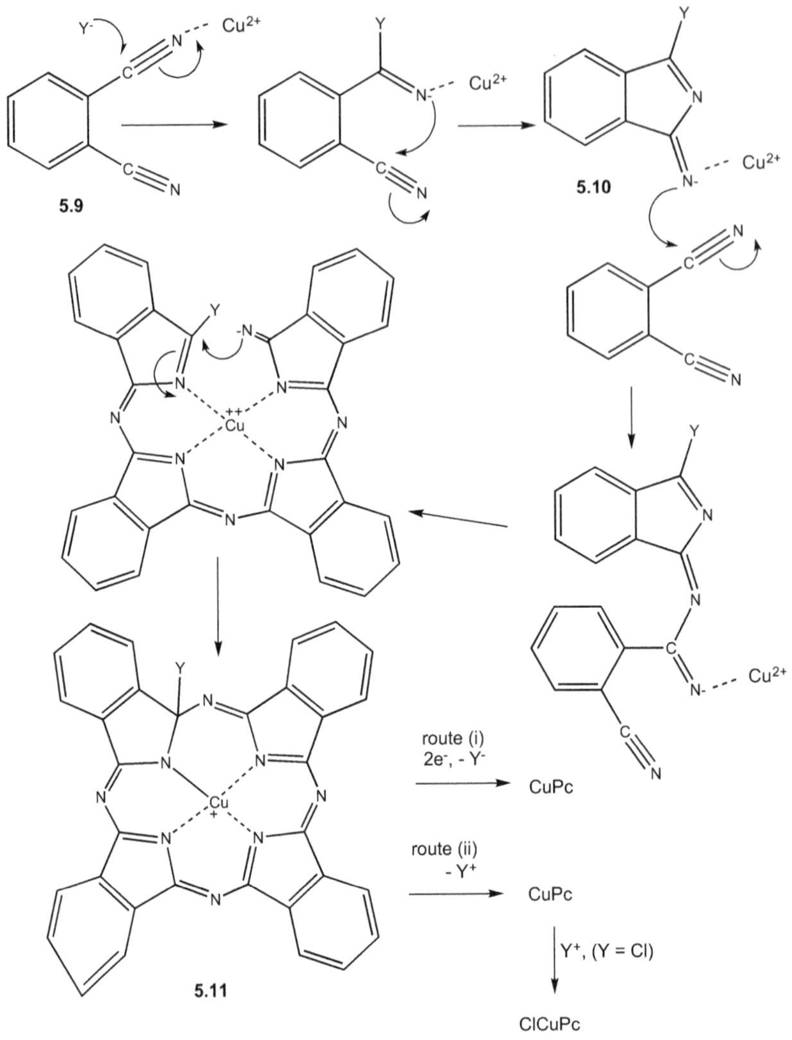

Scheme 5.2 A mechanism for the phthalonitrile route to CuPcs.

for the phthalonitrile route to copper phthalocyanine has been proposed as illustrated in Scheme 5.2.[31] It is suggested that reaction is
initiated by attack by a nucleophile (Y^-), most likely the counteranion
associated with the Cu^{2+} ion, at one of the cyano groups of the
phthalonitrile activated by its coordination with the Cu^{2+} ion. Cyclisation to isoindoline derivative **5.10** then takes place. Attack by
intermediate **5.10** on a further molecule of phthalonitrile then takes
place and, following a series of similar reactions including a cyclisation step, facilitated by the coordinating role of the Cu^{2+}, intermediate **5.11** is formed. When copper metal is the reactant, it is

proposed that two electrons are transferred from the metal, allowing elimination of Y^- to form copper phthalocyanine [route (i)]. Consequently, the Cu(0) is oxidised to Cu(II) as required to participate further in the reaction. When a copper(II) salt is used, it is suggested that Y^+ (the chloronium ion in the case of $CuCl_2$) is eliminated to form CuPc [route (ii)]. The product in this case, rather than copper phthalocyanine itself, is a monochloro derivative, formed by electrophilic attack of Cl^+ on the copper phthalocyanine initially formed. Copper monochlorophthalocyanine is important as it exists exclusively in the α-crystal form, which, unlike unsubstituted CuPc, is stable to solvents. It has been suggested that the single chlorine atom sterically prevents conversion into the β-form.

Both the phthalic anhydride and phthalonitrile routes generally produce a crude blue product, which is of far too large a particle size to be of use as a pigment. The original method developed for particle size reduction used acid pasting, which involves dissolving the crude product in concentrated sulfuric acid, followed by reprecipitation with water. This method gives the α-form of the pigment in a fine particle size form. Mechanical grinding of the crude blue product in the presence of inorganic salts, such as sodium chloride or calcium chloride, produces a mixture of the α- and β-CuPc, which may be converted into pure pigmentary β-CuPc by careful treatment of this mixture with certain organic solvents. Alternatively, grinding the crude material with inorganic salts in the presence of organic solvents can lead directly to β-CuPc in a fine particle size form.

Scheme 5.3 shows an outline of some important substitution reactions of copper phthalocyanine. Synthesis of the phthalocyanine green pigments involves the direct exhaustive halogenation of crude copper phthalocyanine blue with chlorine or bromine or an appropriate mixture of the two halogens, depending on the particular product required, at elevated temperatures in a suitable solvent, commonly an $AlCl_3$/NaCl melt. These reactions are examples of electrophilic substitution, reflecting the aromatic character of the copper phthalocyanine molecule. The crude green products, **5.12**, which are initially formed under these manufacturing conditions, are of large particle size due mainly to a high degree of aggregation. The crude form may be converted into an appropriate pigmentary form either by treatment with suitable organic solvents or by treatment with aqueous surfactant solutions. These processes effect a deaggregation of the product, producing a finer particle size, and in addition they increase the crystallinity of the products. Both of these effects provide a dramatic beneficial effect on their performance as pigments. Treatment of

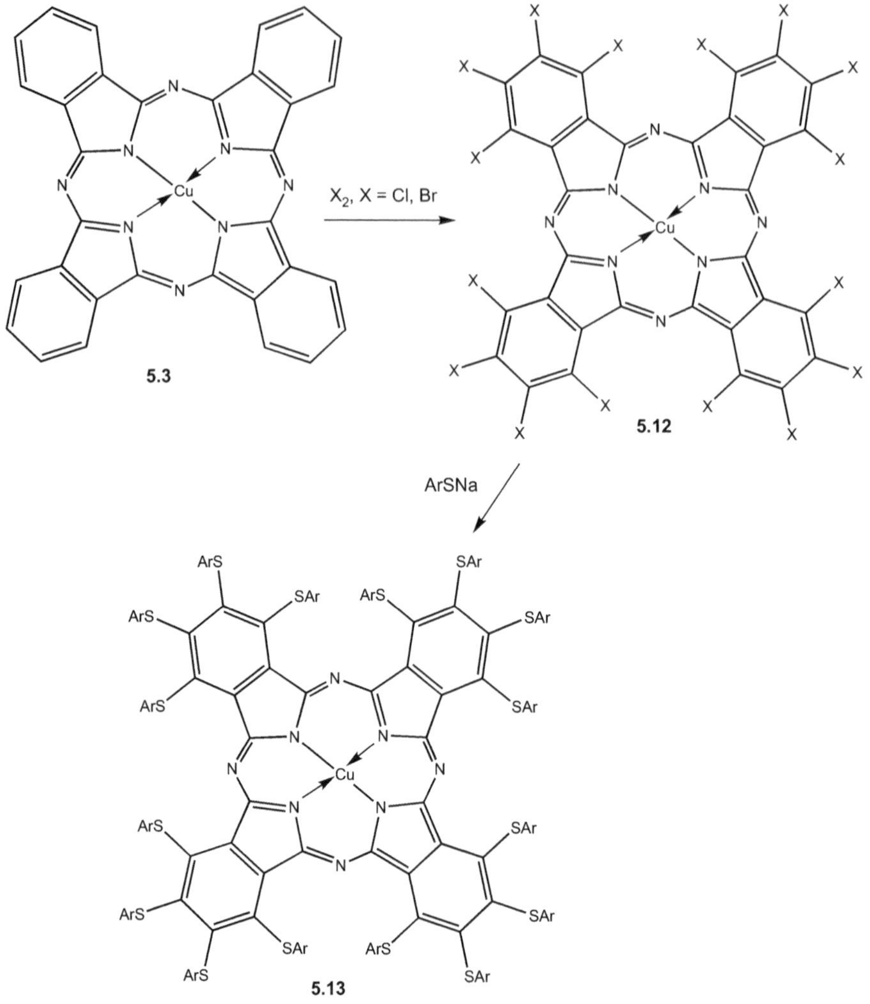

Scheme 5.3   Reactions leading to some substituted CuPc derivatives.

polyhalogenated copper phthalocyanines **5.12** with thiophenols in the presence of alkali at high temperatures in high boiling solvents gives the near-infrared absorbing polyarylthio CuPc derivatives **5.13** as illustrated in Scheme 5.3. This process provides an example of aromatic nucleophilic substitution in the phthalocyanine system. X-Ray structural analysis of these arylthio derivatives demonstrates that the sulfur atoms are located in the plane of the CuPc system while the aryl groups are twisted to accommodate the steric congestion. The disruption of planarity, and hence of the molecular packing, provides these derivatives with solubility in organic solvents.[32]

REFERENCES

1. D. Paterson, in *Colorants and Auxiliaries: Organic Chemistry and Application Properties*, ed. J. Shore, Society of Dyers and Colourists, Bradford, 1990, vol. 1, ch. 2.
2. H. Zollinger, *Color Chemistry: Syntheses, Properties and Applications of Organic Dyes and Pigments*, Wiley-VCH Verlag GmbH, Weinheim, 3rd edn, 2003, ch. 5.
3. P. F. Gordon and P. Gregory, *Organic Chemistry in Colour*, Springer-Verlag, New York, 1983, ch. 5.
4. A. von Braun and J. Tscherniak, *Ber. Dtsch. Chem. Ges*, 1907, **40**, 2709.
5. H. de Diesbach and E. von der Weid, *Helv. Chim Acta*, 1927, **10**, 886.
6. F. H. Moser and A. L. Thomas, *Phthalocyanine Compounds*, Reinhold Publishing Corporation, New York, 1963.
7. F. H. Moser and A. L. Thomas, *The Phthalocyanines*, CRC Press, Boca Raton, Florida, 1983, vol. I and II.
8. C. C. Leznoff and A. B. P. Lever, *Phthalocyanines: Properties and Applications*, VCH, Weinheim, 1989.
9. N. B. McKeown, *Phthalocyanine Materials: Synthesis, Structure and Function*, Cambridge University Press, 1998.
10. G. T. Bourne, R. P. Linstead and A. R. Lowe, *J. Chem. Soc.*, 1934, 1017.
11. R. P. Linstead and A. R. Lowe, *J. Chem. Soc.*, 1934, 1022.
12. C. E. Dent and R. P. Linstead, *J. Chem. Soc.*, 1934, 1027.
13. C. E. Dent, R. P. Linstead and A. R. Lowe, *J. Chem. Soc.*, 1934, 1033.
14. P. A. Barrett, C. E. Dent, R. P. Linstead and A. R. Lowe, *J. Chem. Soc.*, 1936, 1719.
15. J. M. Robertson, *J. Chem. Soc.*, 1935, 613.
16. J. M. Robertson, *J. Chem. Soc.*, 1936, 1195.
17. J. M. Robertson and I. Woodward, *J. Chem. Soc.*, 1937, 219.
18. R. M. Christie and B. G. Freer, *Dyes Pigments*, 1994, **24**, 113.
19. N. V. Tverdova, G. V. Girichev and N. I. Giricheva, *Struct. Chem.*, 2011, **22**, 319.
20. F. Li, Q. Zheng, G. Yang, N. Da and L. Peixiang, *Phys. B-Condensed Matter*, 2008, **403**, 1704.
21. V. G. Maslov, *Opt. Spectrosc.*, 2006, **101**, 853.
22. R. M. Christie and B. G. Freer, *Dyes Pigments*, 1994, **24**, 259.
23. R. M. Christie, *Dyes Pigments*, 1995, **27**, 35.
24. J. R. Fryer, R. B. McKay, R. R. Mather and K. S. W. Sing, *J. Chem. Technol. Biotechnol.*, 1981, **31**, 371.

25. J. W. Owens and M. Robins, *J. Porphyrins Phthalocyanines*, 2001, **5**, 460.
26. H. Ali and J. E. van Lier, *Chem. Rev.*, 1999, **99**, 2379.
27. S. B. Brown and T. G. Truscott, *Chem. Soc. Rev.*, 1995, **29**, 955.
28. G. Guillard, J. Simon and J. P. Germain, *Coord. Chem. Rev.*, 1998, **180**, 1433.
29. D. Woehrle, G. Schnurpfeil, S. Makarov and O. Suvorova, *Chem. Z.*, 2012, **46**, 12.
30. A. M. Paoletti, G. Pennesi, G. Rossi, A. Generosi, B. Paci and V. R. Albertini, *Sensors*, 2009, **9**, 5277.
31. R. M. Christie and D. D. Deans, *J Chem. Soc., Perkin Trans. II*, 1989, 193.
32. P. Gregory in *Colour Chemistry: The Design and Synthesis of Organic Dyes and Pigments*, ed. A. T. Peters and H. S. Freeman, Elsevier Applied Science, London and New York, 1991, ch. 9.

CHAPTER 6

Miscellaneous Chemical Classes of Organic Dyes and Pigments

6.1 INTRODUCTION

The chemistry of the three chemical classes of organic colorants that are industrially most important, the azo, carbonyl and phthalocyanine classes, has been dealt with individually in Chapters 3–5, respectively. In this chapter, the chemistry of some further chemical classes that are of importance for specific applications is discussed. The classes discussed are the polymethines, arylcarbonium ion colorants, dioxazines, sulfur dyes and nitro dyes. A section of this chapter is devoted to each of these, the individual sections containing a description of the principal structural features that characterise the particular colorant type, together with an outline of the chemistry of the main synthetic routes. There are several other chemical compound types that are capable of providing colour but that do not fall into the categories previously mentioned, but which are neglected in this text either because they are commercially of little importance or because they have been less extensively investigated.

6.2 POLYENE AND POLYMETHINE DYES

Polyene and polymethine dyes are two structurally related groups of dyes that contain as their essential structural feature one or more methine (–CH=) groups.[1–3] Polyene dyes contain a series of conjugated double bonds, usually in an *s-trans*-orientation and terminating

Colour Chemistry, 2nd edition
By Robert M Christie
© R M Christie 2015
Published by the Royal Society of Chemistry, www.rsc.org

in aliphatic or alicyclic groups. They owe their colour therefore simply to the presence of the conjugated system. In the case of polymethine dyes, electron donor and electron acceptor groups terminate either end of the polymethine chain, so that they may be considered as typical donor–acceptor dyes.

   The best-known group of polyene dyes is the carotenoids, which are widely encountered as natural colorants. β-Carotene, **6.1**, is given here as an important example. This dye is a hydrocarbon containing eleven conjugated double bonds, which illustrates the length of chain that is necessary to shift the absorption from the UV into the visible range. β-Carotene shows absorption maxima at 450 and 478 nm. Naturally occurring carotenoids not only provide the attractive yellow, orange and red colours that are encountered in many fruit and vegetables, including carrots, sweet potato, mangoes and papayas, but also serve important biochemical functions. For example, they may protect cells and organisms against some of the harmful effects of exposure to light as a result of their action as efficient deactivators of singlet oxygen, and they are also important in certain biological energy transfer processes. The advice from nutritionists encouraging us to include fresh fruit and vegetables in our diet has been emphasized further as evidence has emerged for the potential therapeutic benefits of carotenoids and other related natural colorants, including the possibility that they may provide protection against cancer on the basis, at least in part, of their antioxidant activity by a mechanism that involves the ability to quench singlet oxygen.[4,5] They are also thought to decrease the risk of certain eye diseases. However, it is thought that while there are benefits obtained from diets high in fruit and vegetables containing these nutrients, issues associated with the efficacy and safety of supplements containing the materials have not been addressed fully.[6,7] Another carotenoid that is of some considerable functional biological importance is retinal, which is the chromogenic part of the molecules of rhodopsin and iodopsin, the pigments in the eye that are responsible for colour vision (Chapter 2). There are no true synthetic polyene dyes of any real commercial importance. However, the phthalocyanines (Chapter 5) may, in a sense, be considered structurally as aza analogues of a cyclic polyene system.

6.1

In polymethine dyes, electron donor (D) and acceptor (A) groups terminate the polymethine chain as illustrated by the general structure given in Figure 6.1. They embrace a wide variety of structural types, which may be subdivided into three broad categories as cationic $(z = +1)$, anionic $(z = -1)$ or neutral $(z = 0)$ types, depending on the precise nature of A and D. Dyes of a similar structure in which one or more of the methine carbon atoms are replaced by aza nitrogen atoms are also conveniently considered as polymethines. The presence of the donor–acceptor system means that polymethine dyes can provide strong visible absorption with a much shorter conjugated system than is required with polyenes. Polymethine dyes are capable of providing a wide range of bright, intense colours but, in general, they have tended to show rather inferior fastness properties compared with other chemical classes. This feature has limited their use on textiles, where they are restricted mainly to some disperse dyes for polyester and cationic dyes for acrylic fibres. They are essentially of no industrial consequence as organic pigments. Polymethine dyes have enjoyed commercial importance in traditional film-based colour photographic processes, although this colour application has declined dramatically as digital photography has gained universal acceptance. The dyes have, however, been extensively studied on the grounds both of fundamental theoretical interest and for functional applications, for example in optical recording, dye lasers and solar cells (Chapter 11). The dyes are sometimes referred to as *styryl* dyes where they are derivatives of styrene $(C_6H_5-CH=CH_2)$.[8]

The most important group of cationic polymethine dyes contains nitrogen atoms in both the donor (D) and acceptor (A) groups. These are further classified as cyanines, **6.2**, hemicyanines, **6.3**, or streptocyanines, **6.4**, depending on whether both, one or neither of the nitrogen atoms is contained in a heterocyclic ring as illustrated in Figure 6.2. The history of this type of dye dates from 1856, the same

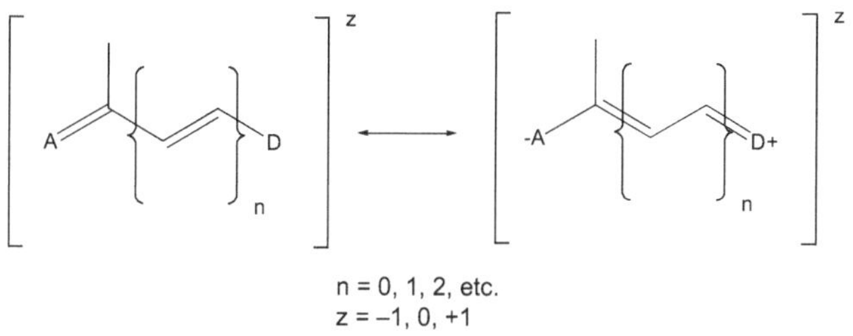

n = 0, 1, 2, etc.
z = –1, 0, +1

Figure 6.1 General structure of polymethine dyes.

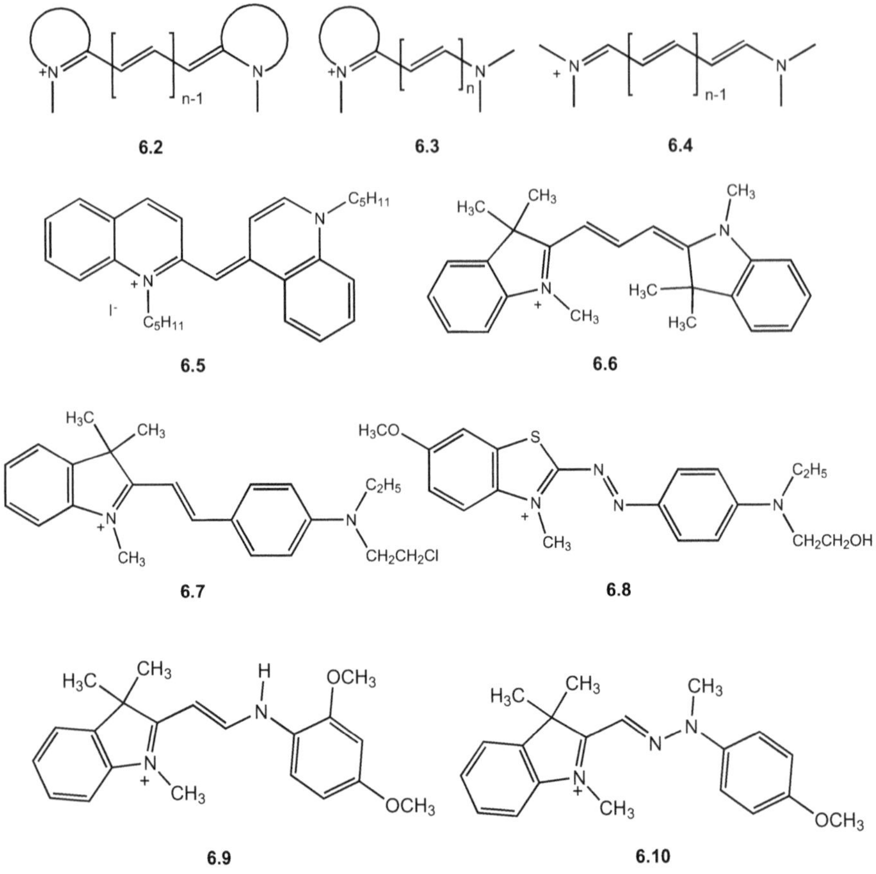

Figure 6.2 Structures of some cationic polymethine dyes.

year in which Perkin discovered Mauveine (Chapter 1). In that year, Williams discovered that the reaction of crude quinoline, which fortuitously contained substantial quantities of 4-methylquinoline, with *iso*-pentyl iodide in the presence of alkali gave a blue dye, which he named *Cyanine*. This dye was eventually identified as having the structure **6.5**. A range of cyanine dyes was subsequently prepared, but these early products proved to be of little use for textiles because of their poor lightfastness. However, considerable interest in their chemistry led ultimately to the discovery that certain of the dyes had photosensitising properties and this gave rise to their application in colour photography. A new era for cationic polymethine dyes was heralded by the introduction in the 1950s of textile fibres based on polyacrylonitrile.[9] It was found that such dyes when applied to acrylic fibres containing anionic sites were capable of giving bright, strong shades with good fastness properties (Chapter 7). Examples of dyes of this type are the cyanine CI Basic Red 12, **6.6**, the hemicyanine CI Basic Violet 7, **6.7**, the diazahemicyanine CI Basic Blue 41, **6.8** (which may also be considered as a member of the azo dye class), the aza-cyanine CI Basic Yellow 11, **6.9**, and the diazacyanine CI Basic Yellow 28, **6.10**. In this series, the cyanines and their aza analogues generally give yellow through to red dyes, while the corresponding hemi-cyanines are more bathochromic, giving reds, violets and blues.

The structures of several neutral and anionic polymethine dyes are illustrated in Figure 6.3. There are many types of neutral polymethines utilising a wide range of electron donor and acceptor groups. For example, CI Disperse Yellow 99, **6.11**, and CI Disperse Blue 354, **6.12**, are important dyes for polyester (Chapter 7). Merocyanines in which the amino group is the donor and the carbonyl group is the acceptor, as represented by the general structure **6.13**, are well known. Because of their generally inferior lightfastness properties, they are not used on textiles, but have found wider use in colour photography. Merocyanines are also formed when certain colourless photochromic compounds, notably the spirooxazines, are irradiated with light.[10] The reversible colour change given by these compounds may be used in a range of applications, for example in ophthalmics, security printing and optical data storage (Chapter 11). Anionic polymethines, such as

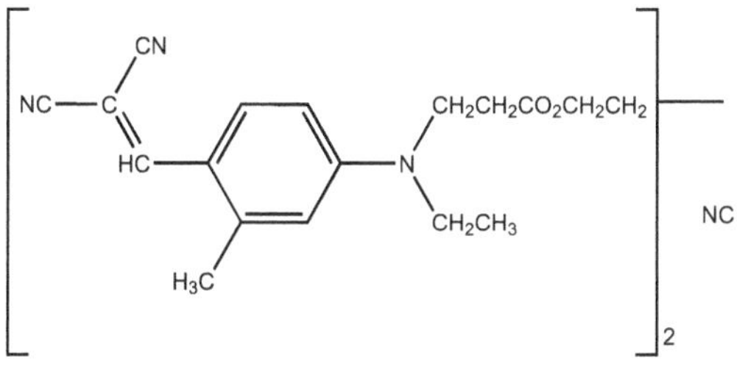

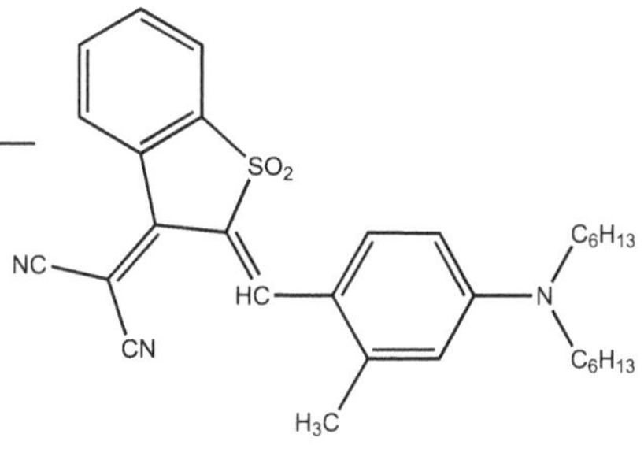

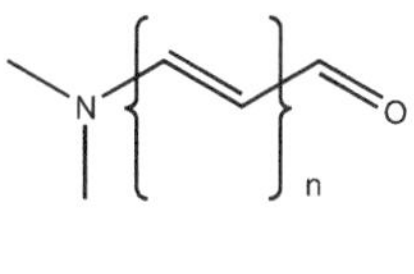

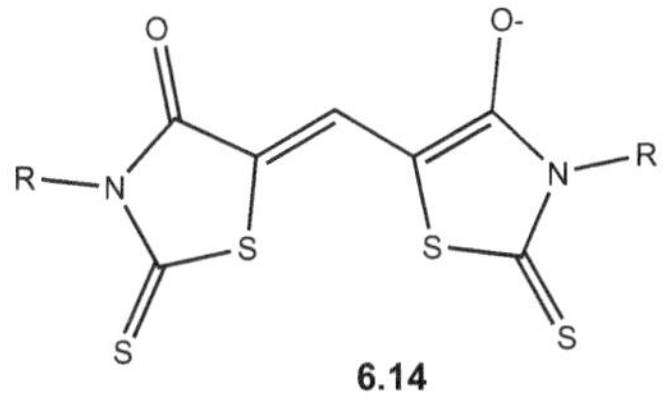

Figure 6.3 Structures of some neutral and anionic polymethine dyes.

the oxonol **6.14**, are usually rather unstable components and so they have not been so extensively investigated.

Symmetrical cyanine dyes, because of the resonance shown in Figure 6.4, in which the two contributing structures are exactly equivalent, are symmetrical molecules.[11,12] X-Ray crystal structure determinations and NMR spectroscopic analysis have demonstrated that the dyes are essentially planar and that the carbon–carbon bond lengths in the polymethine chain are uniform. The colour of cyanine dyes depends mainly on the nature of the terminal groups and on the length of the polymethine chain. The bathochromicity of the dyes is found to increase with the electron-releasing power of the terminal donor group. More bathochromic dyes are also obtained by incorporating the terminal group into a heterocyclic ring system and by extending the conjugation. Cyanine dyes display a highly allowed HOMO → LUMO transition, and hence they show high molar extinction coefficients, i.e., high colour strength. Another important feature of cyanine dyes is the narrowness of the absorption bands, which means that they are capable of providing exceptionally bright colours. One factor that influences the absorption bandwidth of a dye is how closely the geometry of the first excited state of the molecule resembles that of the ground state. In the case of cyanine dyes these two states exhibit very similar geometry and hence the absorption bands are narrow. The results of molecular orbital calculations are generally in good agreement with the experimental spectral data for a wide range of cyanine dyes. In addition, the calculations confirm that there is no major redistribution of π-electron charge densities or π-bond orders on excitation.[13]

There has been some interest in extending the absorption range of cyanine dyes to longer wavelengths into the near-infrared region of the spectrum. A consideration of the spectral data for thiazole derivatives **6.15–6.17** is of some interest in this respect. Cyanine dye **6.15** shows the characteristic visible absorption spectrum for a dye of this type, giving a narrow band with a λ_{max} value of 651 nm in acetonitrile. As the length of the conjugated polymethine chain of cyanine dye **6.15** is extended further, the absorption band is shifted

Figure 6.4   Valence-bond (resonance) approach to cyanine dyes.

bathochromically by about 100 nm for each additional –CH=CH– group. However, the absorption curves become increasingly broad as the chain is extended and there is a consequent reduction in the molar extinction coefficient. It has been suggested that this may be due to trans–cis isomerism, which becomes more facile as the length of the chain increases, and also to an increase in the participation of higher vibrational states of the first excited state as the molecular flexibility is increased. As an alternative to simple extension of the conjugation to provide bathochromicity some structurally related systems have been investigated. For example, the squarylium dye **6.16** absorbs at 663 nm and gives a considerably narrower bandwidth than compound **6.15**, while the croconium derivative **6.17** gives a narrow absorption band at 771 nm, spectral data which are consistent with the results of PPP molecular orbital calculations.[14] Dyes of these structural types are of some interest on the basis of their potential to provide intense narrow absorption bands in the near-infrared region (NIR) with low absorption in the visible region (see Chapter 11 for a discussion of their application in optical data storage).[15]

6.15

6.16

6.17

The strategies used in the synthesis of polymethine dyes are illustrated for a series of indoline derivatives in Scheme 6.1. There is an even wider range of synthetic routes to polymethine dyes than is described here, but they are based for the most part on a similar set of principles. The starting material for the synthesis of this group of polymethine dyes is invariably 2-methylene-1,3,3-trimethylindolenine, **6.18**, known universally as Fischer's base. As illustrated in the scheme, compound **6.18** may be converted by formylation using phosphoryl chloride and dimethylformamide into compound **6.19**, referred to as Fischer's aldehyde, which is also a useful starting material for this series of polymethine dyes. When compound **6.18** (2 moles) is heated with triethylorthoformate (1 mole) in the presence of a base such as pyridine, the symmetrical cyanine dye, CI Basic Red 12, **6.6**, is formed. The synthesis of some hemicyanines may be achieved by heating Fischer's base with an aldehyde, as illustrated for the case of CI Basic Violet 7, **6.7**. The azacyanine, CI Basic Yellow 11,

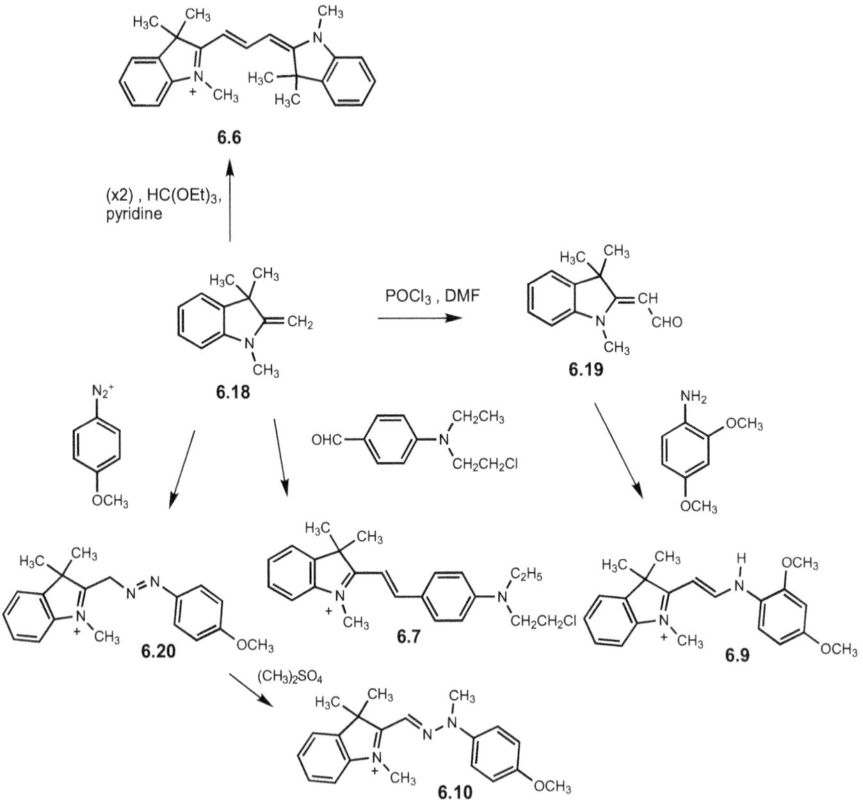

Scheme 6.1   Synthetic approach to some indolenine-based polymethine dyes.

6.9, is synthesised by the condensation reaction of Fischer's aldehyde, **6.19,** with 2,4-dimethoxyaniline. In the synthesis of diazacyanine dye, CI Basic Yellow 28, **6.10,** an important golden yellow dye for application to acrylic fibres, Fischer's base, **6.18,** is treated with 4-methoxybenzenediazonium chloride and undergoes an azo coupling reaction at the reactive methylene group to give the azo dye **6.20.** Methylation of this dye with dimethyl sulfate gives the diazacyanine dye **6.10.**

6.3   ARYLCARBONIUM ION COLORANTS

Arylcarbonium ion colorants were historically the first group of synthetic dyes developed for textile applications. In fact, Mauveine, the first commercial synthetic dye, belonged to this group (Chapter 1).[16] Most of the arylcarbonium ion colorants still in use today were discovered in the late nineteenth and early twentieth centuries. As a group they are used considerably less than in former times, but many are still of some importance, particularly for use as basic (cationic) dyes for the coloration of acrylic fibres and paper, and as pigments.[9] Structurally, they are closely related to the polymethine dyes, especially the cyanine types, and they tend to show similar properties. For example, they provide extremely intense, bright colours covering virtually the complete shade range, but they are generally inferior in technical properties compared with the azo, carbonyl and phthalocyanine chemical classes and as a result their importance has declined over the years.

Arylcarbonium ion dyes encompass a diversity of structural types. Most of the dyes are cationic but there are some neutral and anionic derivatives. The best-known arylcarbonium ion dyes are the diarylmethines, such as Auramine O, CI Basic Yellow 2, **6.21,** and the triarylmethines, the simplest of which is Malachite Green, CI Basic Green 4, **6.22.** The essential structural feature of these two groups is a central carbon atom attached to either two or three aromatic rings. Of these two groups, the triarylmethines are generally the most stable and thus the most useful. Commonly, they are referred to as triaryl *methanes* (or triphenylmethanes), but the name triaryl*methines* more correctly indicates that the carbon atom to which the aromatic rings are attached is sp^2, rather than sp^3 hybridised. Aza analogues of these dyes in which the central carbon is replaced by a nitrogen atom are also conveniently included in this class. There are also a number of derivatives obtained by bridging the di- and triarylmethines and their aza analogues across the *ortho-ortho'* positions of two of the aromatic rings with a heteroatom. Examples of these types of heterocyclic systems, which may be represented by the general structure **6.23**, are illustrated in Table 1.

6.21

6.22

6.23

Mauveine, the original synthetic dye, was of the azine type, its principal component being compound **6.24**.[15] This particular group of dyes is now essentially only of historic interest. Xanthene dyes, such as Rhodamine B, CI Basic Violet 10, **6.25**, Rhodamine 6G, CI

Table 6.1 Heterocyclic arylcarbonium ion dyes and their aza analogues (**6.23**).

X	Y	Type
–C(Ar)=	–O–	Xanthene
–C(Ar)=	–S–	Thioxanthene
–C(Ar)=	–NR–	Acridine
–N=	–O–	Oxazine
–N=	–S–	Thiazine
–N=	–NR–	Azine

Basic Red 1, **6.26**, and Fluorescein, **6.27**, are relatively inexpensive fluorescent dyes. Arylcarbonium dyes are not as a rule highly fluorescent, but the xanthenes exhibit strong fluorescence that is associated with the enhanced molecular rigidity due to the oxygen bridge. Fluorescein is no longer significant as a textile dye but is used in the tracing of water currents and in analytical and biological applications. However, the Rhodamines are of major commercial importance as intensely fluorescing red to violet materials used in a wide variety of traditional and functional applications. Rhodamine 6G is an important red dye for daylight fluorescent pigments and was one of the first to be used in dye lasers[17] (Chapter 11).

6.24

6.25

6.26

6.27

The valence bond (resonance) description of the triphenylmethine dye, Malachite Green, **6.22**, is illustrated in Figure 6.5. A comparison with Figure 6.4 reveals their structural similarity compared with cyanine dyes. Formally, the dye contains a carbonium ion centre, as a result of a contribution from resonance form II. The molecule is stabilised by resonance involving delocalisation of the positive charge on to the *p*-amino nitrogen atoms as illustrated by forms I and III. Because of the steric constraints imposed by the presence of the three rings, triarylmethine dyes cannot adopt a planar conformation. The three rings are twisted out of the molecular plane, adopting a shape like a three-bladed propeller.[18] Malachite Green shows two absorption bands in the visible region with λ_{max} values of 621 and 428 nm. Hence, its observed green colour is due to the addition of blue and yellow components. The long wavelength band is polarised along the *x*-axis and the short wavelength band along the *y*-axis.

The synthesis of arylcarbonium ion dyes and pigments generally follow a similar set of principles. A few selected examples are shown in Scheme 6.2 to illustrate these principles. Essentially, the molecules are constructed from aromatic substitution reactions. In general, a C_1 electrophile, for example phosgene ($COCl_2$), formaldehyde, chloroform or carbon tetrachloride reacts with an aromatic system, which is activated to electrophilic attack by the presence of a strongly electron-releasing group such as the amino (primary, secondary or tertiary) or hydroxyl group. Symmetrical diarylmethines and triarylmethines may be synthesised in one operation, for example by reaction of one mole of phosgene with either two or three moles of the appropriate aromatic compound. Depending on the particular electrophile used, an oxidation may be required at some point in the reaction sequence to generate the final product. Two methods of synthesis of the diarylmethine, Auramine O, **6.21**, are shown in Scheme 6.2. Formerly, this yellow dye was prepared from Mischler's ketone, **6.28**, which may be obtained from the reaction of dimethylaniline (2 moles) with

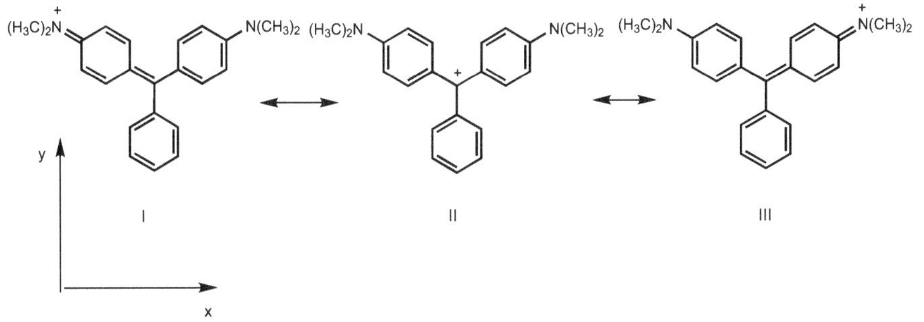

Figure 6.5 Valence-bond (resonance) approach to Malachite Green, **6.22**.

phosgene (1 mole). In the currently preferred method, dimethylaniline (2 moles) is reacted with formaldehyde (1 mole) to give the diaryl compound **6.29**. This compound is then heated with sulfur and ammonium chloride in a stream of ammonia at 200 °C. The dye **6.21** is formed via the thiobenzophenone **6.30** as an intermediate. The synthesis of Malachite Green, **6.22**, is given in Scheme 6.2 to illustrate how an unsymmetrically substituted triarylmethine derivative may be prepared. Dimethylaniline is reacted with benzaldehyde under acidic conditions to give the alcohol **6.31**. This compound is then treated with a further equivalent of dimethylaniline to give the *leuco* base **6.32**, which is subsequently oxidised to the carbinol base **6.33**. Acidification of this compound leads to Malachite Green, **6.22**. For many years, lead dioxide (PbO_2) was the agent of choice for oxidation reactions of this type. Lead-free processes, for example using air or chloranil in the presence of various transition metal catalysts, are now preferred for toxicological and environmental reasons. By using different aromatic aldehydes and aromatic amines as starting materials, this method may be adapted to produce a wide range of triarylmethine dyes.

As an example of a heterocyclic arylcarbonium ion dye, the method of synthesis of Rhodamine B, **6.25**, is shown in Scheme 6.3. The starting materials in this case are phthalic anhydride and 3-*N*,*N*-diethylaminophenol.

Each of the products whose synthesis is illustrated in Schemes 6.2 and 6.3 is a coloured cationic species. When the counteranion is chloride, the products are water-soluble and useful as basic (cationic) dyes for the coloration of acrylic fibres. Precipitation of cationic dyes of these types from aqueous solution using large polymeric counteranions, notably phosphomolybdates, phosphotungstates and phosphomolybdotungstates, leads to a range of highly insoluble red,

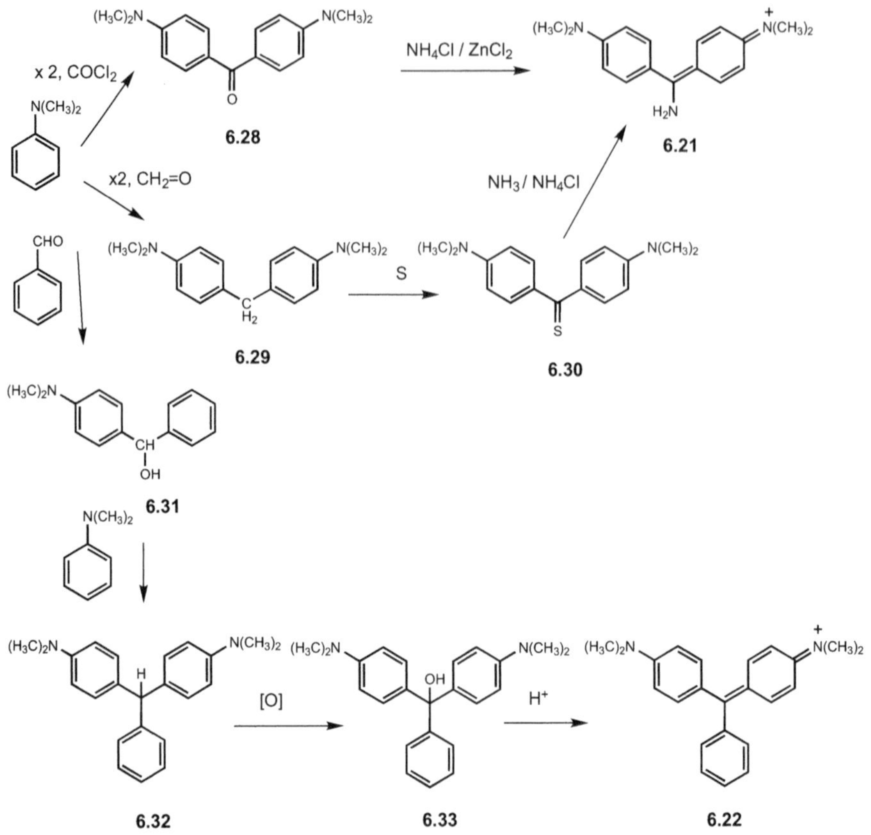

Scheme 6.2 Synthetic approach to some arylcarbonium ion dyes.

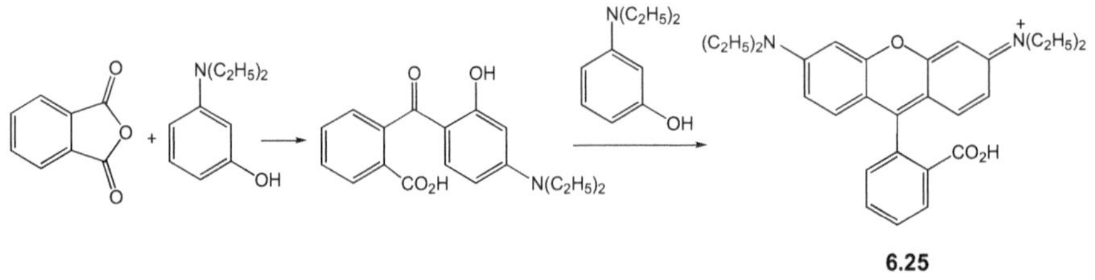

Scheme 6.3 Synthesis of Rhodamine B, **6.25**.

violet, blue and green pigments. These pigments exhibit high brilliance and intensity of colour and high transparency, and are thus well suited to some printing ink applications (Chapter 9).[19–21]

6.4 DIOXAZINES

Dioxazine colorants, as the name implies, contain two oxazine ring systems as the chromophoric grouping. They are relatively few in number and generally restricted to violet to blue shades. Probably the most important dioxazine colorant is CI Pigment Violet 23, **6.34**.[20,21] This product is usually referred to as Carbazole Violet and is the most significant violet pigment for high performance applications (Chapter 9). Compound **6.34** has been shown to have an angular structure as illustrated in Scheme 6.4 rather than the linear structure that is shown in most older texts.[22] The pigment is characterised by a brilliant intense reddish-violet colour, very good lightfastness and resistance to heat and solvents. Its synthesis is illustrated in Scheme 6.4.

In the synthetic scheme, 3-amino-9-ethylcarbazole, **6.36** (2 moles), is condensed with chloranil, **6.35** (1 mole), to form the intermediate **6.37**. This intermediate is then converted into the dioxazine pigment **6.34** by oxidative cyclisation at around 180 °C in an aromatic solvent and in the presence of a catalyst such as aluminium(III) chloride or

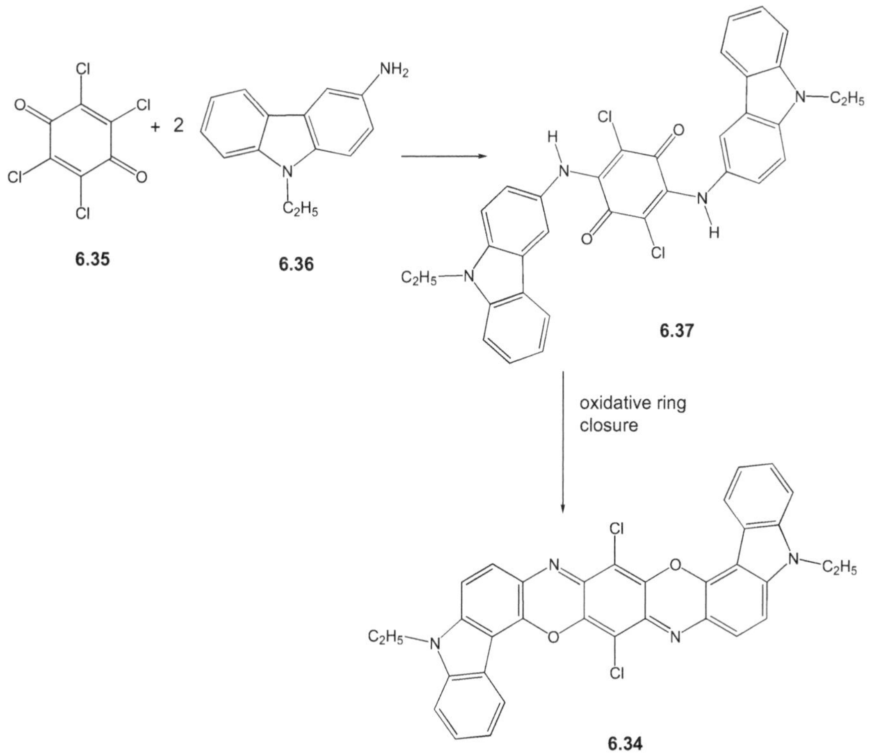

Scheme 6.4   Synthesis of Carbazole Violet, **6.34**.

benzenesulfonyl chloride. Sulfonation of this pigment gives rise to water-soluble dyes that may be used as direct dyes for cotton. An example is CI Direct Blue 108 which contains 3–4 sulfonic acid groups. The dioxazine system is also used as the chromophoric group in some reactive dyes (Chapter 8).[23]

6.5 SULFUR DYES

Sulfur dyes are a group of low cost dyes used in the coloration of cellulosic fibres.[24,25] The dyes currently available commercially are fairly small in number although some of the individual products are manufactured in very large quantities. They are capable of providing a wide range of hues although they tend to give rise to rather dull colours, and thus they are of particular importance as blacks, navy blues, browns and olive greens. Sulfur dyes are treated here as a chemical class of dye, although they may equally be considered as a separate textile application class (Chapter 7). Despite the fact that these products have been known for many years, the chemical structures of sulfur dyes are by no means completely established, but this is completely understandable because the structures are so complex. They are generally known to be complex mixtures of molecular species containing a large proportion of sulfur in the form of sulfide (–S–), disulfide (–S–S–) and polysulfide (–S$_n$–) links and in heterocyclic rings, especially the benzothiazole, thiazone and thianthrene ring systems. It has been proposed that some sulfur dyes are based structurally on the phenothiazonethioanthrone chromophoric system shown in Figure 6.6.

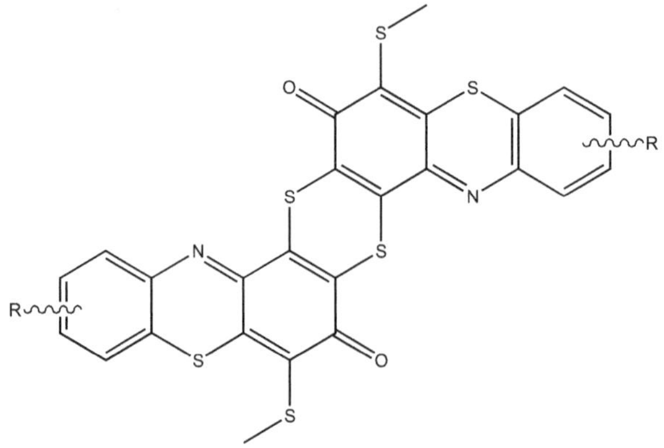

Figure 6.6 Phenothiazonethioanthrone chromophoric system proposed as a constituent of some sulfur dyes.

Traditional sulfur dyes are products of high insolubility in water. They are applied to cellulosic fibres after conversion into a water-soluble *leuco* form by treatment with an aqueous alkaline solution of sodium sulfide. The chemistry of this process, which shows certain similarities to vat dyeing (Chapter 7), is thought to involve mainly reductive cleavage, by the sulfide anion, of disulfide (–S–S–) and polysulfide linkages, leading to monomeric *leuco* species containing alkali-soluble thiol (–SH) groups. After application of the *leuco* form, which is absorbed into the fibre, the insoluble structure of the dye is regenerated by oxidation in air or with an agent such as potassium dichromate in the presence of acetic acid, and aggregates become trapped within the fibre. A second group of sulfur dyes are pre-formed leuco dyes which are 'ready-to-use' in concentrated aqueous solutions. A third group contains thiosulfate ($-S-SO_3^-Na^+$) water-solubilising groups. These dyes, referred to as *Bunte salts*, are applied to the fibres together with sodium sulfide. During their application the thiosulfate groups are reduced to form insoluble dimeric and polymeric species as a result of disulfide bond formation. The particular advantage of sulfur dyes as a class of dyes for cellulosic fibres is that they provide reasonable technical performance at low cost. However, traditional sulfur dyes present significant environmental problems, largely associated with residues of sulfide and other sulfur-containing species, electrolytes and unfixed dye, which contaminate the dyehouse effluent and it is conceivable that this feature may cause their use to decline in years to come. However, there has been considerable effort in recent years to develop more ecologically-friendly processing conditions for the application of sulfur dyes, for example using reducing sugars or other sulfur-free reducing agents, with the aim also to maximise dye exhaustion on to the fibres, thus minimising the effluent load. There has also been some experimentation with their application using electrochemical reduction processes.[26]

The manufacture of sulfur dyes involves *sulfurization* processes, the chemistry of which remains rather mysterious and may arguably be considered still to be in the realms of alchemy! The processes involve heating elemental sulfur or sodium polysulfide, or both, with aromatic amines, phenols or aminophenols. These reactions may be carried out either as a dry bake process at temperatures between 180 and 350 °C or in solvents such as water or aliphatic alcohols at reflux or at even higher temperatures under pressure. CI Sulfur Black 1, for example, is prepared by heating 2,4-dinitrophenol with sodium polysulfide. CI Sulfur Black 1 is by far the most important product in the series. In fact, it may well be the individual dye of all chemical types that has the largest production volume worldwide.

6.6 NITRO DYES

The nitro group is commonly encountered as a substituent in dyes and pigments of most chemical classes, but it acts as the essential chromophore in only a few dyes.[27] Nitro dyes constitute a small group of dyes of some importance as disperse dyes for polyester (Chapter 7) and as semi-permanent hair dyes (Chapter 10). Picric acid, **6.38**, was historically probably the first nitro dye, although it was never really commercially important due to its poor dyeing properties, its toxicity and its potential explosive properties (Chapter 1). The nitro dyes used today have relatively simple aromatic structures, some examples of which are shown in Figure 6.7. They contain at least one nitro (NO_2) group as the chromophore and electron acceptor, and one or more electron-releasing amino groups complete the donor–acceptor system. Nitro dyes are capable of providing bright yellow, orange and red shades, but the colours are amongst the weakest provided by the common commercial chromophores. Nitrodiphenylamines are nevertheless of some importance as yellow disperse dyes, such as CI

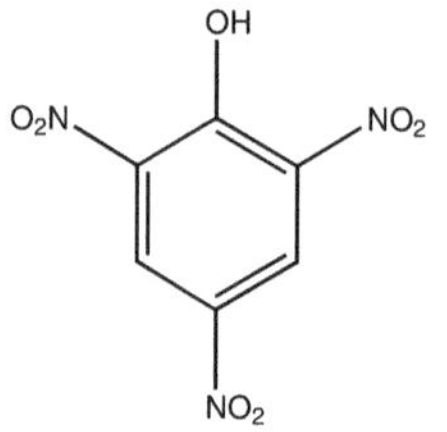

6.38

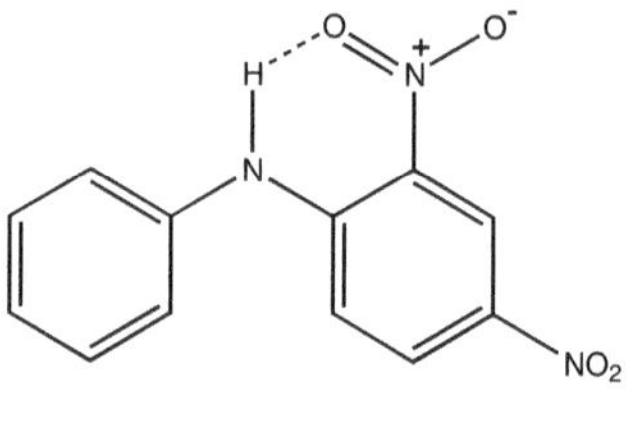

6.39

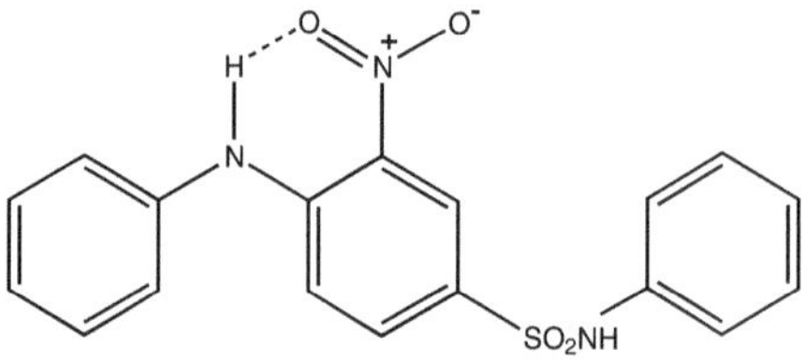

6.40

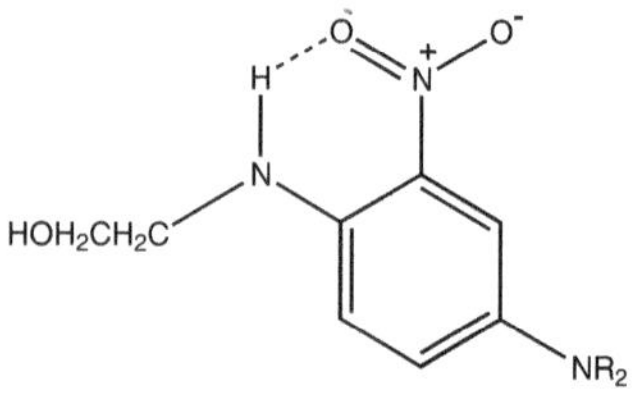

6.41a: R = H
6.41b: R = CH_2CH_2OH

Figure 6.7 Structures of some nitro dyes.

Disperse Yellows 14 and 42, **6.39** and **6.40**, respectively, because of their low cost and their good lightfastness. The good lightfastness of these dyes is attributed to the intramolecular hydrogen bonding between the *o*-nitro group and the amino group while the electron withdrawing (nitro or sulfonamide) group in the *para*-position is also important as it gives rise to an increase in tinctorial strength. A wider range of hues is provided by some nitro semi-permanent hair dyes, such as compounds **6.41a**, which is red, and **6.41b**, which is violet (Chapter 10).[28] The bathochromicity of dye **6.41b** may be explained on the basis of the stronger electron-donating power in the donor–acceptor system.

The synthesis of nitro dyes is relatively simple, a feature that accounts to a certain extent for their low cost. The synthesis, illustrated in Scheme 6.5 for compounds **6.39** and **6.40**, generally involves a nucleophilic substitution reaction between an aromatic amine and a chloronitroaromatic compound. The synthesis of CI Disperse Yellow 14, **6.39**, involves the reaction of aniline with 1-chloro-2,4-dinitroaniline while compound **6.40** is prepared by reacting aniline (2 moles) with the chlorosulfonyl derivative **6.42** (1 mole).

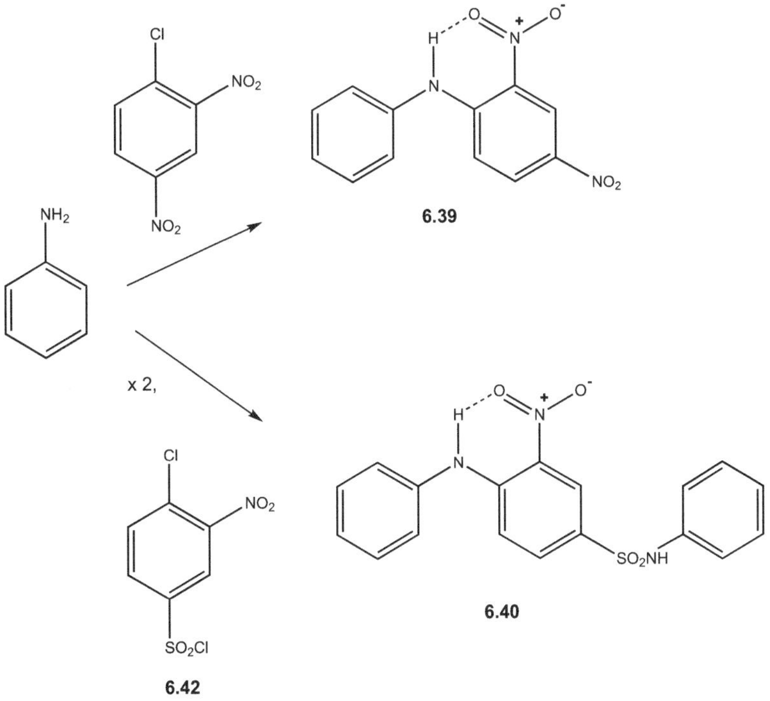

Scheme 6.5 Synthesis of nitro dyes **6.39** and **6.40**.

REFERENCES

1. G. Hallas, in *Colorants and Auxiliaries: Organic Chemistry and Application Properties*, ed. J. Shore, Society of Dyers and Colourists, Bradford, 1990, vol. 1, ch. 6.
2. H. Zollinger, *Color Chemistry: Syntheses, Properties and Applications of Organic Dyes and Pigments*, Wiley-VCH Verlag GmbH, Weinheim, 3rd edn, 2003, ch. 3.
3. P. F. Gordon and P. Gregory, *Organic Chemistry in Colour*, Springer-Verlag, New York, 1983, ch. 5.
4. H. Li, R. Tsao and Z. Deng, *Can. J. Plant. Sci.*, 2012, **92**, 1101.
5. H. Nishino, M. Murakosh, T. Li, M. Takemura, M. Kuchide, M. Kanazawa, X. Y. Mou, M. Masuda, Y. Ohsaka, S. Yogosawa, Y. Satomi and K. Jinno, *Cancer Metastasis Rev.*, 2002, **21**, 257.
6. C. L. Rock, *Pure Appl. Chem.*, 2002, **74**, 1451.
7. E. J. Johnson, *Nutr. Clin. Care*, 2002, **5**, 56.
8. T. Deligeorgiev, A. Vasilev, S. Kaloyanova and J. J. Vaquero, *Color Technol.*, 2010, **126**, 55.
9. R. Raue, *Rev. Prog. Color.*, 1984, **14**, 187.
10. R. M. Christie, L. J. Chi, R. A. Spark, K. M. Morgan, A. S. F. Boyd and A. Lycka, *J. Photochem. Photobiol. A: Chem.*, 2005, **169**, 37.
11. F. M. Hamer, *The Cyanines and Related Compounds*, Interscience, New York, 1964.
12. A. Mishra, R. K. Behera, P. K. Behera, B. K. Mishra and G. B. Behera, *Chem. Rev.*, 2000, **100**, 1973.
13. A. D. Kachkovski and M. L. Dekhtyar, *Dyes Pigments*, 1996, **30**, 43.
14. S. Yasui, M. Matsuoka and T. Kitao, *Dyes Pigments*, 1988, **10**, 13.
15. J. Fabian, H. Nakazumi and M. Matsuoka, *Chem. Rev.*, 1992, **92**, 1197.
16. O. Methcohn and M. Smith, *J. Chem. Soc., Perkin Trans. I*, 1994, 5.
17. R. M. Christie, *Rev. Prog. Color.*, 1993, **23**, 1.
18. H. Zollinger, *Color Chemistry: Syntheses, Properties and Applications of Organic Dyes and Pigments*, Wiley-VCH Verlag GmbH, Weinheim, 3rd edn, 2003, ch. 4.
19. A. Daviddson and B. Norden, *Chem. Scr.*, 1977, **11**, 68.
20. W. Herbst and K. Hunger, *Industrial Organic Pigments. Production, Properties, Applications*, Wiley-VCH Verlag GmbH, Weinheim, 3rd edn, 2004.
21. R. M. Christie, *The Organic and Inorganic Chemistry of Pigments*, Oil & Colour Chemists Association, London, 2002.
22. E. Dietz, *Chimia*, 1991, **45**, 13.
23. A. H. M. Renfrew, *Rev. Prog. Color.*, 1985, **15**, 15.

24. J. N. Chakraborty, Sulphur dyes, in *Handbook of Textile and Industrial Dyeing: Principles, Processes and Types of Dyes*, ed. M. Clark, Woodhead Publishing, Cambridge, 2011, vol. 1, ch. 14.
25. W. E. Wood, *Rev. Prog. Color.*, 1971–75, 7, 80.
26. M. Bozic and V. Kokol, *Dyes Pigments*, 2008, **76**, 299.
27. H. Zollinger, *Color Chemistry: Syntheses, Properties and Applications of Organic Dyes and Pigments,* Wiley-VCH Verlag GmbH, Weinheim, 3rd edn, 2003, ch. 6.
28. R. M. Christie and O. J. X. Morel, The coloration of human hair, in *The Coloration of Wool and other Keratin Fibres*, ed. D. M. Lewis and J. A. Rippon, John Wiley & Sons Ltd, Chichester, 2013, ch. 11.

CHAPTER 7

Textile Dyes (excluding Reactive Dyes)

7.1 INTRODUCTION

In Chapters 3–6, the commercially important chemical classes of dyes and pigments are discussed in terms of their essential structural features and the principles of their synthesis. The reader will encounter further examples of these individual chemical classes of colorants throughout Chapters 7–11 which, as a complement to the content of the earlier chapters, deal with the chemistry of their application. Chapters 7 and 8 are concerned essentially with the application of dyes to textiles, whereas Chapter 9 is devoted to pigments and their applications. The distinction between these two types of colorants has been made previously in Chapter 2. In contrast to the mechanism of action of pigments in their application, which involves mechanical anchoring as discrete solid particles in a polymeric matrix, dyeing relies on equilibrium processes involving diffusion or sorption of dye molecules or ions within the substrate. Dyes are used in the coloration of a wide range of substrates, including paper, leather and plastics, but by far their most important outlet is on textiles. Textiles are ubiquitous materials that are used in a wide variety of products, including clothing of all types, soft furnishings, such as curtains, upholstery and carpets, as well as towels and bedding.[1,2] Textiles are also used in a range of arguably less familiar technical applications, for example as used in transport, healthcare and construction.[3] This chapter deals with the chemical principles underlying the main application classes of dyes that may be applied to textile

Colour Chemistry, 2nd edition
By Robert M Christie
© R M Christie 2015
Published by the Royal Society of Chemistry, www.rsc.org

fibres, with the exception of the reactive dye class, which is dealt with exclusively in Chapter 8.

Textile fibres may be classified into three broad groups: natural, semi-synthetic and synthetic.[1,2] Unlike the commercial range of dyes and pigments, which are now almost entirely synthetic in origin, natural fibres continue to play a prominent part in textile applications. The most important natural fibres are either of animal origin, for example the protein fibres, wool and silk, or of vegetable origin, such as cotton, which is a cellulosic fibre.[4,5] Semi-synthetic fibres are derived from natural sources although their production involves chemical processing. The most significant semi-synthetic fibres used today are derived from cellulose as the starting material. Viscose, modal and lyocell are regenerated cellulosic fibres.[6] Viscose is a long-established textile fibre that is manufactured by reacting cellulose, *e.g.*, from wood pulp, with carbon disulfide in alkali to give its water-soluble xanthate derivative. This is followed by regeneration of the cellulose in fibrous form using sulfuric acid. Lyocell is a more recent development, introduced in 1994 as a regenerated cellulosic fibre with strong environmental credentials, commercialized under the trade name *Tencel*.[7,8] Lyocell is manufactured from wood pulp, from sustainably-farmed eucalyptus, dissolved in an organic solvent, *N*-methylmorpholine *N*-oxide (NMMO), which is reported to be non-toxic. The fibres are prepared from the concentrated solution by spinning into a water bath, which regenerates the cellulose. The manufacture involves a closed loop process from which the solvent is virtually completely recovered and recycled. Cellulose acetate is a chemically modified cellulose derivative, manufactured by acetylation and partial hydrolysis of cotton. The most important completely synthetic fibres are polyester, polyamides (nylon), acrylic fibres and polypropylene.[9] Textile fibres share the common feature that they are made up of polymeric organic molecules. However, the physical and chemical natures of the polymers involved vary widely and this explains why each type of fibre essentially requires its own 'tailor-made' application classes of dyes. Textile fibres are characterised by extreme fineness. Most natural fibres exist as *staple* fibres, roughly 2–50 cm long and 10–40 μm in diameter, and are converted by a spinning process into yarns. In contrast, synthetic fibres are produced as continuous filament yarn, as is silk (by the silkworm).

Dye molecules are designed to ensure that they have a set of properties that are appropriate to their particular applications.[10,11] The most obvious requirement for a dye is that it must possess the desired colour, in terms of hue, strength and brightness. The relationships between colour and molecular constitution of dyes have been discussed principally in Chapter 2, although the reader will find specific aspects relating to particular chemical classes in Chapters 3–6. A further feature of dye molecules, which is of some practical importance, is their ability to dissolve in water. Since textile dyes are almost always applied from an aqueous dyebath solution, they are required to be soluble in water, or, alternatively, to be capable of conversion into a water-soluble form suitable for application. Many dye application classes, including acid, mordant, premetallised, direct, reactive and cationic dyes, are readily water-soluble. Disperse dyes for polyester are, in contrast, only sparingly soluble in water, but they have sufficient solubility for their application at the high temperatures employed in their application. A few groups of dyes, including vat and sulfur dyes for cellulosic fibres, are initially insoluble in water and are thus essentially pigments. However, they may be converted chemically into a water-soluble form and in this form they can be applied to the fibre, after which the process is reversed and the insoluble form is regenerated in the fibre.

Dyes must be firmly attached to the textile fibres to which they are applied in order to resist removal, for example by washing.[12–14] This may be achieved in several ways. The molecules of many dye application classes are designed to provide forces of attraction for the polymer molecules that constitute the fibre.[15] In the case of reactive dyeing, the dye molecules combine chemically with the polymer molecules, forming covalent bonds (Chapter 8). In further cases, for example vat, sulfur and azoic dyes for cellulosic fibres, an insoluble pigment is generated within the fibres and is retained by mechanical entrapment. In other cases, a set of dye–fibre intermolecular forces operate, varying according to the particular dye–fibre system. These interactions commonly involve a combination of ionic, dipolar, van der Waals forces and hydrogen bonding. An additional feature of textile dyeing is that the dye must distribute itself evenly throughout the material to give a uniform colour, referred to as a *level* dyeing. Finally, the dye must provide an appropriate range of fastness properties, for example to light, washing, heat, rubbing, *etc.* This chapter provides an overview of the most important application classes of dyes for textiles, with a range of selected examples that illustrate how the dye molecules are designed to suit their particular application. The discussion is organised according to fibre type in three sections: dyes for protein fibres, dyes for cellulosic fibres and dyes for synthetic fibres. In each case there is a description of the structure of the polymer, followed by a discussion of the structural features of the dyes that determine their suitability for application to the particular type of fibre. The subject of reactive dyes is considered to be of sufficient interest and importance to warrant separate treatment in Chapter 8.

Large quantities of textile fabrics are also coloured by printing to produce multicolour patterns and images.[16,17] The most important technique used industrially in the manufacture of printed textiles is screen printing, mostly by continuous rotary screen printing, which is an especially economic method for long print runs. Screen printing is in effect a stencilling process in which the image is produced on top of a woven polyester mesh using a photopolymerisation process. The polymer forms the non-image areas which block the passage of ink while the image is contained in the open areas. This image is transferred on to the fabric by pressing the ink through the open areas. Textile printing may be considered essentially as localized dyeing. This feature means that the dye classes appropriate to the fibre in question are used and that the principles of the dye application and its interactions with the fibre are the same as in dyeing. In screen printing, the dye is incorporated into a highly concentrated printing paste containing a range of auxiliary ingredients. The most important ingredient is a thickening agent whose role is to hold the design in place while the print is subjected to fixation conditions after it has been transferred to the fabric. Urea (NH_2CONH_2) is another common ingredient in printing pastes. Its function is to swell and open up the fibre structure, due mainly to its ability to participate in hydrogen bonding, thus facilitating penetration of the dye. The paste also contains ingredients which ensure that the pH is maintained as appropriate for dye fixation, for example, acidic conditions for acid dyeing of protein fibres and alkaline conditions for reactive dyeing of cellulosic fibres (Chapter 8). Generally, the fixation process involves treating the printed fabric at a high temperature, using either steam or dry heat, under which conditions the dyes diffuse into the fibres and the dye/fibre intermolecular forces are formed. In addition to selection from the appropriate application classes, dyes for printing should have high water solubility, because of the relatively low volumes of water used in the print pastes, and any unfixed dye should be relatively easily removed in the final washing process so that there is no staining of non-image areas of the fabric. Pigments may also be used in screen printing on to textiles. Their use requires the use of a binder to contain the colorant and to ensure that the design adheres to the fabric. The use of pigments in screen printing offers certain advantages over dyes. The use of a binder means that they can be applied to any fabric, whereas dyes are fibre-specific, and also the process requires minimal wash off after printing. A disadvantage in the use of a binder in pigment printing, however, is that the handle or feel of the fabric may be adversely affected. While rotary screen printing remains the dominant process used industrially for textile printing, the use of inkjet printing, in which the digital image is transferred directly from the computer on to the fabric, is growing in importance.[18] The principles of inkjet printing are dealt with in Chapter 11.

7.2 DYES FOR PROTEIN FIBRES

Protein fibres are natural fibres derived from animal hair sources. The most important protein fibre used commercially is wool (from sheep), although luxury fibres such as silk (from the silkworm), cashmere and mohair (both from goats) are important high value products. Human hair is also a protein fibre. The chemical principles of hair coloration are considered in Chapter 10. Both the physical and chemical structures of protein fibres are highly complex and there is considerable variation depending on the source. The principal component of the fibres is the protein keratin, the molecular structure of which is illustrated in outline in Figure 7.1. The protein molecules consist of a long polypeptide chain constructed from the eighteen commonly encountered amino acids that are found in most naturally-occurring proteins. The structures of these amino acids are well documented in general chemical and biochemical textbooks and so they are not reproduced here. As a result of the diverse chemical nature of these amino acids, the protein side-chains (R^1, R^2, *etc.* in Figure 7.1) are of widely varying character, containing functionality that includes, for example, amino and imino, hydroxy, carboxylic acid, thiol and alkyl groups and heterocyclic functionality. At intervals, the polypeptide chains are linked together by disulfide (-S-S-) bridges derived from the amino acid cystine. There are also ionic links between the protonated amino ($-NH_3^+$) and carboxylate ($-CO_2^-$) groups, which are located on

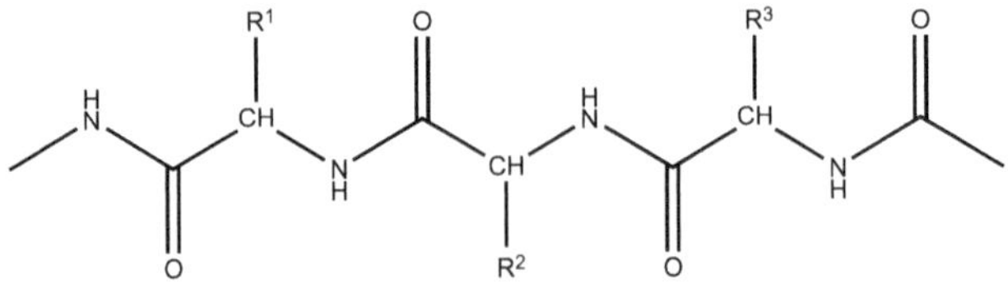

Figure 7.1 Structure of the protein keratin.

the amino acid side-groups and at the end of the polypeptide chains. Many of the functional groups on the wool fibre play some part in the forces of attraction involved when dyes are applied to the fibres. Protein fibres may be dyed using a number of application classes of dyes, the most important of which are acid, mordant and pre-metallised dyes, the structural features of which are discussed in the rest of this section, and reactive dyes which are considered separately in Chapter 8. Protein fibres may be degraded chemically under aqueous alkaline conditions, but they are relatively stable to acidic conditions. Thus, most protein dyeing processes are carried out by applying the dyes under mildly acidic to neutral aqueous conditions, usually at elevated temperatures.[19,20]

Acid dyes derive their name historically from the fact that they are applied to protein fibres such as wool under acidic conditions.[13,14] They are also used to a certain extent to dye polyamide fibres such as nylon. Acid dyes may be conveniently classified as either acid levelling or acid milling types. Acid-levelling dyes are a group of dyes that show only moderate affinity for the wool fibres. Because the intermolecular forces between the dye and the fibre molecules are not strong, these dyes are capable of migrating through the fibre and thus produce a level dyeing, as the name implies. Acid milling dyes are a group of dyes that show much stronger affinity for the wool fibres. Because of the superior strength of the intermolecular forces between the dye and the fibre molecules, the dyes are less capable of migration through the fibre and appropriate processing conditions have to be adopted to ensure that level dyeing is achieved. However, they give superior fastness to washing. Indeed, their name is derived historically from their ability to provide superior resistance to the wet treatments, such as milling, a mechanical wet process used traditionally to introduce bulk and a degree of dimensional stability to wool fabrics. There is also a group of acid dyes referred to as acid supermilling because they are capable of producing superior washfastness properties, while at the same time presenting difficulties in achieving levelness.

A characteristic feature of acid dyes for protein and polyamide fibres is the presence of one or more sulfonate ($-SO_3^-$) groups, usually present as the sodium (Na^+) salt. These groups have a dual role. Firstly, they provide solubility in water, the medium from which the dyes are applied to the fibre. Secondly, they ensure that the dyes carry a negative charge (*i.e.*, they are anionic). When acid conditions are used in the dyeing process, the protein molecules acquire a positive charge. This is due mainly to the fact that, under acidic conditions, the amino (-NH$_2$) and imino (=NH) groups on the amino acid side-chains are protonated as -NH$_3{}^+$ and =NH$_2{}^+$ groups, respectively, while ionisation of the carboxylic acid groups is suppressed. The positive charge on the polymer attracts the acid dye anions by ionic forces, and these displace the counteranions within the fibre by an ion exchange process. As well as these ionic forces of attraction, van der Waals forces, dipolar forces and hydrogen-bonding between appropriate functionality of the dye and fibre molecules also play a part in the affinity of acid dyes for protein fibres. The molecular size and shape is very commonly a critical feature in the design of dyes for application to specific substrates. In this context, acid levelling dyes may be described as a small to medium-sized planar molecules. These features allow the dyes to penetrate easily into the fibre and also permit a degree of movement or migration within the fibre as the ionic bonds between the dye and the fibre are capable of breaking and then re-forming, thus producing a level or uniform colour. However, as the dye is not very strongly bonded to the fibre, it may show only moderate fastness towards wet-treatments such as washing. Acid-milling dyes are significantly larger molecules than acid levelling dyes and they show enhanced affinity for the fibre, and hence improved fastness to washing, as a result of a more extensive set of van der Waals forces, dipolar forces and hydrogen bonding. Molecular planarity is important in maximizing the efficiency of the intermolecular interactions. Acid milling dyes, because of their higher affinity for the protein fibres, do not require as strongly acidic application conditions as acid levelling dyes for complete dye absorption.

Most acid dyes, especially yellow, oranges and reds, belong to the azo chemical class while blues and greens are often provided by carbonyl dyes, especially anthraquinones, and to a certain extent by arylcarbonium ion types. Figure 7.2 illustrates some typical acid dye structures. A notable aspect of the structure of dyes **7.2–7.5** is the strong intramolecular hydrogen-bonding that exists contained in six-membered rings, a feature that enhances the stability of the compounds and, in particular, confers good lightfastness properties. One explanation that has been proposed is that the hydrogen bonding leads to a reduction in electron density at the chromophore, and that this in turn reduces the sensitivity of the dye towards photo-chemically-induced oxidation. It has also been suggested that intra-molecular proton transfer within the excited state of the dye molecules can lead to enhanced photostability.[21] Whatever the exact mechanism might be, intramolecular hydrogen-bonding is ubiqui-tously encountered in the structures of a wide range of dyes and

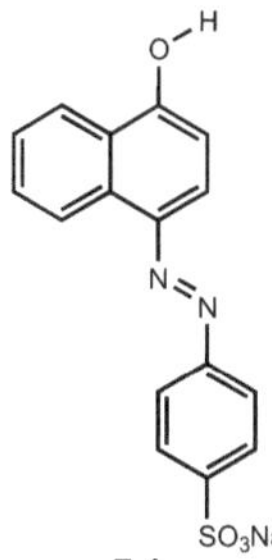

7.1

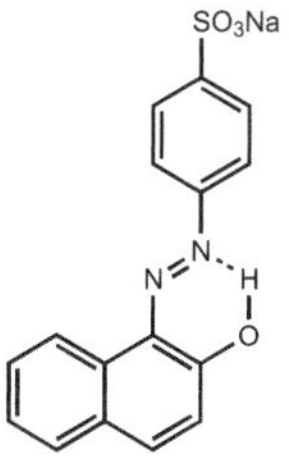

7.2

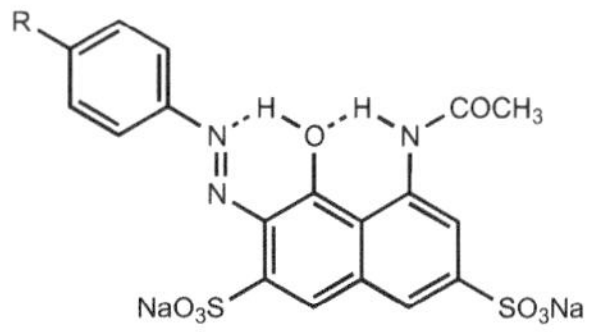

7.3a: R = H; **7.3b**: R = $C_{12}H_{25}$

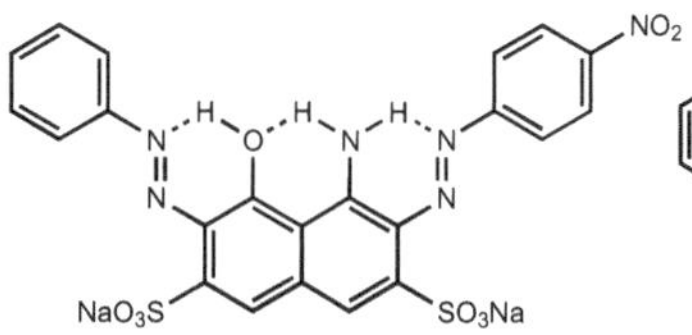

7.4

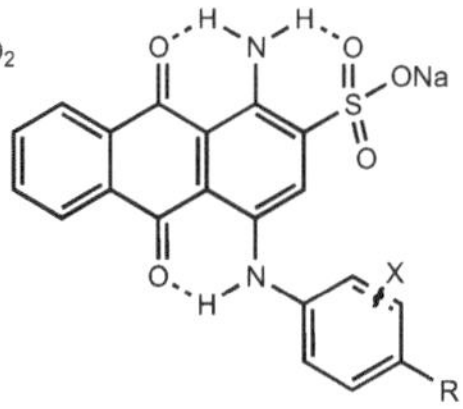

7.5a: R = H, X = H;
7.5b: R = $C_{12}H_{25}$, X = SO_3Na

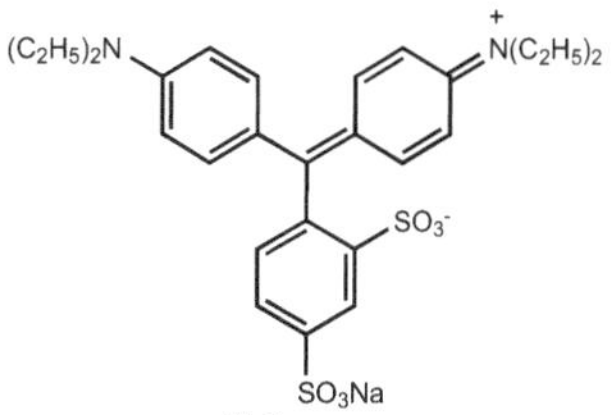

7.6

Figure 7.2 Structures of some typical acid dyes for protein fibres.

pigments because of its ability to enhance stability. Intramolecular hydrogen-bonding also reduces the acidity of the hydroxyl group which is commonly encountered in dye molecules and thus can lead to improved resistance towards alkaline conditions, such as those commonly used in domestic laundering. A comparison between the two isomeric monoazo acid dyes CI Acid Orange 20, **7.1** and CI Acid Orange 7, **7.2**, illustrates the effect of intramolecular hydrogen bonding. Dye **7.2** shows significantly improved fastness to alkaline washing and lightfastness compared with dye **7.1** in which intramolecular hydrogen bonding is not possible. A comparison of the structurally-related monoazo dyes **7.3a** (CI Acid Red 1) and **7.3b** (CI Acid Red 138), and of the anthraquinone acid dyes **7.5a** (CI Acid Blue 25) and **7.5b** (CI Acid Blue 138), illustrates the distinction between acid levelling and acid milling dyes. Dyes **7.3b** and **7.5b** show excellent resistance to washing as a result of the presence of the long alkyl chain substituent ($C_{12}H_{25}$), which is attracted to hydrophobic or non-polar parts of the protein fibre molecules by van der Waals forces. Because of the extremely strong dye/fibre affinity, dyes of this type are of the acid supermilling type. Dyes **7.3a**, **7.5a**, CI Acid Black 1, **7.4**, a typical disazo acid dye, and CI Acid Blue 1, **7.6**, an example of a triphenylmethine acid dye, are acid-levelling dyes. In the case of dye **7.6**, it is important to note that while the nitrogen atoms carry a formal single delocalised positive charge, the presence of two sulfonate groups ensures that the dye is overall anionic.

The ability of transition metal ions, and especially chromium (as Cr^{3+}), to form highly stable metal complexes may be used to produce dyeings on protein fibres with superior fastness properties, especially towards washing and light. The chemistry of transition metal complex formation with azo dyes is discussed in some detail in Chapter 3. There are two application classes of dyes in which this feature is utilised, mordant dyes and premetallised dyes, differing significantly in application technology but involving similar chemistry.[13,14,22]

Chrome mordant dyes generally have the characteristics of acid dyes but with the ability in addition to form a stable complex with chromium in its Cr(III) oxidation state. Most commonly, this feature takes the form of two hydroxyl groups on either side of (*ortho, ortho'* to) the azo group of a monoazo dye, as illustrated for the case of CI Mordant Black 1, 7.7. In the most important method for application of mordant dyes, the so-called afterchrome process, the dye is applied to the fibre as an acid dye and then the dyed fibres are treated with a source of chromium, most frequently sodium dichromate ($Na_2Cr_2O_7$) in which the chromium exists in oxidation state Cr(VI). During the process, the chromium(VI) undergoes reduction by functional groups (which are consequently oxidized) on the wool fibre, for example the cysteine thiol groups, and a chromium(III) complex of the dye is formed within the fibre by a process such as that illustrated in Figure 7.3. A dye of this type acts as a tridentate ligand, the chromium bonding with two oxygen atoms derived from the hydroxyl groups and with one nitrogen atom of the azo group. Complexes of Cr(III) are invariably six-coordinate with octahedral geometry. It has not been established with certainty how the remaining three valencies of the chromium are satisfied in the mordant dyeing of protein fibres. There are several possibilities, which include bonding with water molecules, with coordinating groups (-OH, -SH, -NH_2, -CO_2H, *etc.*) on the amino acid side chains on the fibre, or with another dye molecule. Chrome mordanting has been traditionally of particular importance in the dyeing of loose wool in very dark (especially black) shades to provide very high levels of fastness. However, there are serious ecological problems associated with the use of chrome mordant dyes, associated with the severe toxicity of Cr(VI) and its presence in dyehouse effluent. Consequently, the process is little used nowadays and continues to decrease in importance.

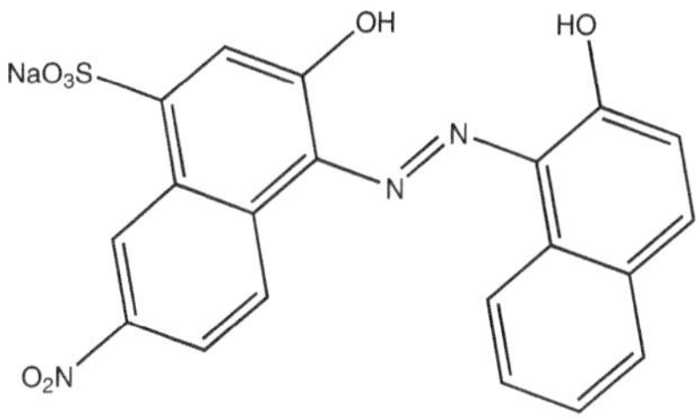

7.7

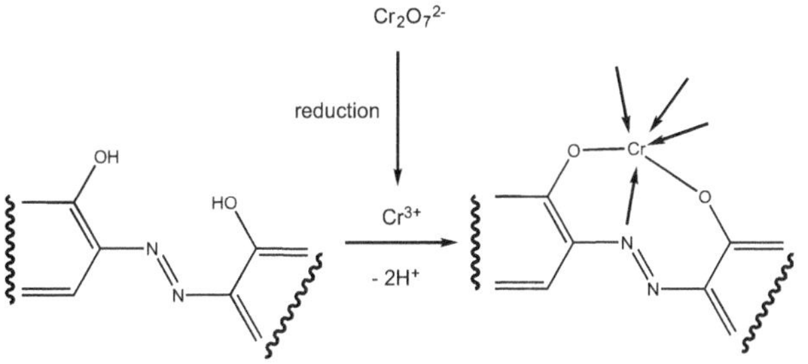

Figure 7.3 Chemistry of chrome mordanting.

Premetallised dyes, as the name implies, are pre-formed metal complex dyes.[22,23] They are usually six-coordinate complexes of chromium(III) with octahedral geometry, as exemplified for example by CI Acid Violet 78, **7.8**, although some complexes of cobalt(III) are also used. Most premetallised dyes are azo dyes, with one nitrogen atom of the azo group playing a part in complexing with the central metal ion. Since in this case there are two azo dye molecules co-ordinated with one chromium atom, compound **7.8** is referred to as a 2:1 complex. 1:1 Complexes are also used commercially, but to a lesser extent as they have generally inferior properties. Premetallised dyes of this type, like traditional acid dyes, are anionic in nature even though, as is the case with complex **7.8**, they may not contain sulfonate groups. Indeed, the presence of sulfonate groups can cause the dye anions to be too strongly attracted to the fibre, which leads in turn to levelness problems. The purpose of the sulfone group in dye **7.8** is to enhance the hydrophilic character of the molecule and hence its water solubility, without increasing the charge on the dye anion. Premetallised dyes are designated in the Colour Index as acid dyes, and they are applied to protein fibres in a similar way as traditional acid dyes. Because of the special stability of chromium(III) complexes, which is attributed to the d^3 transition metal configuration, premetallised dyes provide dyeings with excellent lightfastness.

Protein fibres may also be dyed with certain chemical types of reactive dye, as discussed in Chapter 8.

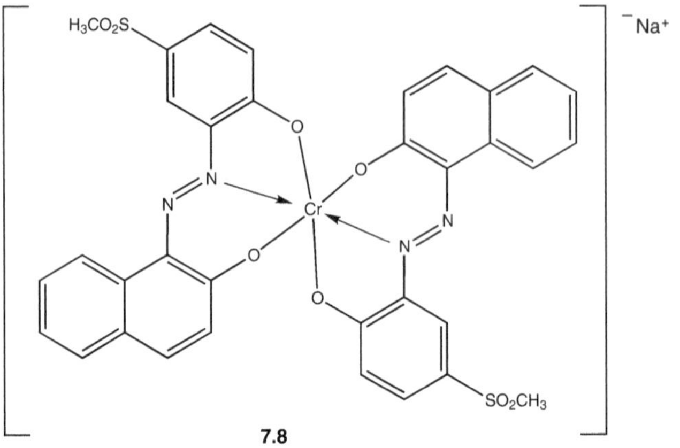

7.8

7.3 DYES FOR CELLULOSIC FIBRES

Cellulosic fibres provide the most important natural fibres and are derived from plant sources. The most important cellulosic fibre used in textiles is cotton, but there are many others, including linen, jute, hemp and flax, and there is also interest in cellulosic fibres derived from bamboo and nettles. The principal component of the cotton fibre is cellulose, the structure of which is shown in Figure 7.4. Cotton is in fact almost pure cellulose (up to 95%). Cellulose is a polysaccharide. It is a high molecular weight polymer consisting of long chains of repeating glucose units, with up to around 1300 such units in each molecule. Cellulose has a fairly open structure, which allows large dye molecules to penetrate relatively easily into the fibre. Each glucose unit contains three hydroxyl groups, two of which are secondary and one primary, and these give the cellulose molecule a considerable degree of polar character. The presence of the hydroxyl groups is of considerable importance in the dyeing of cotton. The hydrophilicity that they confer means that the fibre is

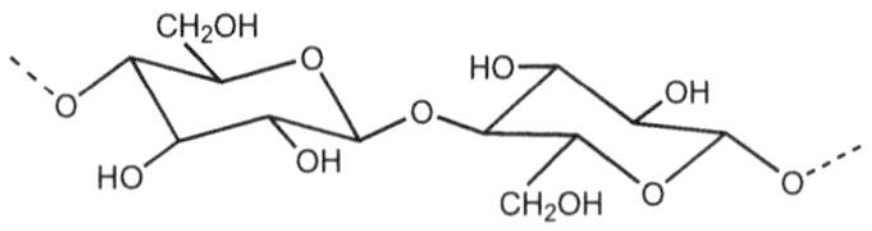

Figure 7.4   Structure of cellulose.

capable of absorbing significant quantities of water, the medium from which the dyes are applied. For example, the ability of the hydroxyl groups to form intermolecular hydrogen bonds is thought to be of some importance in direct dyeing, while reactive dyeing (Chapter 8) involves a chemical reaction of the hydroxyl groups with the dye to form dye–fibre covalent bonds. The tendency of the hydroxyl groups to ionise to a certain extent (to $-O^-$) means that the fibres may carry a small negative charge. There are a larger number of application classes of dyes that may be used to dye cellulosic fibres such as cotton than for any other fibre. These dye application classes include direct, vat, sulfur, azoic and reactive dyes. The chemical principles and structural features of direct, vat and azoic dyes are considered in this section while the discussion of reactive dyes is continued separately in Chapter 8. An outline of the chemistry of sulfur dyes, which is not well-established, is presented in Chapter 6. Viscose, modal and lyocell are regenerated cellulosic fibres that may be dyed in essentially the same way, and with the same dye classes, as natural cellulosic fibres. Paper is also derived from cellulose.

Cellulosic materials are generally sensitive to degradation under acidic conditions but are quite resistant to alkaline conditions. This contrasts with the behaviour of protein fibres where the opposite is the case. Cellulosic fibres are thus commonly wet-processed under alkaline conditions and this includes dyeing.[13,14,24] It is vital that cotton, in particular, is subject to a series of preparatory processes prior to dyeing in order to ensure satisfactory performance. The most important of these processes is *scouring*, using a surfactant often in the presence of alkali, which is essentially a cleaning process to remove impurities such as fats, oils, waxes and particulate materials that would otherwise inhibit dye uptake, and it also improves the wettability of the fibres. *Desizing* is an enzymatic process used to remove *sizes*, which are starch-like materials applied to cotton yarns to provide a degree of mechanical protection during the process of weaving into fabrics. *Mercerisation* involves the treatment of cotton with hot aqueous sodium hydroxide, applied commonly to enhance a range of properties of the cotton fabrics. In particular, mercerisation leads to significantly improved dye uptake as a result of swelling of the fibres and also reducing the degree of crystallinity within the cellulose.

Direct dyes are a long-established class of dyes for cellulosic fibres.[25] They derive their name historically from the fact that they were the first application class to be developed that could be applied directly to these fibres without the need for a fixation process such as mordanting. For this reason, they are also commonly referred to as *substantive* dyes. In some ways, direct dye molecules are structurally similar to acid dye molecules used for protein fibres. For example, they are anionic dyes as a result of the presence of sulfonate ($-SO_3^-$) groups. However, the role of the sulfonate groups in the case of direct dyes is simply to provide water-solubility. In contrast to the acid dyeing of protein fibres, ionic attraction to the fibre is not involved in the direct dyeing of cellulosic fibres. In fact, the anionic nature of the dyes can reduce affinity for the fibres because cellulosic fibres may carry a small negative charge. For this reason, usually only as many sulfonate groups as are required to give adequate solubility in water are present in direct dyes and, in addition, the groups are distributed evenly throughout the molecules. Direct dyes are usually applied close to the boiling point of water and in the presence of an electrolyte, sodium chloride or sulfate, which, by a 'salting-out' effect, enhances dye exhaustion on to the fibre and is also suggested to promote a degree of dye aggregation within the fibre.

Arguably the most important features of direct dye molecules that influence their application properties are associated with their size and shape. They are, in general, large molecules and in shape they are long, narrow and planar. Direct dyes show affinity for cellulose by a combination of van der Waals, dipolar and hydrogen-bonding intermolecular forces. Individually, these forces are rather weak. The long, thin and flat molecular geometry allows the dye molecules to align with the long polymeric cellulose fibre molecules and hence maximise the overall effect of the combined set of intermolecular forces. Chemically, direct dyes, of which compound **7.9** (CI Direct Orange 25) is a typical example, are almost invariably azo dyes, commonly containing two or more azo groups. The long, flat, linear shape of compound **7.9** allows groups such as the -OH, -NHCO (amide), and $-N=N-$ groups in principle to form hydrogen bonds with OH groups on cellulose as it lines up with the cellulose molecule. There are only two sulfonate groups in compound **7.9** and these are well separated from one another. This is sufficient to give adequate water-solubility for their application. In addition, it may be argued that the sulfonate groups are on the opposite side of the molecule from groups that may be participating in hydrogen-bonding with the fibre and this means that they will be oriented away from the cellulose molecule, thus minimising any negative charge repulsion effects.

7.9

Direct dyes provide a full range of hues although frequently not with a high level of brightness, and in comparison with other dye classes for cellulosic fibres, particularly vat and reactive dyes, provide only moderate washfastness. They are, however, inexpensive and are therefore commonly the dyes of choice for applications, for example in the coloration of paper, where cost is of prime concern and fastness to wet treatments is of lesser importance. For textile applications, the washfastness may in certain cases be improved by specific chemical aftertreatments. For example, a group of products referred to as *direct and developed* dyes contain free aromatic amino (-NH$_2$) groups. Treatment of the dyed fabric with aqueous sodium nitrite under acidic conditions results in diazotisation of these groups and the resulting diazonium salts may be reacted with a variety of coupling components, such as 2-naphthol. The larger azo dye molecule, which is thus formed (see Chapter 3 for a discussion of the chemistry involved in azo dye formation), is more strongly attracted to the fibre and less soluble in water, both features leading to improved washfastness. However, the process is complex, requiring careful control, and has thus suffered a decline in commercial importance. Alternatively, in a process referred to as *after-coppering,* some *o,o'-*dihydroxyazo direct dyes may be treated with copper(II) salts to form square-planar metal complexes that show improved washfastness properties and may also enhance lightfastness. Fastness to wet-treatments may also be achieved by aftertreatment with organic cationic auxiliary materials, such as quaternary ammonium salts. The large organic cations replace the sodium ions associated with the dye anions, forming much less soluble salts that are more resistant to removal from the fibre.

Vat dyes are a group of totally water-insoluble dyes, and in this respect they are essentially pigments.[26] In fact, some vat dyes, after conversion into a suitable physical form, may be used as pigments (Chapters 4 and 9). For application to cellulosic fibres, vat dyes are initially converted into a water-soluble *leuco* form by an alkaline reduction process. Generally, this conversion is carried out using a reducing agent such as sodium dithionite ($Na_2S_2O_4$), also referred to as hydrosulfite, in the presence of sodium hydroxide. In the leuco form, the dyes have affinity for the cellulosic fibre and are consequently taken up by the fibre. Subsequently, they are converted back into the insoluble pigment by an oxidation reaction. Hydrogen peroxide is frequently used as the oxidising agent for this process, although atmospheric oxygen can also effect the oxidation under appropriate conditions. The pigment or molecular aggregate thus produced becomes trapped mechanically within the fibre and its insolubility gives rise to the excellent washfastness properties which are characteristic of vat dyes. The lightfastness of cellulosic fibres dyed with vat dyes is also excellent, typical of the levels given by some of the higher performing pigments. The vat dyeing process is usually completed with a high temperature aqueous surfactant treatment, or 'soaping', which enhances molecular aggregation and develops crystallinity within the entrapped particles, in addition to removing particles that are only loosely held at the fibre surface. The chemistry of the reversible reduction/oxidation process involved in vat dyeing, which is illustrated in Figure 7.5, requires the presence in the vat dye of two carbonyl groups linked *via* a conjugated system. The carbonyl (C=O) groups are reduced under alkaline conditions to enolate ($-O^-Na^+$) groups which give the leuco form water-solubility. Investigations have been carried out into vat dyeing with alternative reducing agents, such as hydroxyacetone, a biodegradable organic material, or glucose, with a view to avoiding issues associated with residual sulfur-containing species in textile dyeing effluent when sodium dithionite is used.[27] These explorations have demonstrated

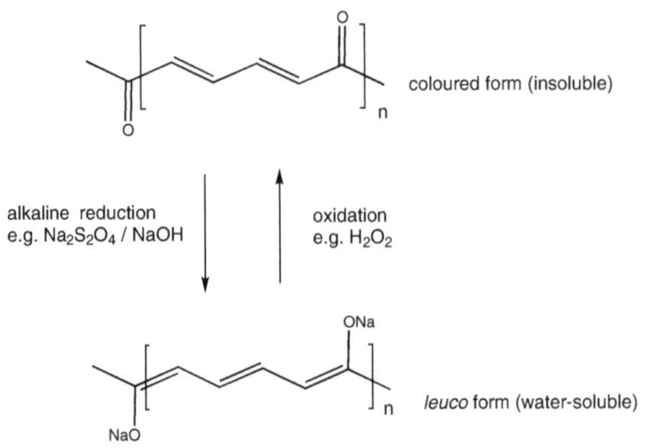

Figure 7.5   Chemistry of vat dyeing.

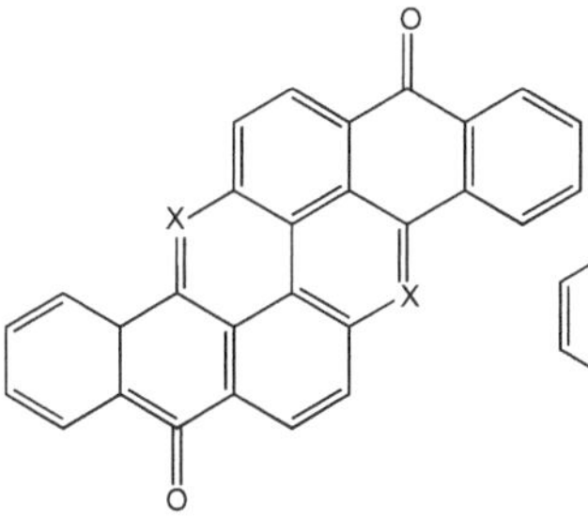

7.10a: X=CH
7.10b: X=N

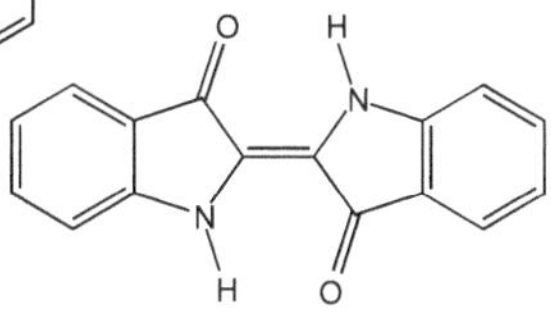

7.11

Figure 7.6 Structures of some typical vat dyes.

some success on a laboratory scale although it is not clear that there has as yet been significant commercial exploitation.

Vat dyes are exclusively of the carbonyl chemical class (Chapter 4). Dyes of other chemical classes, including azo dyes, are generally inappropriate as vat dyes because they undergo a reduction that cannot be reversed. Vat dyes, illustrated with selected examples in Figure 7.6, are generally large planar molecules, often containing multiple ring systems, to provide the leuco form of the dyes with some affinity for the cellulosic fibre as a result of van der Waals and dipolar forces. There is usually a notable absence of other functional groups in the molecules, because such groups can be sensitive to the reduction and oxidation reactions. However, halogen substituents (Cl, Br) are encountered on occasions. Pyranthrone, **7.10a** (CI Vat Orange 9) and flavanthrone, **7.10b** (CI Vat Yellow 1) provide examples, respectively, of carbocyclic and heterocyclic anthraquinone vat dyes. Indigo, **7.11**, is a further notable example of a vat dye. Natural indigo dyeing (Chapter 1) is carried out directly in a vat where the fermentation of composted leaves from the *Indigofera tinctoria* plant is performed in the presence of alkali from wood ash or limestone to produce precursors that are oxidised in air on the fibre to give indigo. Synthetic indigo is long-established as the vat dye used in the dyeing of denim fabric.

Solubilised vat dyes (Figure 7.7) are a group of water-soluble, prereduced forms of vat dyes, stabilised as the sodium salts of sulfate esters, which are prepared *via* the leuco form by chlorosulfonation. These dyes may be applied directly to the cellulosic fibres and subsequently generate the vat dye within the fibres by acid hydrolysis and oxidation with hydrogen peroxide.

Azoic dyeing of cellulosic fibres is a long-established process although it is used only to a limited extent currently. The synthesis of

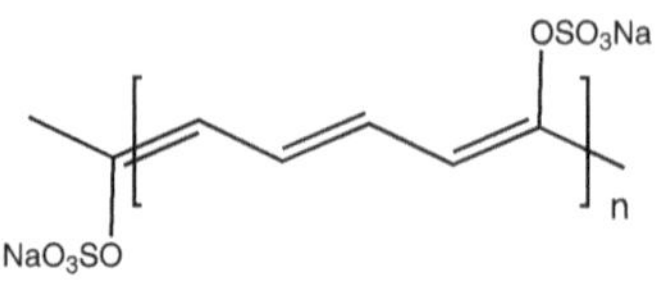

Figure 7.7 Generalised structure of a solubilised vat dye.

azo dyes and pigments, the most important chemical class of organic colorants, by the two-stage reaction sequence involving diazotization and azo coupling has been dealt with in detail in Chapter 3. In the azoic dyeing process, an azo pigment is formed by this sequence of chemical reactions within the fibre. The cotton fibres are first impregnated with an appropriate coupling component such as the anilide of 3-hydroxy-2-naphthoic acid, **7.12**, under aqueous alkaline conditions. The fibre prepared in this way is then treated with a solution of a diazonium salt, either prepared directly by diazotization of the appropriate aromatic amine or in a stabilised form in which the counteranion is, for example, tetrafluoroborate (BF_4^-) or tetrachlorozincate $(ZnCl_4^{2-})$. As a consequence of the azo coupling reaction which ensues, insoluble azo pigment aggregates are formed and these become trapped mechanically within the fibres. Azoic dyeing provides good fastness properties, although the range of colours is restricted to those typical of azo pigments, namely yellow, orange, red, Bordeaux and browns. The main disadvantages of azoic dyeing are that the processes are complex and lack the versatility provided by other dye classes, notably reactive dyes.

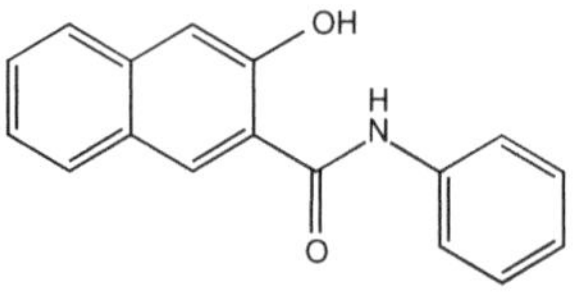

7.12

7.4 DYES FOR SYNTHETIC FIBRES

The three most important types of synthetic fibres used commonly as textiles are polyester, polyamides (nylon) and acrylic fibres.[28] Polyester and the semi-synthetic fibre cellulose acetate are dyed almost exclusively using disperse dyes. Polyamide fibres may be coloured using either acid dyes, the principles of which have been discussed in

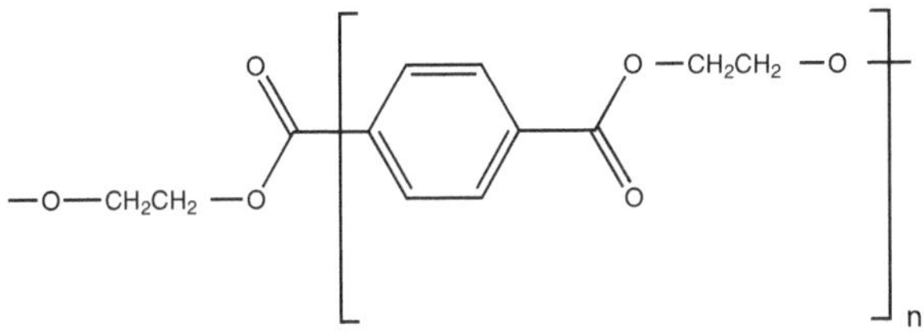

Figure 7.8 Structure of polyester (PET, poly(ethylene terephthalate)).

the section on protein fibres in this chapter, or with disperse dyes. Acrylic fibres are dyed mainly with basic (cationic) dyes.

Polyesters are polymers whose monomeric units are joined through ester (-COO-) linkages. The most important polyester fibre by far is poly(ethylene terephthalate), PET, whose production far outstrips that of any other synthetic fibre. PET, commonly referred to simply as polyester, has the chemical structure shown in Figure 7.8. Polyester is relatively hydrophobic (non-polar) in character, certainly in comparison to the natural protein and cellulosic fibres, largely as a result of the prominence of the benzene rings and the $-CH_2CH_2-$ groups. However, the ester groups do confer a degree of polarity on the molecule, so that the fibres are not as hydrophobic as, for example, hydrocarbon polymers such as polyethylene and polypropylene. Nevertheless, it is not surprising that relatively hydrophobic fibres such as polyester and cellulose acetate show little affinity for dyes that are ionic in character. This feature means that dye application classes containing ionic water-solubilising groups such as the sulfonate ($-SO_3^-$) group, for example acid and direct dyes, are inappropriate for application to polyester. The inevitable consequence is that dyes for polyester and cellulose acetate cannot be expected to have substantial water-solubility. A further feature of polyester is that its physical structure is highly crystalline, consisting of tightly packed, highly-ordered polymer molecules, and the polymer is below its glass transition temperature at normal textile dyeing temperatures. As a result, it is relatively inaccessible even to small molecules. Further, its hydrophobic nature means that it has little affinity for water, which is the usual dyeing medium. Polyester is thus a difficult fibre to dye. Disperse dyes are an application class of dyes with non-ionic molecular structures providing relatively low water-solubility. They may be applied to these relatively hydrophobic synthetic fibres as a fine dispersion of solid particles in water, prepared with the assistance of specific dispersing agents. Disperse dyes were originally developed for application to cellulose acetate, but they have assumed much greater importance for application to polyester, and to a lesser extent to polyamides.

   Disperse dyes are required to be relatively small, planar molecules to allow the dyes to penetrate between the polymer chains and into the bulk of the fibre.[29-33] These dyes are commonly applied to the fibre as a fine aqueous dispersion at temperatures of around 130 °C under pressure. They may also be applied by padding the dispersion on to the fabric followed by *thermofixation*, which involves treatment at around 210 °C for a short time. At elevated temperatures, the tight physical structure of the polymer is loosened by thermal agitation, which reduces the strength of the intermolecular interactions, thus enhancing segmental mobility of the chains and facilitating entry of the dye molecules. After dyeing with disperse dyes, especially when applied at higher concentrations, small dye aggregates that have not penetrated into the polyester often remain on the fibre surface. These aggregates are removed by a washing process known as *reduction clearing*, generally involving treatment with aqueous alkaline sodium dithionite which destroys the surface dye chemically by reduction.[34] Disperse dyes are non-ionic molecules that effectively dissolve in the polyester. In the solid solutions formed, the dye–fibre affinity is generally considered to involve a combination of van der Waals and dipolar forces and hydrogen-bonding, while π–π interactions between aromatic groups on the dye and polyester molecules may also be involved. A general feature of disperse dye molecules is that they possess several polar, though not ionic, groups. Frequently encountered polar groups include the nitro (NO_2) cyano (CN), hydroxyl, amino, ester, amide (NHCO) and sulfone (SO_2) groups. These polar groups can be considered as making a number of contributions to the application properties of disperse dyes. One of their roles is to provide an adequate degree of water-solubility at the high temperatures at which the dyes are applied from the aqueous dyebath. A second function of the polar groups is to enhance affinity as a result of dipolar intermolecular forces with the ester groups of the polyester molecule. Disperse dyes can therefore be considered as compromise molecules possessing a balanced degree of polar (hydrophobic) and non-polar (hydrophilic) character, similar to that of the polyester molecule. The polar groups also have an important influence on the colour of the dye molecules according to the principles discussed in Chapter 2.

   A consequence of the fact that disperse dyes are relatively small, non-polar molecules is that they may have a tendency to be volatile, and hence prone to sublimation out of the fibre at high temperatures.

This can lead to a loss of colour and the possibility of staining of adjacent fabrics when dyed polyester is subjected to high temperatures, *e.g.*, in heat setting and ironing. Increasing the size and/or the polarity of the dye molecules enhances fastness to sublimation. However, as a consequence, the compromise arises that these larger, more polar dye molecules will require more forcing conditions, such as higher temperatures and pressures, to enable the dyes to penetrate into the fibre.

Figure 7.9 illustrates the chemical structures of some typical disperse dyes and demonstrates that a wider range of chemical types are encountered in the range of commercial disperse dyes than is the case with any other dye application class.[29,30] Numerically, azo dyes form by far the most important chemical class of disperse dyes. Azo disperse dyes may be classified into four broad groupings. The most numerous of these are the aminoazobenzenes, which provide important orange, red, violet and blue disperse dyes. They are exemplified by CI Disperse Orange 25, **7.13**, CI Disperse Red 90, **7.14**, and CI Disperse Blue 165, **7.15**. A comparison of these three aminoazobenzene dyes provides an illustration of the bathochromic shift provided by increasing the number of electron-accepting and electron-donating groups in appropriate parts of the molecules, in accordance with the principles of the relationship between colour and molecular structure as discussed in Chapter 2. There are two further groups of disperse dyes that are heterocyclic analogues of the aminoazobenzenes. Derivatives based on heterocyclic diazo components provide bright intense colours and are bathochromically shifted so that they serve the purpose of extending the range of blue azo disperse dyes available. An example of such a product is CI Disperse Blue 339, **7.16**. Derivatives based on heterocyclic coupling components are useful for their ability to provide bright intense yellow azo disperse dyes. An example is CI Disperse Yellow 119, **7.17**, which, as illustrated in Figure 7.9, exists as the ketohydrazone tautomer. The fourth group are disazo dyes of relatively simple structures, for example CI Disperse Yellow 23, **7.18**. Carbonyl disperse dyes, especially anthraquinones, are next in importance to the azo dyes and there are also a few products belonging to the nitro and polymethine chemical classes. CI Disperse Red 60, **7.19**, and CI Disperse Green 5, **7.20** are examples of typical anthraquinone disperse dyes, while compound **7.21** is an example of the more recently-introduced benzodifuranone carbonyl type. CI Disperse Yellow 42, **7.22** and CI Disperse Blue 354, **7.23**, respectively, provide commercially relevant examples of the nitro and polymethine chemical classes.

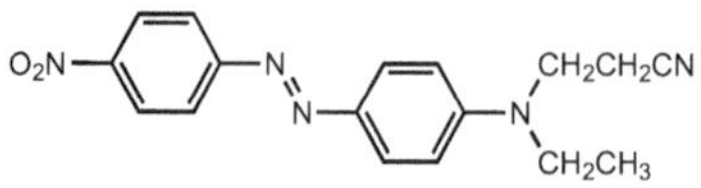

7.13

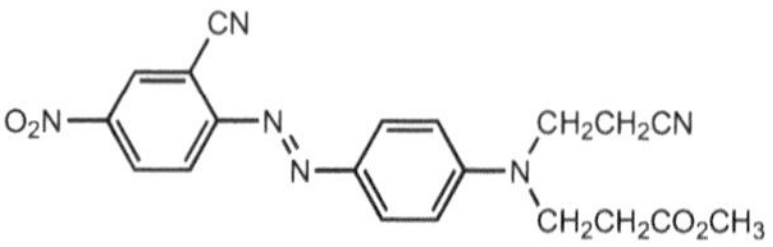

7.14

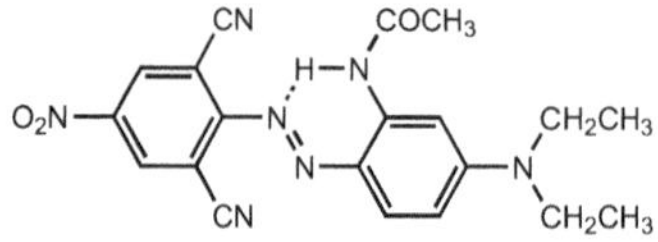

7.15

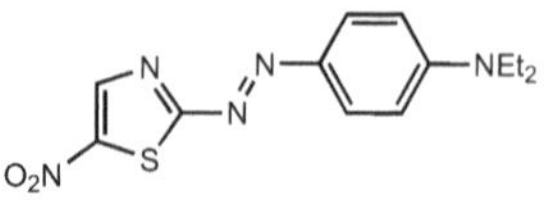

7.16

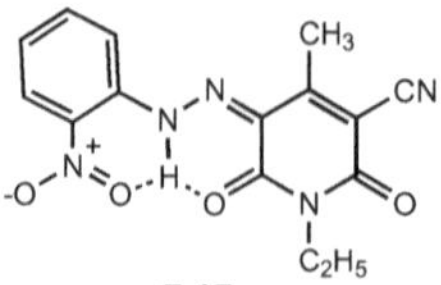

7.17

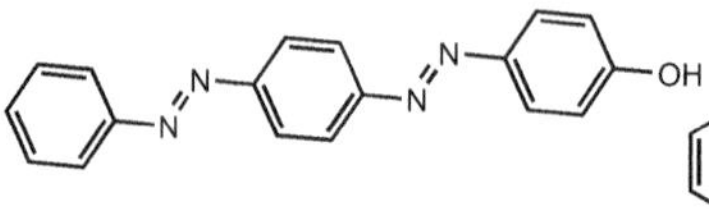

7.18

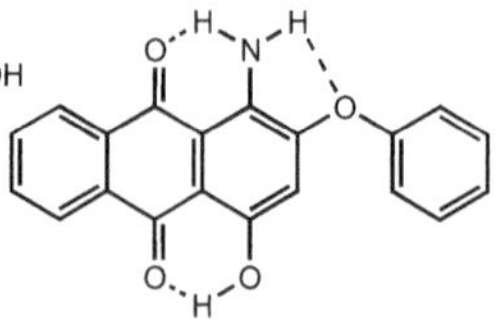

7.19

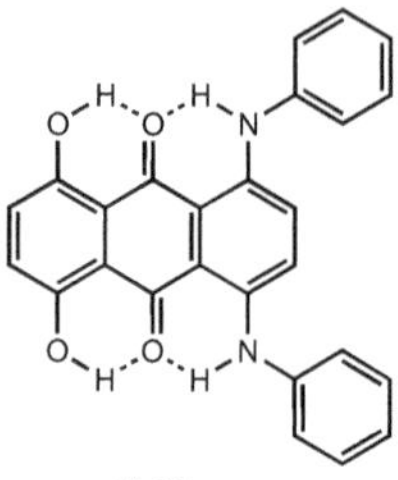

7.20

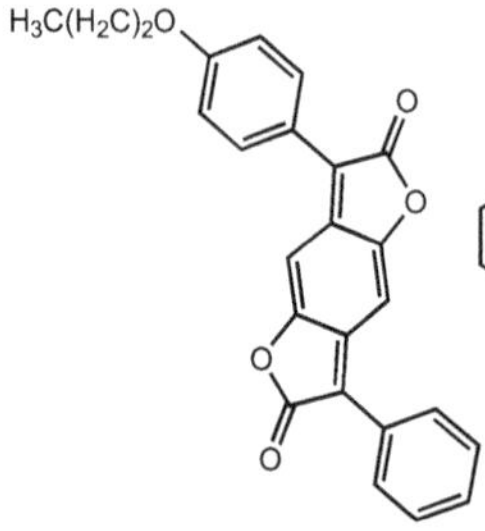

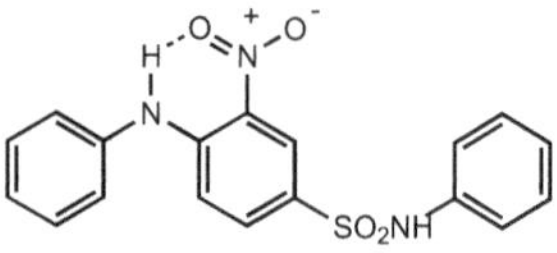

7.22

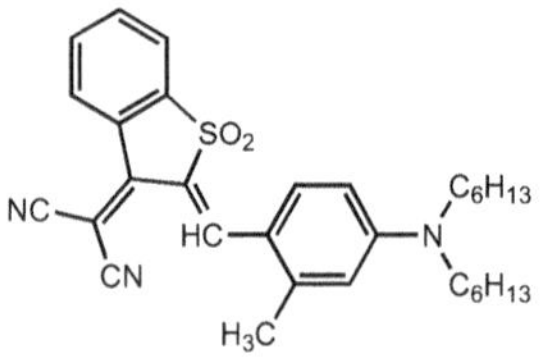

7.21 **7.23**

Figure 7.9 Some examples of chemical structures of disperse dyes.

Polylactic acid (PLA), the structure of which is shown in Figure 7.10, is a polyester fibre in which there has been recent interest because of its environmental credentials. PLA may be derived from renewable resources, such as cornstarch, and it is biodegradable. PLA may be coloured using certain disperse dyes, although the dyes do not exhaust as well as on PET, mainly because of its aliphatic character.[35,36]

Acrylic fibres are synthetic fibres based essentially on the addition polymer polyacrylonitrile, the essential structure of which is illustrated in Figure 7.11. However, most acrylic fibres are rather more complex and contain within their structure anionic groups, most commonly sulfonate ($-SO_3^-$), but also carboxylate ($-CO_2^-$) groups either as a result of the incorporation of co-polymerised monomers in

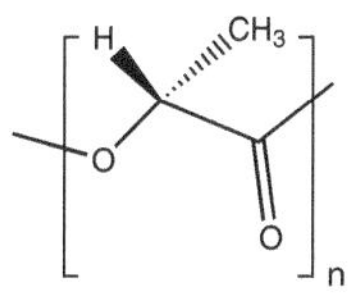

Figure 7.10 Structure of polylactic acid (PLA).

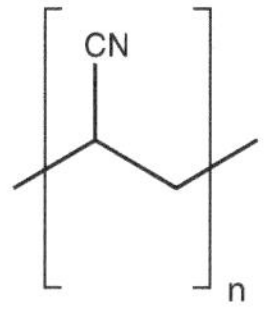

Figure 7.11 Structure of polyacrylonitrile.

which these groups are present or due to the presence of residual amounts of anionic polymerisation inhibitors. Co-polymerisation with appropriate monomers is also used to reduce the glass transition temperature and crystallinity of the acrylic fibres in order to enhance the accessibility of dye molecules. The anionic character of these acrylic fibres provides the explanation as to why cationic dyes provide the principal application class used for their coloration.[37,38] These dyes are classified by the Colour Index as basic dyes, a term that originated from their use, now largely obsolete, to dye protein fibres, such as wool, from a basic or alkaline dyebath under which conditions the protein molecules acquired a negative charge. Cationic dyes are generally found to exhibit rather poor fastness properties, especially lightfastness, when applied to natural fibres but they are observed to give much better performance on acrylic fibres.

Some examples of the structures of cationic dyes used to dye acrylic fibres are shown in Figure 7.12. These include the azo dye, CI Basic Red 18, **7.24**, the arylcarbonium ion (triphenylmethine) dye, CI Basic Green 4, **7.25**, and the methine derivative, CI Basic Yellow 11, **7.26**. As the name implies, the dyes are coloured cationic species, generally as a result of the presence of positively-charged quaternary nitrogen atoms (as $-NR_3^+$, or $=NR_2^+$). In the case of dyes **7.24** and **7.26**, the positive charge is localised on the nitrogen atom, whereas in dye **7.25** it is delocalised by resonance. These groups serve two purposes. Firstly, they provide the water-solubility necessary for the application of the dyes, due to their ionic character. Secondly, they provide affinity for the acrylic fibres as a result of ionic attraction between the dye

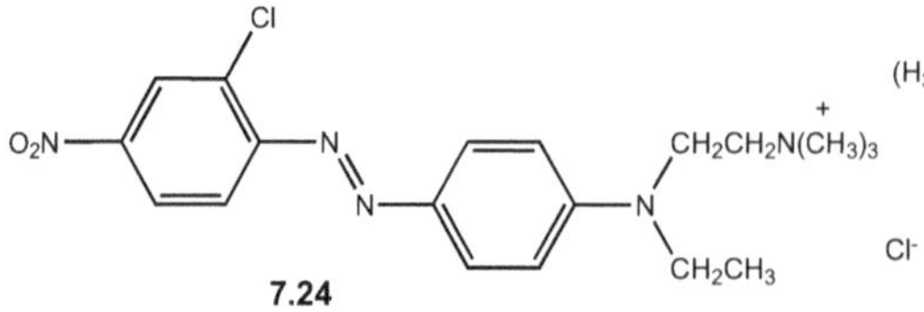

Figure 7.12   Some examples of chemical structures of cationic dyes for acrylic fibres.

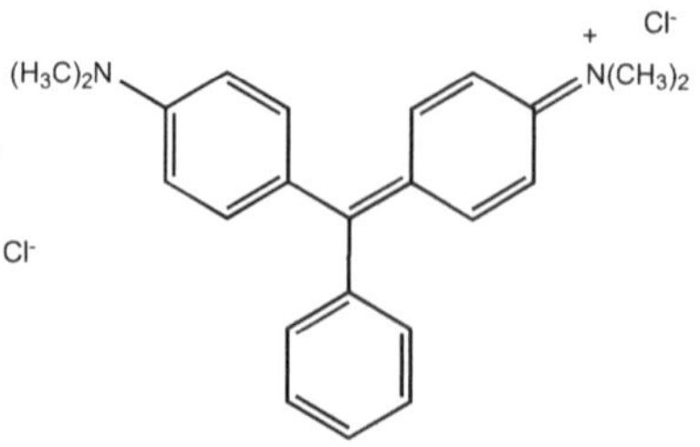

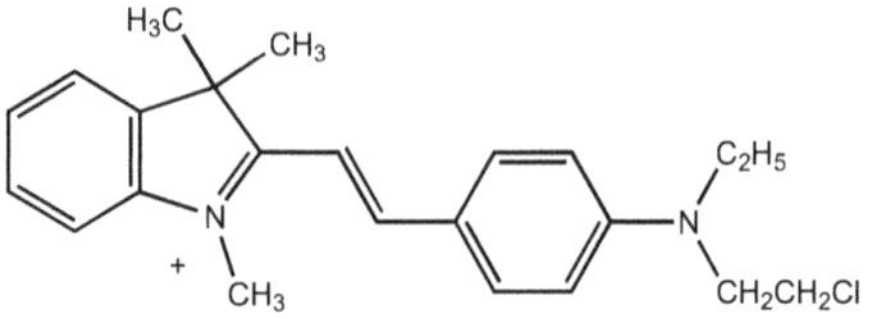

cations and the anionic groups ($-SO_3^-$ and $-CO_2^-$) that are present in the acrylic fibre polymer molecules. In a sense, the means of attachment of cationic dyes to acrylic fibres may be considered as the converse of that involved in the acid dyeing of protein fibres, discussed previously in this chapter, which involves the attraction of dye anions to cationic sites on the fibre.

Polyamides are polymers whose monomeric units are joined through secondary amide (-NHCO-) linkages. The two most important polyamide fibres are based on nylon 6.6, **7.27**, and nylon 6, **7.28**, whose structures, both completely aliphatic in character, are illustrated in Figure 7.13. A comparison with Figure 7.1 will reveal the structural analogy between natural protein fibres such as wool and polyamide fibres, and thus they may be dyed by some of the application classes suitable for protein fibres. Polyamides are commonly dyed using acid dyes. These are attracted to the fibres by a mechanism similar to the acid dyeing of wool, involving attraction of the dye anions to amino groups, which are present for example at the end of the polyamide chains, and which are protonated under acidic conditions. Alternatively, because the molecular structures of polyamide fibres are relatively hydrophobic, similar in this respect to polyester, they may be dyed using the range of disperse dyes by a mechanism analogous to the dyeing of polyester discussed earlier in this chapter.

Fibres based on polyolefins (polyalkenes), such as polyethylene and polypropylene, find some applications in textiles. As hydrocarbons,

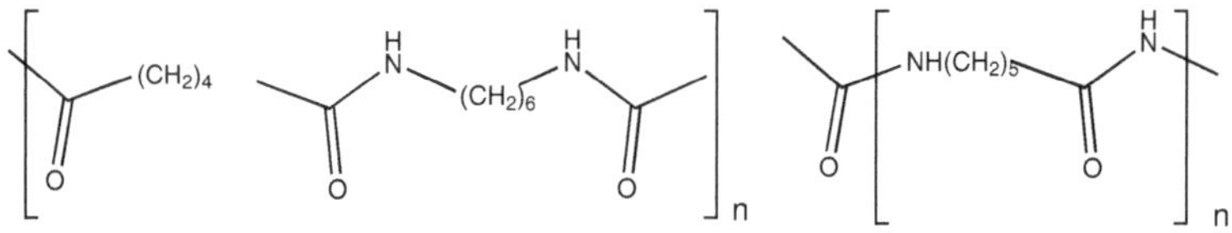

7.27: nylon 6,6   **7.28**: nylon 6

Figure 7.13   Structure of polyamide fibres.

these fibres are highly hydrophobic. They also commonly exist in a highly crystalline form and are consequently extremely difficult to dye. Colour is generally introduced into synthetic fibres of this type using pigments by *mass coloration* in which the pigment is dispersed intimately into the molten polymer before it is extruded into a fibrous form (Chapter 9). Other synthetic fibres, including polyester, may also be coloured by mass pigmentation in this way. Coloration at such an early stage of the textile manufacturing process offers the advantage that it ensures levelness, as further processing can average out differences. However, it requires careful planning of the amounts of coloured fabrics required commercially, whereas dyeing at later stages of the production process provides a much more versatile approach to the production of the appropriate range of colours.

REFERENCES

1. R. R. Mather and R. H. Wardman, *The Chemistry of Textile Fibres*, Royal Society of Chemistry, Cambridge, 2011.
2. W. E. Morton and J. W. S. Hearle, *Physical Properties of Textile Fibres*, Woodhead Publishing, Cambridge, UK, 2008.
3. A. R. Horrocks and S. C. Anand (ed.), *Handbook of Technical Textiles*, Woodhead Publishing, Cambridge, UK, 2000.
4. R. Kozlowski (ed.), *Handbook of Natural Fibres: Types, Properties and Factors affecting Breeding and Cultivation*, Woodhead Publishing, Cambridge, UK, 2012, vol. 1.
5. R. Kozlowski (ed.), *Handbook of Natural Fibres: Processing and Applications*, Woodhead Publishing, Cambridge, UK, 2012, vol. 2.
6. C. Woodings (ed.), *Regenerated Cellulose Fibres*, Woodhead Publishing, Cambridge, UK, 2001.
7. R. S. Blackburn (ed.), *Biodegradable and Sustainable Fibres*, Woodhead Publishing, Cambridge, UK, 2005, pp. 157–188.
8. C. R. Woodings, *Int. J. Biol. Macromol.*, 1995, **17**, 305.

9. J. E. McIntyre (ed.), *Biodegradable and Sustainable Fibres*, Woodhead Publishing, Cambridge, UK, 2004.
10. H. Zollinger, *Color Chemistry: Syntheses, Properties and Applications of Organic Dyes and Pigments*, Wiley-VCH Verlag GmbH, Weinheim, 3rd edn, 2003, ch. 11.
11. P. F. Gordon and P. Gregory, *Organic Chemistry in Colour*, Springer-Verlag, New York, 1983, ch. 6.
12. A. D. Broadbent, *Basic Principles of Textile Coloration*, Society of Dyers and Colourists, Bradford, UK, 2001.
13. J. N. Chakraborty, *Fundamentals and Practices in Colouration of Textiles*, Woodhead Publishing India Pvt Ltd, New Delhi, 2012.
14. M. Clark (ed.), *Handbook of Textile and Industrial Dyeing: Principles, Processes and types of Dyes*, Woodhead Publishing, Cambridge, 2011, vols 1 and 2.
15. D. M. Lewis, *Rev. Prog. Color.*, 1998, **28**, 12.
16. L. W. C. Miles, *Textile Printing*, Society of Dyers and Colourists, Bradford, UK, 2nd edn, 2003.
17. T. L. Dawson and C. J. Hawkyard, *Rev. Prog. Color.*, 2000, **30**, 7.
18. H. Ujiie (ed.), *Digital Printing of Textiles*, Woodhead Publishing, Cambridge, UK, 2006.
19. D. M. Lewis and J. A. Rippon (ed.), *The Coloration of Wool and other Keratin Fibres*, John Wiley & Sons Ltd, Chichester, 2013, in press.
20. D. M. Lewis (ed.), *Wool Dyeing*, Society of Dyers and Colourists, Bradford, UK, 1992.
21. T. P. Smith, K. A. Zaklika, K. Thakur, G. C. Walker, K. Tominaga and P. F. Barbara, *J. Phys. Chem.*, 1991, **95**, 10465.
22. F. Beffa and G. Back, *Rev. Prog. Color.*, 1984, **14**, 33.
23. M. Szymczyk and H. S. Freeman, *Rev. Prog. Color.*, 2004, **34**, 39.
24. J. Shore (ed.), *Cellulosics Dyeing*, Society of Dyers and Colourists, Bradford, UK, 1995.
25. J. Shore, *Rev. Prog. Color.*, 1991, **21**, 23.
26. U. Baumgarte, *Rev. Prog. Color.*, 1987, **17**, 29.
27. M. Bozic and V. Kokol, *Dyes Pigments*, 2008, **76**, 299.
28. S. M. Burkinshaw, *Chemical Principles of Synthetic Fibre Dyeing*, Blackie Academic and Professional, Glasgow, 1995.
29. J. F. Dawson, *J. Soc. Dyers Colourists*, 1991, **107**, 395.
30. J. F. Dawson, *J. Soc. Dyers Colourists*, 1983, **99**, 183.
31. P. W. Leadbetter and A. T. Leaver, *Rev. Prog. Color.*, 1989, **19**, 33.
32. O. Annen, R. Egli, R. Hasler, B. Henzi, H. Jakob and P. Matzinger, *Rev. Prog. Color.*, 1987, **17**, 72.

33. R. Egli, in *Colour Chemistry: The Design and Synthesis of Organic Dyes and Pigments*, ed. A. T. Peters and H. S. Freeman, Elsevier Applied Science, London and New York, 1991, ch. 1.
34. J. R. Aspland, *Textile Chem. Colorist.*, 1992, **24**, 18.
35. D. Karst, D. Nama and Y. Yang, *J. Colloid Interface Sci.*, 2007, **310**, 106.
36. L. E. Scheyer and A. Chiweshe, *AATCC Rev.*, 2001, **1**, 44.
37. R. Raue, *Rev. Prog. Color.*, 1984, **14**, 187.
38. I. Holme, *Rev. Prog. Color.*, 1983, **13**, 10.

CHAPTER 8

Reactive Dyes for Textile Fibres

8.1 INTRODUCTION

It is probable that history will judge the development of reactive dyes to have been the most significant innovation in textile dyeing technology of the twentieth century. As a consequence of their particular industrial importance, and also because they make use of some interesting organic chemistry, this chapter is devoted entirely to a consideration of the chemical principles involved in the application of reactive dyes, and their synthesis.[1–6] Reactive dyes, after application to the textile fibre, are induced to react chemically to form a covalent bond between the dye and the fibre. This covalent bond is formed between a carbon atom of the dye molecule and an oxygen, nitrogen or sulfur atom of a hydroxyl, amino or thiol group on the polymer that constitutes the fibre. Because of the strength of the covalent bond, reactive dyes once applied to the textile material resist removal and as a consequence show outstanding washfastness properties. Initially, reactive dyes were introduced commercially for application to cellulosic fibres, and this remains by far their most important use, although dyes of specific types have also been developed for application to protein and polyamide fibres. The potential to apply reactive dyes to other fibre types, including polyester and polypropylene, has been demonstrated technically although processes of this type are not as yet a commercial reality.

The concept of linking dye molecules covalently to fibre molecules in order to produce colours with superior fastness to washing had

Colour Chemistry, 2nd edition
By Robert M Christie
© R M Christie 2015
Published by the Royal Society of Chemistry, www.rsc.org

been envisaged long before the 1950s when the major breakthrough took place. The early studies were mostly directed towards attaching colour covalently to cellulosic fibres, such as cotton and viscose, because of the modest wet-fastness properties provided by direct dyes which were at the time the most commonly used dye class applied to those fibre types. There was also significant early research into covalent attachment of colour to protein fibres such as wool and silk, although there was less commercial urgency because high washfastness was achievable on those fibres using chrome mordant dyes. In general, the early attempts at dye fixation employed rather vigorous conditions under which serious fibre degradation occurred. The landmark discovery was reported in 1954, when Rattee and Stephen demonstrated that dyes containing the 1,3,5-triazinyl group with chlorine substituents were capable of reaction with cellulosic fibres under mildly alkaline conditions, and under these conditions no significant degradation of the fibre was observed.[7] This discovery led, within two years, to the commercial launch by ICI of the first series of water-soluble reactive dyes for cellulosic fibres, named Procion MX, based on this reactive system.[8,9] The new application class of dyes proved to be an almost immediate success. The dyes were capable of providing superior washfastness compared with direct dyes, a much wider range of brilliant colours than were available from vat and azoic dyes, and they were capable of both continuous and batchwise application.

This initial success provided the impetus for the intense body of research conducted by most other dye manufacturers that followed into this promising area, to a large extent involving the development of alternative types of fibre-reactive group and of multifunctional dyes with more than one reactive group. Subsequently, reactive dyes have grown steadily in importance to become the most significant industrial class of dyes for application to cellulosic fibres. A notable trend in recent years has been the progressive decline in the percentage of textiles derived from cotton produced worldwide, which has been accompanied by a corresponding increase in the share of synthetic fibres. Indeed, in the first decade of the twenty-first century, the global production of polyester overtook that of cotton. Nevertheless, the production of reactive dyes for cellulose has continued to increase, largely because they have been gaining market share at the expense of other dye classes, such as azoic, sulfur and direct dyes, and this trend has been projected to continue.[3] Growth rates for cellulosic fibres may be limited in the future by the availability of land suitable for growing cotton, although this may be augmented

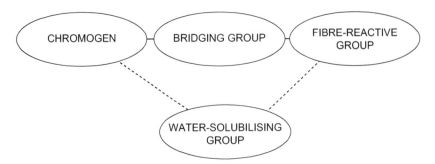

Figure 8.1   General structure of a reactive dye.

by the development of more intensive agricultural methods and the increased production of regenerated cellulosic fibres, such as lyocell.

A general, if rather simplistic, schematic representation of the structure of a reactive dye is illustrated in Figure 8.1. In the figure, four important structural features of the molecules may be identified separately: the chromogen, the water-solubilising group, the bridging group and the fibre-reactive group. The *chromogen* is that part of the molecule which essentially gives the molecule its colour and may contribute to other features of the dye, such as its lightfastness. As encountered in most other application classes of textile dyes, these chromogens typically belong to the azo, carbonyl or phthalocyanine chemical classes. Specific examples are used to illustrate this characteristic later in the chapter. Reactive dyes are required to be water-soluble for their application to the textile fibres, and they invariably contain for this purpose one or more ionic groups, most commonly the sulfonate group as the sodium salt. They are thus anionic in nature and, as such, show many of the chemical features of acid dyes for protein fibres (Chapter 7). Most commonly, the *water-solubilising group* is located in the chromogenic part of the reactive dye molecule, although it may also be more closely associated with the fibre-reactive group. The essential structural characteristic of a reactive dye is a specific functional group that is capable of reacting chemically with the fibre. This feature is, for obvious reasons, termed the *fibre-reactive group* and the organic chemistry underlying the reaction of these groups with functionality on the fibre forms a substantial part of this chapter. Commonly, the term *bridging group* is used to identify the group of atoms that is used to link the chromogenic part of the molecule to the fibre-reactive group. In many dyes, the bridging group is the amino (–NH–) group, often for reasons of synthetic convenience. There are, however, certain types of reactive dyes in which no obvious bridging functionality may be clearly identified.

8.2 FIBRE-REACTIVE GROUPS

Cotton fibres are based on cellulose, a polysaccharide, whose structure is given in Figure 7.4 (Chapter 7). Most reactive dyes for cotton utilise the ability of the many hydroxyl (OH) groups present in the cellulose molecule to act as nucleophiles. The dyes are commonly induced to react with the cellulose in aqueous alkali under which conditions deprotonation of the hydroxyl groups (Cell-OH) takes place. This produces the more powerfully nucleophilic cellulosate anions (Cell-O$^-$), which are generally regarded as the active nucleophiles in the reactive dyeing of cellulose. In the reactive dyeing of protein fibres, such as wool or silk, the nucleophilic group on the fibre may be the amino (–NH$_2$), hydroxyl (–OH) or thiol (–SH) groups present in the amino side-chains of the polypeptide (see Figure 7.1, Chapter 7). The most common types of fibre-reactive dyes for cellulosic fibres react either by aromatic nucleophilic substitution or by nucleophilic addition to activated alkenes.[1,2]

8.2.1 Fibre-Reactive Groups Reacting by Nucleophilic Substitution

A characteristic feature of the chemistry of aromatic ring systems is their tendency to undergo substitution reactions in which the aromatic character of the ring is retained. Since most aromatic systems are electron-rich in nature, by virtue of the system of π-electrons, the most frequently encountered reactions of aromatic compounds are electrophilic substitution reactions. Nucleophilic aromatic substitution reactions are less commonly encountered, although they are not rare. For reaction with nucleophiles under mild conditions to take place, aromatic systems require the presence of structural features that reduce the electron density of the π-system and thus facilitate nucleophilic attack. To illustrate this by way of example, chlorobenzene undergoes nucleophilic substitution (*e.g.*, of Cl by OH) only under vigorous conditions. In contrast, 1-chloro-2,4-dinitrobenzene undergoes nucleophilic substitution very readily because of the activating effect of the two powerfully electron-withdrawing nitro groups. Reactions of this type are used in the synthesis of some nitrodiphenylamine dyes (see Scheme 6.5 in Chapter 6). Another feature of nucleophilic substitution reactions is the requirement for a good, stable leaving group, such as the chloride anion.

By far the most important fibre-reactive groups that react by nucleophilic substitution contain six-membered aromatic nitrogen-containing heterocyclic rings with halogen substituents. The first group of commercial reactive dyes, the Procion dyes, which remain among the most important in use today, contain the 1,3,5-triazine (or *s*-triazine) ring, a six-membered ring of alternate carbon and nitrogen atoms, containing at least one chlorine (or fluorine) substituent, and with the amino (–NH–) group as the bridging group. The reaction of a chlorotriazinyl dye **8.1** with the cellulosate anion, incorporating an outline mechanism for the reaction which is characteristic of aromatic nucleophilic substitution, is illustrated in Scheme 8.1. Several features may be identified as responsible for facilitating the nucleophilic substitution reaction. Firstly, the electron-withdrawing nature of the heterocyclic nitrogen atoms and, to a lesser extent, of the

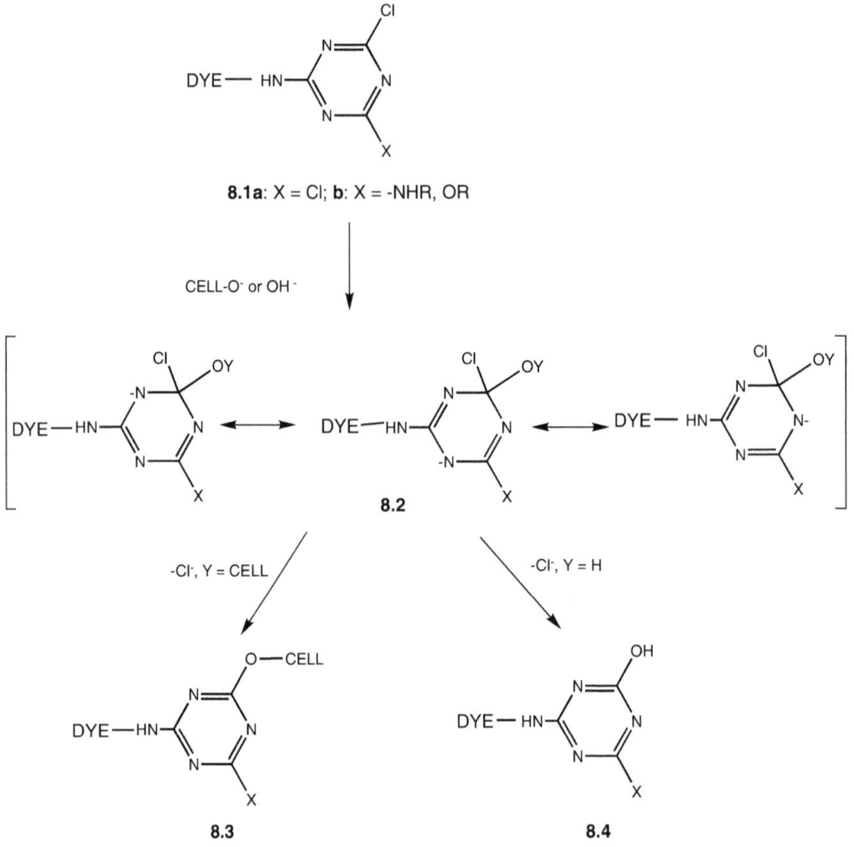

Scheme 8.1   Reaction of Procion dyes with cellulose.

chlorine substituent reduces the electron density in the aromatic ring, which is thereby activated towards nucleophilic attack. An important feature is the extensive resonance stabilisation of the anionic intermediate **8.2**. In particular, it is of special importance that the delocalised negative charge is favourably accommodated on the electronegative heterocyclic nitrogen atoms in each of the three contributing resonance forms of the intermediate. In the final step of the sequence, the chloride ion, a particularly good leaving group, is eliminated. As a result of the reaction, a dye–fibre covalent (C–O) bond is formed, as illustrated for structure **8.3**, and it is to this bond that the excellent fastness of the dyed textile fibres to wet treatments may be attributed.

Both dichloro- and monochlorotriazinyl dyes have been commercialised. Dichlorotriazinyl dyes, **8.1a**, (marketed as Procion M dyes), are more reactive than the monochlorotriazinyl (Procion H) types, **8.1b**. This feature may be explained by the fact that in the monochloro dyes **8.1b** the group X is electron-releasing, commonly amino (either primary or secondary) or methoxy, replacing the electron-withdrawing chloro group in the dichloro derivatives **8.1a**. There is therefore an increase in the electron density of the triazine ring in the monochlorotriazinyl dyes, which reduces their reactivity towards nucleophilic attack. Procion M dyes are capable therefore of reaction with cellulosic fibres at lower temperatures than is required for the Procion H dyes. Procion M dyes are thus applied at around 40–60 °C whereas Procion H dyes are applied at around 80–90 °C.

A feature of the chemistry of triazinyl reactive dyes, which is in fact common to all reactive dye systems, is that they undergo, to a certain extent a hydrolysis reaction, involving reaction of the dye with hydroxide (⁻OH) anions present in the aqueous alkaline dyebath in competition with the dye–fibre reaction. The hydrolysis reaction is also illustrated in Scheme 8.1. Reactive dye hydrolysis is a highly undesirable feature of reactive dyeing for various reasons. In the first instance, the hydrolysed dye **8.4** that is formed is no longer capable of reacting with the fibre and so must be washed out of the fibre after dyeing is complete, to ensure the excellent fastness to washing that is characteristic of this dye application class. A second reason is that a degree of fixation less than 100% and the need for the wash-off after treatment considerably reduces the cost-effectiveness of the process. Finally, the hydrolysed dye inevitably ends up in the dyehouse effluent, and this presents significant environmental concerns to which society is becoming increasingly sensitive (Chapter 12). In view of the fact that the dyes are generally applied to the fibres in such a way that water is in vast excess, it might be expected intuitively that hydrolysis would overwhelmingly predominate over reaction with the fibre. However, there are several factors that oppose this. Commonly, the dye is applied to the fibre at around neutral pH in the presence of a salt, such as sodium sulfate, which induces the dye to exhaust in the fibre. The dye molecules are initially adsorbed on to the fibre and are located in an environment that is dominated by the presence of cellulose hydroxyl groups, thus favouring dye–fibre reaction. To cause dye–fibre reaction, the pH is then adjusted to alkaline conditions. The ionization of cellulose means that the fibre acquires a negatively charge and this causes hydroxide ions to be expelled from the interior of the fibres. Factors such as these combine to cause the dye–fibre reaction to predominate over hydrolysis, giving a level of fixation that is variable, but typically around 70%. The final stage in the dyeing process is a very thorough wash-off, to ensure that any hydrolysed or unreacted dye is completely removed, thus ensuring excellent wash-fastness properties because only covalently-bound dye remains.

Monochlorotriazinyl dyes are of particular importance for textile printing, using both screen printing and digital inkjet printing techniques.[10] These dyes are relatively inert towards alkali at low temperature, so that reasonably stable screen print pastes may be formulated that incorporate both the dye and the alkali required for fixation. After printing on to the fabric, the dye–fibre reaction is initiated by raising the temperature to above $100\,^{\circ}C$, normally by steaming. The relatively low substantivity of these dye molecules for the fibre is an advantage in this case in that it facilitates the wash-off of unreacted or hydrolysed dye at the end of the process and ensures that there is no backstaining in white areas of the fabric. In the case of inkjet inks, however, the alkali (sodium carbonate or bicarbonate) is applied to the fabric as a pre-treatment as its presence in the ink would lower the shelf-life to an unacceptable level. Unlike screen print pastes, which are generally used soon after formulation, commercial inkjet inks are required to have a shelf-life of several months.

The initial commercial success of the reactive dyes based on the triazine ring system was immediately followed by intense industrial research activity into the possibilities offered by other related electron-deficient nitrogen-containing heterocyclic systems. Numerous systems have been patented as fibre-reactive groups although only a few of these have enjoyed significant commercial success. Some examples are illustrated in Figure 8.2. They include the trichloropyrimidines, **8.5**, the dichloroquinoxalines, **8.6**, the dichloropyridazines, **8.7** and the chlorobenzothiazolyl dyes, **8.8**.

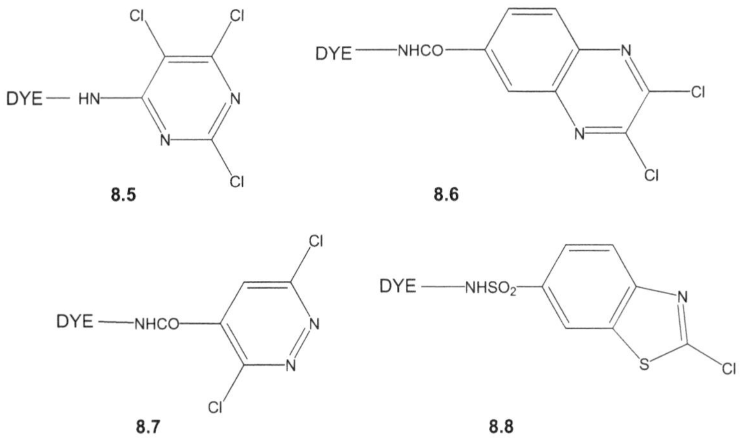

Figure 8.2 Structures of some other heterocyclic reactive dye systems.

Perhaps the best-known of these related groups of reactive dyes are those that utilise the 2,4,5-trichloropyrimidinyl (2,4,5-trichloro-1,3-diazinyl) system, **8.5**, the basis of the Drimarene dyes marketed originally in the 1960s by Sandoz. These dyes react with cellulosate anions by nucleophilic displacement of either the 2- or 4-chlorine atoms according to a mechanism analogous to that shown in Scheme 8.1 for triazine dyes. In contrast, the chlorine atom at the 5-position is not readily substituted. The reasons for this are immediately apparent from examination of the resonance forms of the intermediates derived formally from attack of the nucleophile at the relevant carbon atom as illustrated in Figure 8.3. There are three contributing resonance forms in each case. In the anionic intermediate **8.9**, which is formed by attack at the 4-position, two of the three contributing resonance forms have the negative charge accommodated favourably on the electronegative nitrogen atoms. The reader might like, as an exercise, to demonstrate that a similar situation arises from attack at the 2-position. In contrast, however, attack at the 5-position gives an intermediate **8.10** in which the negative charge is localised on carbon atoms in each of the three contributing forms. The instability of this intermediate explains the lack of reactivity at the 5-position. The 5-chloro substituent, although unreactive, is nevertheless of technological importance. Since the fibre reactive group has only two activating heterocyclic nitrogen atoms (compared with three in the Procion dyes), the 5-chloro substituent serves to enhance the reactivity of the ring towards nucleophilic attack as a result of its electron-withdrawing nature and ensure that the dyes are capable of

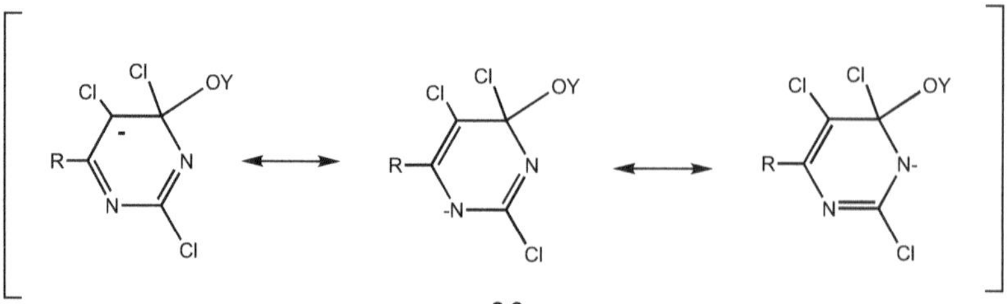

8.9

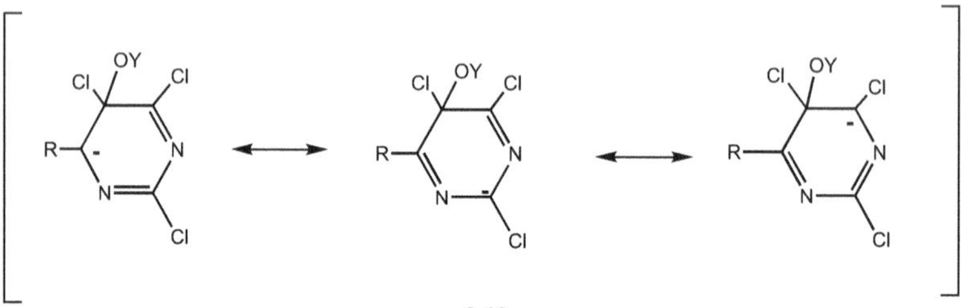

8.10

Figure 8.3 Structure of the intermediates formed from the reaction of tri-chloropyrimidinyl dyes with nucleophiles.

reacting with cellulose at reasonable temperatures. Interestingly, in the reactive dye systems **8.6–8.8**, the bridging group is electron-withdrawing (carbonyl or sulfonyl) and no doubt plays a decisive part in activating the system to nucleophilic attack, ensuring that dyes have appropriate reactivity for the fibre.

Another method whereby the reactivity of reactive dyes may be modified involves the use of alternative leaving groups other than the chloride ion. Some examples of reactive dyes using this modification are illustrated in Figure 8.4. One of the earliest ventures in this field involved the use of quaternary amino groups, as for example in the triazine derivative, **8.11**. Dyes using systems of this type were found to have enhanced reactivity towards cellulosic fibres, but their commercial development was inhibited by a degree of nervousness concerning the liberation of tertiary amines, pyridine in this case, into the dyebath. Some commercial success has been achieved using fluorinated heterocycles, such as those found in the mono-fluorotriazine **8.12** (Cibacron F dyes) and the difluoropyrimidine **8.13** (Levafix EA dyes). Because of the high electronegativity of the fluorine atoms, these dyes show particularly high reactivity and are therefore capable of fixation to cellulosic fibres under mild conditions, for example at relatively low temperatures. Another leaving group that has been utilised to produce highly activated dyes is the methylsulfonyl

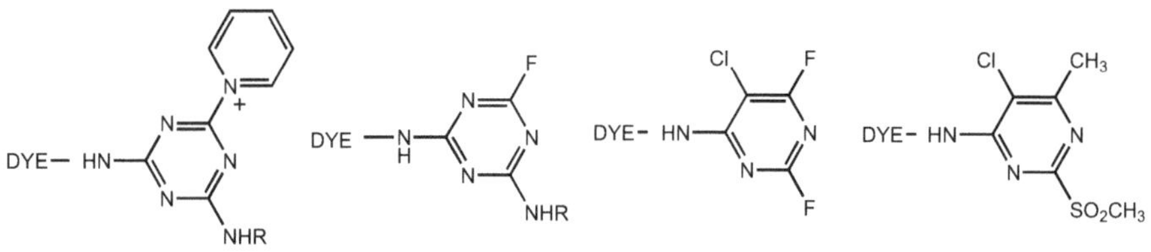

8.11　　　　　　　8.12　　　　　　　8.13　　　　8.14

Figure 8.4 Structures of some reactive dye systems using alternative leaving groups.

group ($-SO_2CH_3$) used in Levafix P dyes (Bayer), which are based on the pyrimidine structure **8.14**.

8.2.2 Fibre-Reactive Groups Reacting by Nucleophilic Addition

Alkenes are generally regarded as relatively reactive compounds, their reactivity being attributable to the presence of the C=C double bond. Characteristically, alkenes readily undergo addition reactions, most commonly of the electrophilic type because of the electron-rich nature of the π-bond. Nucleophilic addition reactions of alkenes are less commonly encountered but can take place when there are strongly electron-withdrawing groups attached to the double bond, thereby reducing the electron density and thus facilitating nucleophilic attack. Nucleophilic addition to substituted alkenes of this type is alternatively referred to as either conjugate addition or Michael addition.

The most important reactive dyes in commercial use for application to cellulosic fibres in which the fibre-reactive groups react by nucleophilic addition are the Remazol reactive dyes. These dyes, based on the vinylsulfone reactive group, were introduced by Hoechst soon after the launch of the Procion dyes based on the triazine system by ICI.[1,11] The chemistry of the process in which vinylsulfone dyes react with cellulose under alkaline conditions is illustrated in Scheme 8.2. The dye is supplied by the manufacturers as the β-sulfatoethylsulfone **8.15**, which is not itself fibre-reactive and is commonly referred to as the stable storage form. The water-solubilising sulfonate group, as a sulfate ester, is in the case of these dyes attached to the latent fibre-reactive group. As illustrated mechanistically in the scheme, the reaction of compound **8.15** with aqueous alkali causes its conversion by an elimination reaction into the highly reactive vinylsulfone, **8.16**. The presence of the powerfully electron-withdrawing sulfone group activates the double bond towards nucleophilic attack. Attack by the cellulosate anion on the vinylsulfone initially leads to

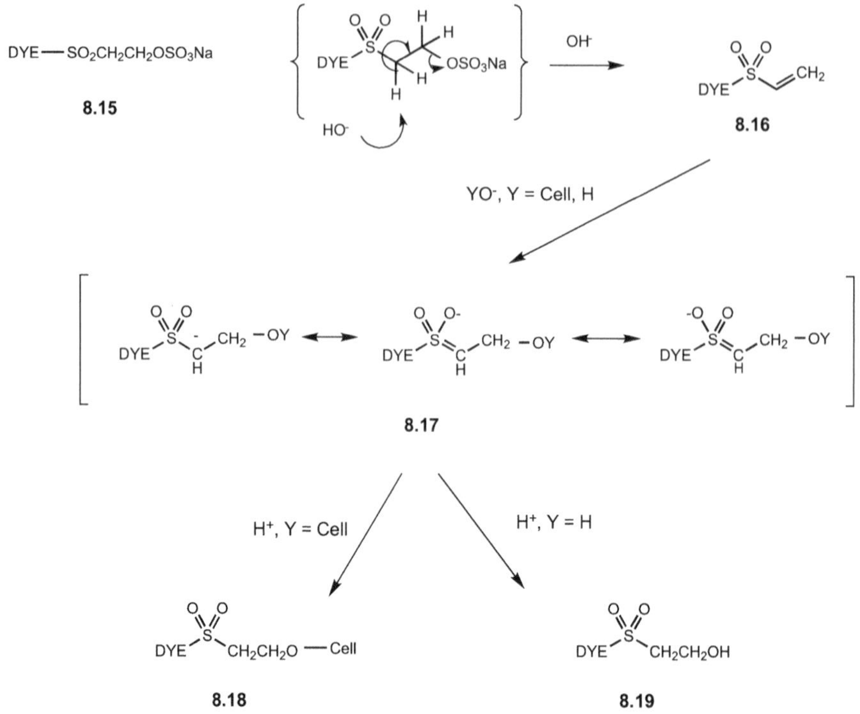

Scheme 8.2   Reaction of vinylsulfone dyes with cellulosic fibres.

the anionic intermediate **8.17** which is stabilised by resonance, with important contributions from canonical forms in which the negative charge is delocalised on to the electronegative oxygen atoms of the sulfone group. The addition reaction is completed by protonation, and the ultimate outcome is the permanent formation of a covalent dye–fibre (C–O) bond as illustrated by structure **8.18**. In this reactive dyeing process, as with dyes that react by nucleophilic substitution, hydrolysis also takes place in competition with the dye–fibre reaction. This reaction, leading to hydrolysed dye **8.19**, is also illustrated in Scheme 8.2.

There are relatively few other reactive groups reacting by nucleophilic addition that have achieved significant commercial exploitation. However, some systems are worthy of note in reactive dyes for protein fibres, such as wool and silk.[12–14] The protein that constitutes the fibre contains amino and thiol groups, both of which are excellent nucleophiles. Some reactive dyes used for wool are similar chemically to those used for cellulosic fibres, for example based on nucleophilic substitution with monochlorotriazines and chlorodifluoropyrimidines, but modified to conform with the different dyeing requirements of protein fibres. For example, there are usually fewer sulfonate groups on the dye molecule as too high a negative charge reduces the ability of the dye to migrate on the fibre, with a consequent effect on levelness. To allow migration, reactive groups on a wool dye should ideally be relatively slow to react. One appropriate group that reacts by nucleophilic add- ition is the acrylamide group (–NHCO–CH=CH$_2$). The α-bromoacryla- mide group is amongst the most widely used of the reactive groups specifically devised for wool, as found for example in the Lanasol dyes. These dyes are applied to wool initially at pH 4.5–6.5 and 40 °C. The temperature is then raised to the boil slowly to allow migration. The pH is then adjusted to around 8.5 at 80 °C for a short period to com- plete the dyeing process and a final washing process is required to remove hydrolysed and unfixed dye together with dye that may have reacted with soluble peptide material. Dye fixation is high in the case of these dyes, at around 85–95%. A proposed mechanism for the re- action of these dyes with the wool fibre, illustrated for the case of primary amino groups, is shown in Scheme 8.3.[13] Reaction with thiol and imidazole groups on the fibre has also been proposed. The dyes generally exist in the α,β-dibromopropionylamide form, **8.20**, which dehydrobrominates in the presence of primary amino groups to form

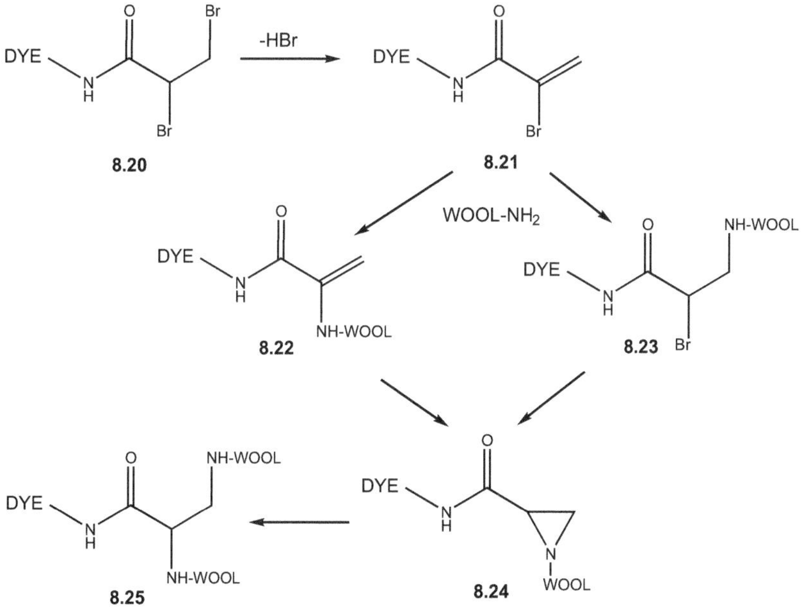

Scheme 8.3   Reaction of α-bromoacrylamide reactive dyes with wool.

the α-bromoacrylamide **8.21**. There remain some incompletely an-
swered issues in the way that the covalent attachment to the fibre takes
place subsequently, although some interesting studies of reactions
with model compounds have shed light on possible mechanisms.[15,16]
Covalent bond formation may result, in principle, from reaction
pathways involving either substitution/addition, *i.e.*, *via* intermediate
8.22, or addition/substitution, *i.e.*, *via* intermediate **8.23**. It is not absolutely certain which of these options operates under practical dyeing
conditions, although the latter is regarded as theoretically more likely.
However, there is convincing evidence from the study of model reactions that cyclisation takes place to form an aziridine ring, as in
structure **8.24**. The possibility of further reaction as shown in Scheme
8.3, leading, in principle, to cross-linking of the fibre as illustrated in
structure **8.25**, has been suggested but this does not appear to have
been experimentally established with certainty.

8.3 POLYFUNCTIONAL REACTIVE DYES

There is no doubt that the major weakness of the reactive dyeing
process is the competing hydrolysis reaction, which reduces the cost
efficiency of the process, and requires a thorough wash-off process,
with the consequence of significant dye residues being discharged to
effluent. The extent to which dye hydrolysis takes place in competition with dye–fibre reaction varies quite markedly within the
range 10–40% depending upon the system in question. A considerable amount of research has therefore been devoted to the search for
reactive dyes with improved fixation. The most successful approach to
addressing this issue has involved the development of dyes with more
than one fibre-reactive group in the molecule, which statistically
improves the chances of dye–fibre bond formation. Examples of
products of this type are the Procion H–E range, which contain two
monochlorotriazinyl reactive groups, for example as represented by
general structure **8.26** and some dyes of the Remazol range that
contain two β-sulfatoethylsulfone groups. These are examples of
homo-bifunctional dyes. Another approach utilises two different types
of reactive groups, the so-called *hetero-bifunctional dyes*, such as for
example structure **8.27**, which contains both a monochlorotriazinyl
and a β-sulfatoethylsulfone group.[17] This particular class of reactive
dye, because of the differing reactivities of the two groups, can offer
greater flexibility in performance, for example reducing sensitivity to
temperature and to pH variation. Indeed, most reactive dye manufacturers now offer at least one range of heterobifunctional dyes,

which are characterized by high fixation efficiency. It would seem logical that the fixation efficiency might be improved further by incorporating even more reactive groups into the dye molecules. Indeed, there are a few commercial examples of trifunctional reactive dyes, for example in the Cibacron and Remazol ranges. However, there are opposing effects associated with the increase in molecular size as more groups are added that can inhibit migration leading to reduced fixation and low colour build up. Thus, although reactive dyes with multiple functionality now offer a significantly improved degree of fixation, the development of a practical reactive dye system that is completely free of the problems associated with hydrolysis or incomplete dye–fibre reaction remains an important target for dye chemists and one which so far has proved elusive despite many years of effort.

8.26

8.27

8.4 CHROMOGENIC GROUPS

A selection of representative chemical structures from the vast range of reactive dyes now available commercially is illustrated in Figure 8.5. Reactive dyes may be prepared, in principle, from any of the chemical classes of colorant by attaching a fibre-reactive group to an appropriate molecule. In common with most application classes of textile dyes and pigments, most reactive dyes belong to the azo chemical class, especially in the yellow, orange and red shade areas. Examples are typified by the structurally related red monoazo reactive dyes, the dichlorotriazine **8.28a**, CI Reactive Red 1, the

monochlorotriazine **8.28b**, CI Reactive Red 3, and the trichloropyrimidine **8.29**, CI Reactive Red 17. Bright blue reactive dyes
are commonly derived from anthraquinones. Examples include the
triazinyl dyes CI Reactive Blue 4, **8.30a**, CI Reactive Blue 5, **8.30b**, and
the structurally related vinylsulfone dye, CI Reactive Blue 19, **8.31**.
Copper or nickel phthalocyanine derivatives are used to achieve bluish to greenish turquoise shades and, in combination with appropriate yellow dyes, to give bright greens. However, despite the high
tinctorial strength that the phthalocyanine chromophore can provide,
the reactive dyes derived from the system do not build up easily to
deep shades on cellulosic fibres. This feature is probably due, at least
in part, to low fixation levels, which may arise from a tendency of the
dyes to form molecular aggregates. Ruby, violet and navy blue dyes
commonly make use of the square-planar copper complexes of
appropriate azo dyes. Reactive dyes based on the dioxazine system are
growing in importance for violet and blue shades, since they are
capable of providing intense colours, much stronger than those based
on the anthraquinone chromophore.[18]

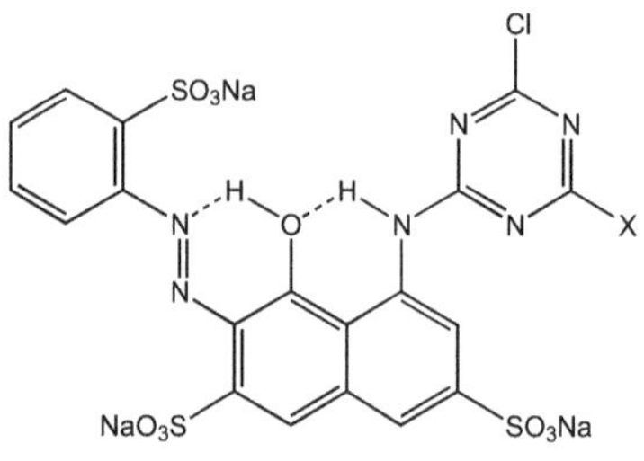

8.28a: X = Cl; **8.28b**: X = NHPh 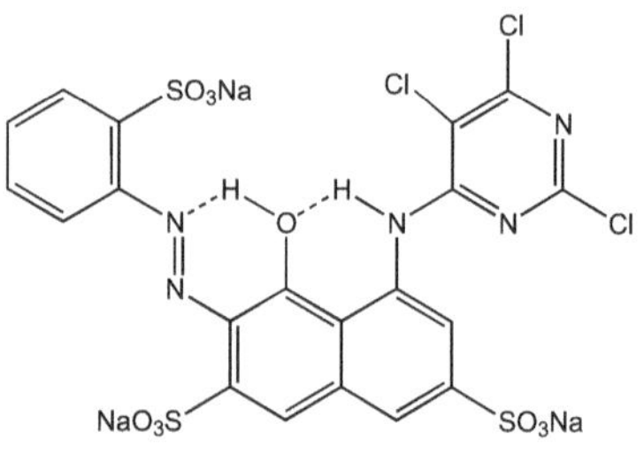   **8.29**

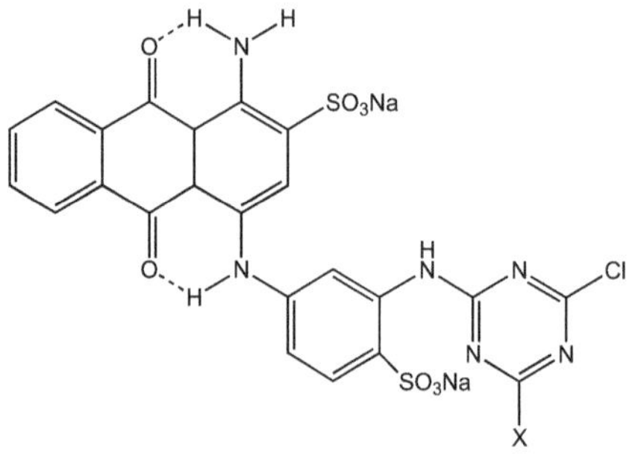

8.30a: X = Cl;
8.30b: X = -NH-C$_6$H$_4$-3-SO$_3$Na

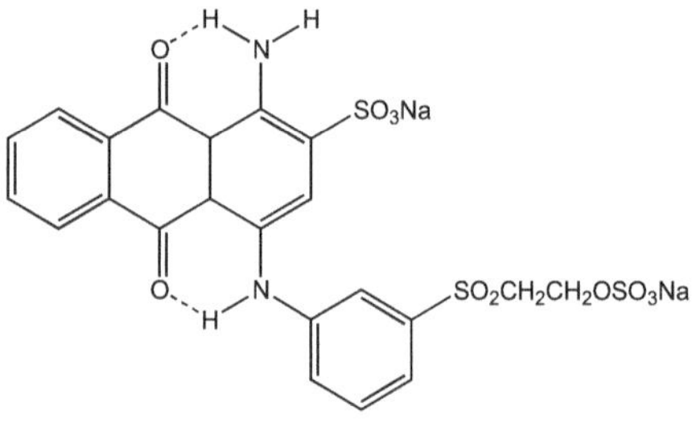

8.31

Figure 8.5 Some typical chemical structures of reactive dyes.

8.5 SYNTHESIS OF REACTIVE DYES

The principles of the synthesis of the important chemical classes of dyes from which reactive dyes may be prepared have been discussed in Chapters 3–6 of this book. This section deals specifically with those aspects of the synthetic sequences that are used to introduce the fibre-reactive group. The starting material for the synthesis of chlorotriazinyl reactive dyes is the highly reactive material cyanuric chloride (2,4,6-trichloro-1,3,5-triazine), **8.32**. The strategy used in the synthesis of these dyes, as illustrated in Scheme 8.4, involves at appropriate stages of the overall reaction scheme sequential nucleophilic substitution of the chlorine atoms of compound **8.32** by reaction with primary amines. As an example, dichlorotriazinyl dye **8.28a** is synthesised by formation of the monoazo dye **8.33** by reaction of diazotised aniline-2-sulfonic acid with H-Acid under alkaline conditions, followed by its condensation with cyanuric chloride, **8.32**. Treatment of dye **8.28a** with aniline under appropriate conditions gives the monochlorotriazinyl dye **8.28b**. Since replacement of an electron-withdrawing chlorine atom in compound **8.32** by an

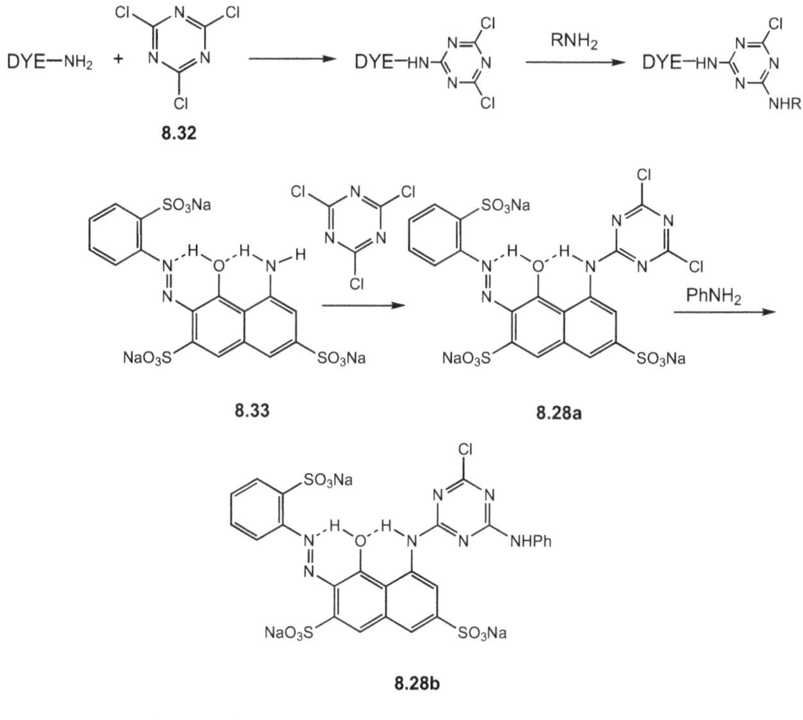

Scheme 8.4   Synthesis of chlorotriazinyl reactive dyes.

electron-releasing amino group deactivates the product of the reaction towards further nucleophilic substitution, replacement of a subsequent chlorine atom requires more vigorous conditions. This is a useful feature of the chemistry of the process since it facilitates the selectivity of the reaction sequence leading to mono and dichlorotriazinyl reactive dyes.

The syntheses of fluorotriazine, trichloropyrimidine and chlorodifluoropyrimidine dyes are completely analogous, using respectively as starting materials cyanuric fluoride, **8.34**, 2,4,5,6-tetrachloropyrimidine, **8.35**, and 5-chloro-2,4,6-trifluoropyrimidine, **8.36**. The fluorinated intermediates are prepared from the corresponding chloro derivatives by an exchange reaction with potassium fluoride.

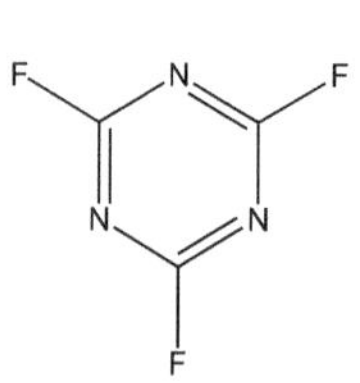

8.34 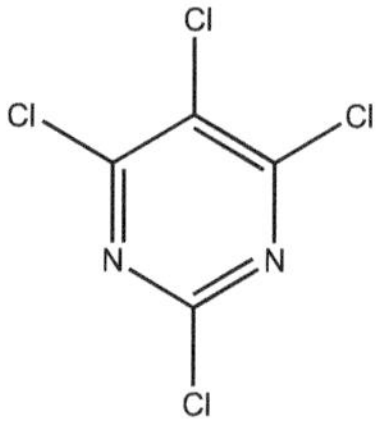   **8.35** 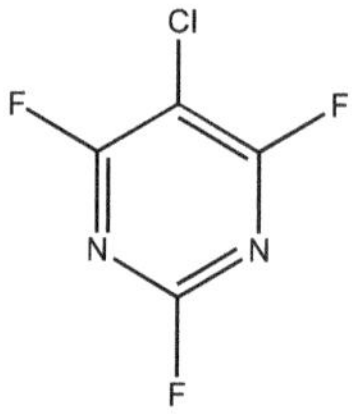   **8.36**

The routes most commonly employed for the preparation of the β-sulfatoethylsulfone group, which is the essential structural feature of vinylsulfone reactive dyes, are illustrated in Scheme 8.5. One method of synthesis involves, initially, the reduction of an aromatic sulfonyl chloride, for example with sodium sulfite, to the corresponding sulfinic acid. Subsequent condensation with either 2-chloroethanol or ethylene oxide gives the β-hydroxyethylsulfone, which is converted into its sulfate ester by treatment with concentrated sulfuric acid at 20–30 °C. An alternative route involves treatment of an aromatic thiol with 2-chloroethanol or ethylene oxide to give the β-hydroxyethylthio compound, which may then be converted by oxidation into the β-hydroxyethylsulfone. The synthesis of the

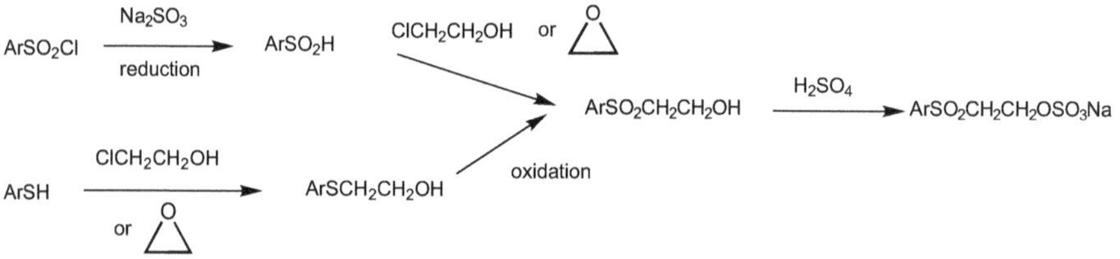

Scheme 8.5   Synthesis of vinylsulfone reactive dyes.

homo- and hetero-bifunctional dyes commonly employs logical extensions of the chemistry described in this section, often involving the monofunctional dyes as intermediates.

REFERENCES

1. A. H. M. Renfrew, *Reactive Dyes for Textile Fibres*, Society of Dyers and Colourists, Bradford, UK, 1999.
2. W. F. Francis, *Fibre-Reactive Dyes*, Logos Press, London, 1970.
3. J. A. Taylor, *Rev. Prog. Color.*, 2000, **30**, 93.
4. A. H. M. Renfrew and J. A. Taylor, *Rev. Prog. Color.*, 1990, **20**, 1.
5. P. Rosenthal, *Rev. Prog. Color.*, 1976, 7, 23.
6. C. V. Stead, Chemistry of reactive dyes, in *Colorants and Auxiliaries: Organic Chemistry and Application Properties*, ed. J. Shore, Society of Dyers and Colourists, Bradford, 1990, vol. 1, ch. 7.
7. I. D. Rattee and W. E. Stephen (ICI), *Br. Pat.*, 1954, 772030.
8. W. E. Stephen and I. D. Rattee, *J. Soc. Dyers Colourists*, 1960, **76**, 6.
9. W. E. Stephen and I. D. Rattee, *Chimia*, 1965, **19**, 261.
10. S. O. Aston, J. R. Provost and H. Masselink, *J. Soc. Dyers Colourists*, 1993, **109**, 147.
11. E. Bohnert, *J. Soc. Dyers Colourists*, 1959, **75**, 581.
12. W. Zuwang, *Rev. Prog. Color.*, 1998, **28**, 32.
13. J. S. Church, A. S. Davie, P. J. Scammells and D. J. Tucker, *Rev. Prog. Color.*, 1999, **29**, 87.
14. M. L. Gulrajani, *Rev. Prog. Color.*, 1993, **23**, 51.
15. J. S. Church, A. S. Davie, P. J. Scammells and D. J. Tucker, *Dyes Pigments*, 1998, **39**, 291.
16. J. S. Church, A. S. Davie, P. J. Scammells and D. J. Tucker, *Dyes Pigments*, 1998, **39**, 313.
17. F. Fujioki and A. Abeta, *Dyes Pigments*, 1982, **3**, 281.
18. A. H. M. Renfrew, *Rev. Prog. Color.*, 1985, **15**, 15.

CHAPTER 9

Pigments

9.1 INTRODUCTION

The distinction between pigments and dyes, which is based on the differences in their solubility characteristics, has been discussed in detail in Chapter 2. A pigment is a finely divided solid colouring material, which is essentially insoluble in its application medium.[1-9] Pigments are used mostly in the coloration of paints, printing inks and plastics, although they are applied to a certain extent in a much wider range of substrates, including paper, textiles, rubber, glass, ceramics, cosmetics, crayons and building materials such as cement and concrete. In most cases, the application of pigments involves their incorporation into a liquid medium, for example a wet paint or ink or a molten thermoplastic material, by a dispersion process in which clusters or agglomerates of pigment particles are broken down into primary particles and small aggregates. The pigmented medium is then allowed or caused to solidify, by solvent evaporation, physical solidification or by polymerisation, and the individual pigment particles become fixed mechanically in the solid polymeric matrix. In contrast to textile dyes where the individual dye molecules are strongly attracted to the individual polymer molecules of the fibres to which they are applied, pigments are considered to have only a weak affinity for their application medium, and only at surface where the pigment particle is in contact with the medium.

Pigments are incorporated in order to modify the optical properties of a substrate, the most obvious effect being to provide colour.

Colour Chemistry, 2nd edition
By Robert M Christie
© R M Christie 2015
Published by the Royal Society of Chemistry, www.rsc.org

However, this is not their only optical function. The pigment may also be required to provide opacity. This feature is most important in paints, which are generally designed to obscure the surface to which they are applied. Alternatively, and in complete contrast, high transparency may be important. For example, in multicolour printing processes–which generally use inks of four colours, the three subtractive primaries, yellow, magenta and cyan, together with black, transparency is essential to ensure that subsequently printed colours do not obscure the optical effect of the first colour printed. A pigment owes its optical properties to a combination of two effects that result from the interaction of the pigment particles with visible light: light *absorption* and light *scattering*. The colour provided by a pigment is determined mainly by the light absorption and, as is the case with dyes, this is primarily dependent on its molecular constitution, as discussed in Chapter 2. However, there are two further factors that influence the colour of pigments as a consequence of the fact that they are used as solid, crystalline particles. These factors are the crystal structure, *i.e.*, the way in which the molecules pack in their crystal lattice, and the particle structure, which includes particle size and shape distribution.[10,11] Opacity is mainly dependent on the degree of light scattering by the pigment particles, which is in turn dependent on the refractive index of the material, with high refractive index materials giving high opacity. This property is also dependent on the crystal structure of the pigment and its particle size and shape. A third optical property that is influenced by pigments is gloss, the mirror-like reflection, which is a desirable optical effect of many coatings. The gloss shown by a painted or printed surface depends on several factors, including the viewing angle and the nature of the film-forming polymer, its refractive index and the extent to which it forms a smooth film. However, the property may also be influenced by the pigment. In essence, gloss is determined by the degree of smoothness of the surface of the paint or ink film and will be reduced by the protrusion of pigment particles through the surface, so that maximum gloss is achieved using pigments of small particle size at low concentrations, very finely dispersed in the coatings formulation.

In addition to providing the appropriate optical properties, pigments must be capable of withstanding the effects of the environment in which they are placed both during processing and in their anticipated useful lifetimes. They are required to be fast to light, weathering (if external exposure is envisaged), heat and chemicals such as acids and alkalis to a degree dependent on the demands of the particular application.

Fastness to light is determined principally by the chemical structure of the pigment and the nature of the crystal packing, and also depends to an extent on the particle size of the pigment, its concentration in the application medium and on the nature of the medium. Lighter shades, especially in combination with white pigments, tend to show poorer lightfastness than deeper shades. Heat stability is critically important in pigments for application in ceramics, which are processed at temperatures in excess of 800°C; only certain inorganic pigments are capable of withstanding such temperatures. Heat stability is also a vital factor in the coloration of thermoplastics, the level required depending not only on the processing temperature for the polymer in question, which can range from 150 to 350°C, but also on the time of exposure. In addition, pigments are required to show solvent resistance, which refers to their ability to resist dissolving in solvents with which they may into contact in their application, in order to minimise problems such as 'bleeding' and migration. A pigment will be selected for a particular application on the basis of its optical and fastness properties, but with due regard also to toxicological and environmental considerations and, inevitably, cost.

In chemical terms, pigments are conveniently classified as either *inorganic* or *organic*.[1,2] These two broad groups of pigments are of roughly comparable importance industrially. In general, inorganic pigments are capable of providing excellent resistance to heat, light, weathering, solvents and chemicals, and in those respects they can often offer technical advantages over most organic pigments. In addition, many inorganic pigments are of significantly lower cost than typical organics. On the other hand, they commonly lack the intensity and brightness of colour of organic pigments. Organic pigments are characterised by high colour strength and brightness although the fastness properties that they offer are somewhat variable. There is, however, a range of high performance organic pigments that offer excellent durability while retaining their superior colour properties, but these products tend to be rather more expensive.[12] The ability either to provide opacity or to ensure transparency in application provides a further contrast between inorganic and organic pigments. In general, inorganic pigments are high refractive index materials that are capable of providing high opacity, while organic pigments are of low refractive index and consequently are transparent.

This chapter provides an overview of the characteristic structural features of the most important commercial pigments. If it seems to the reader that the chapter places considerable emphasis on inorganic pigments, this is to an extent because the various chemical classes of organic pigments are dealt with to a certain extent in Chapters 3–6. In individual cases there is some discussion of structure–property relationships. Such relationships are rather more complex with pigments than with dyes, because of the dependence of the colouristic and technical performance of pigments not only on the molecular structure but also on the crystal structure arrangement and on the nature of the pigment particles, particularly their size and shape distribution. The section on inorganic pigments presents an outline of the synthetic procedures used for their manufacture. Discussion of the chemistry involved in the synthesis of organic pigments is omitted as this is dealt with in relevant earlier chapters concerned with the specific chemical classes. The process by which a pigment is manufactured may be considered as involving a number of stages. The first of these is the synthesis in which the sequence of chemical reactions that form the pigment is carried out. However, simply carrying out this chemistry does not necessarily give the product in a form that is optimised for a particular application. In a second phase of the process, commonly referred to as *conditioning*, the product is developed into an appropriate crystalline form, with a controlled particle size and shape distribution, and the surfaces of the particles may be altered by suitable treatments. Depending on the particular pigment, these two phases may be either distinctly separate or combined into a single process. Finally, the pigment is subjected to *finishing* in which it is converted into a form that is acceptable to the user. The conditions of synthesis of those organic and inorganic pigments that are prepared by precipitation from reaction in water may often be controlled to ensure that the pigments are formed directly in an appropriate particle size range. Commonly, they will precipitate from this aqueous medium in a poorly-defined crystal form, and so they are subsequently treated at elevated temperatures, either in the presence or absence of solvents (water or organic solvents), to develop high crystallinity and ensure that the products show optimum optical properties. Many inorganic pigments are subjected to very high temperatures towards the end of their manufacture in a process referred to as *calcination* (temperatures which organic pigments would not survive), in order to remove tightly bound water and to develop the appropriate crystal form. Some high performance organic pigments are synthesised at high temperatures by reaction in organic solvents, and are produced initially in a large particle size form, which must be reduced to a fine particle size to make them useful pigmentary products. Organic pigments are often prepared in as fine a particle size as is technically feasible in order to provide maximum colour strength and transparency. In contrast, the particle size of many inorganic pigments is controlled carefully (often in the range 0.2–0.3 μm) to provide maximum opacity.

For the most part, pigments are solid, crystalline materials manufactured in a finely divided form. A feature commonly encountered in the crystalline solid state structures of pigments is *polymorphism*. A material exhibits polymorphism if it is capable of existing in forms with identical chemical composition but different molecular or ionic arrangements in the crystal structures. These polymorphic forms may show significantly different colouristic and technical properties, and often the different forms have separate commercial applications. There are numerous examples of pigments that exhibit this phenomenon, including titanium dioxide, iron oxides, lead chromates, cadmium sulfides, copper phthalocyanine, quinacridones and several azo pigments, and reference to this will be found in the discussion of the individual types which follow. Particle size distribution is a further important factor in determining the properties of a pigment. A reasonable generalisation is that organic pigments are prepared in as fine a particle size as is technically feasible in order to give maximum colour strength and transparency. This is especially important for printing ink applications where the particle size is commonly below 0.05 μm. For some applications, notably in industrial surface coatings, a rather larger particle size is desirable for improved opacity and fastness properties, but this is achieved at the expense of colour strength. The particle size of many inorganic pigments is controlled carefully (often in the range 0.2–0.3 μm) to provide maximum opacity.

It is essential that a pigment is well dispersed into its application medium, for example during its introduction into a paint, printing ink or plastic article. Pigment powders consist of clusters of aggregated particles which must be subjected to conditions that break down large aggregates into small aggregates and primary particles to the degree required to give the optimum optical properties. The term *dispersibility* refers to the ease with which the desired level of dispersion is achieved. The dispersibility of the pigments produced commercially is optimised to minimise the energy requirements of the dispersion process. This is usually achieved by surface treatment. The nature of the surfaces of pigment particles is of importance in determining application performance because it is these surfaces that are in contact with the application medium. The particle surfaces that form directly in the synthesis of pigments often have highly polar character and thus they show poor compatibility with many of the non-polar, or hydrophobic, media with which they will be required to interact in use. The pigments are therefore often treated with materials such as organic surface-active agents or resins, either during or after the synthesis phase of their manufacture, in order to change the nature of the surfaces of the particles. When the pigment is treated with the surface-active agent, which contains separate hydrophilic and hydrophobic regions, hydrophilic groups become attached to the polar pigment surface, as a result of dipolar or ion–dipole forces of attraction. As a result, the pigment effectively acquires a new surface that is hydrophobic and hence more compatible with an organic application medium. This surface presents a steric barrier so that the pigment particles are less susceptible to aggregation and also those aggregates that are formed are more easily broken down by the dispersion process. Another way in which the properties of pigments may be improved by surface treatment is the coating of inorganic pigments, such as titanium dioxide and lead chromates, with inorganic oxides such as silica or alumina, in order to stabilise the particles and lead to a marked improvement in lightfastness and chemical stability.

9.2 INORGANIC PIGMENTS

Natural inorganic pigments, derived mainly from mineral sources, have been used as colorants since prehistoric times and a few, notably iron oxides, remain of some significance today. The origin of the synthetic inorganic pigment industry may be traced to the rudimentary products produced by the ancient Egyptians, pre-dating the synthetic organic colorant industry by several centuries (Chapter 1).[13] The range of modern inorganic pigments was developed for the most part during the twentieth century and encompasses white pigments, by far the most important of which is titanium dioxide, black pigments, notably carbon black, and coloured pigments of a variety of chemical types, including oxides (*e.g.*, of iron chromium and the mixed metal oxides), cadmium sulfides, lead chromates and the structurally more complex ultramarine and Prussian blue.

The structural chemistry and properties of the important chemical types of inorganic pigments are dealt with in the sections that follow, together with an outline of the most important synthetic methods. The colour of inorganic pigments arises from electronic transitions that are quite diverse in nature and broadly different from those responsible for the colour of organic colorants. For example, they may involve charge transfer transitions, either ligand–metal (*e.g.* in lead chromates) or between two metals in different oxidation states (in Prussian blue). The colours of cadmium sulfide and sulfoselenide pigments have been explained on the basis of semiconductor properties. In ultramarines, the colour is due to radical anions trapped in the crystal lattice. Inorganic pigments generally exhibit high inherent opacity, a property that may be attributed to the high refractive index resulting from the compact atomic or ionic arrangement in their crystal structure. A wide variety of synthetic methods are employed in the manufacture of inorganic pigments. Frequently, the chemistry is carried out in aqueous solution from which the pigments can precipitate directly in a suitable physical form. In some cases, high temperature solid state reactions are used (*e.g.*, mixed phase oxides, ultramarines), while gas-phase processes, because of their suitability for continuous large-scale manufacture, are of importance for the manufacture of the two largest tonnage pigments, *viz.*, titanium dioxide and carbon black.

9.2.1   Titanium Dioxide and other White Pigments

White pigments are conveniently classified as either hiding or non-hiding types, depending on their ability to provide opacity. By far the most important white opaque pigment is titanium dioxide (TiO_2, CI Pigment White 6).[14-16] It finds widespread use in paints, plastics, printing inks, rubber, paper, synthetic fibres, ceramics, textiles and cosmetics. It owes its dominant industrial position to its ability to provide a high degree of opacity and whiteness (maximum light scattering with minimum light absorption) and to its excellent durability and non-toxicity. The pigment is manufactured in two polymorphic forms, the *rutile* and the *anatase* forms, the former being far more important commercially. In the crystal structures of both forms, each Ti^{4+} ion is surrounded octahedrally by six O^{2-} ions. The rutile form has a higher refractive index (2.70) than the anatase form (2.55), a feature that is attributed to the particularly compact atomic arrangement in its crystal structure, and it is therefore more opaque. Rutile is also the more stable form thermodynamically. Anatase converts thermally into rutile at temperatures above around 700 °C, although for most applications this phase conversion does not present a problem. Anatase does offer a few advantages over rutile, which suits it to certain applications. It shows lower absorption than rutile at low visible wavelengths (380–420 nm), giving a less yellowish shade, and it tends to be softer in texture. For these reasons, it is often appropriate for paper and synthetic fibre applications. Pigmentary TiO$_2$ is generally manufactured to give a primary particle size of around 0.25 μm, at which opacity is maximized. Fine primary particle size grades (<0.1 μm), often referred to as 'nano TiO$_2$', are transparent and show high UV absorption. They are consequently of little value as traditional pigments but have niche applications, for example in sun-screening preparations.

Two processes are used in the manufacture of titanium dioxide pigments: the *sulfate* process and the *chloride* process. Both processes use as starting materials natural deposits of titanium dioxide, which vary in composition from the rather crude *ilmenite* ores to the much purer, but less abundant and hence more expensive, *rutile* ores. Both manufacturing methods use chemical conversion processes to prepare pigmentary titanium dioxide with carefully controlled particle characteristics. The chemistry of the sulfate process, the longer established of the two methods, is illustrated schematically in Scheme 9.1. In this process, crude ilmenite ore, which contains titanium dioxide together with substantial quantities of oxides of iron, is digested with concentrated sulfuric acid, giving a solution containing the sulfates of Ti(IV), Fe(III) and Fe(II). Treatment of this solution with iron metal then effects reduction of the Fe(III) ensuring that the iron in solution exists exclusively in the Fe(II) oxidation state. The solution at this stage is then concentrated, thereby depositing crystals of FeSO$_4 \cdot$7H$_2$O (copperas), a major by-product of the process that is removed by filtration. Subsequently, in the critical step, the solution is boiled, leading to a precipitate of hydrated titanium dioxide as a result of hydrolysis of the aqueous titanium (IV) sulfate. The hydrated oxide formed is finally calcined at 800–1000 °C to remove water and

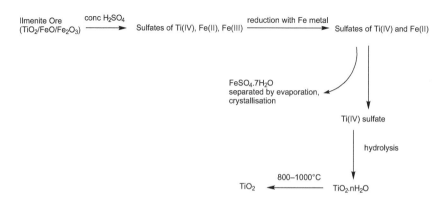

Scheme 9.1   Sulfate process for the manufacture of TiO$_2$.

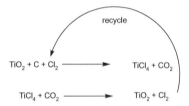

Scheme 9.2 Chloride process for the manufacture of TiO_2.

residual sulfate (as H_2SO_4) and this leads to the formation of an-hydrous titanium dioxide. The sulfate process may be adapted to prepare either the rutile or anatase form of the pigment, by using a 'seed' of the appropriate material at the precipitation stage.

In the chloride process (Scheme 9.2), rutile titanium dioxide ore is initially treated with chlorine in the presence of carbon as a reducing agent at 800–1000°C to form titanium tetrachloride. After purification by distillation, the tetrachloride is subjected to gas-phase oxidation at 1500°C with air or oxygen to yield a high purity, fine particle size rutile titanium dioxide pigment. Chlorine is generated at this stage and may be recycled. The two manufacturing processes are of roughly comparable importance on a worldwide basis. However, the chloride process offers certain inherent advantages over the sulfate route. These include suitability for continuous operation, excellent control of pigment properties and fewer by-products, which in the case of the sulfate process can lead to waste disposal issues. Titanium dioxide pigments are commonly subjected to surface aftertreatments to improve application performance. Inorganic surface treatments, primarily with oxides of silicon, aluminium or zirconium, are used to enhance weathering properties and to provide a barrier that inhibits the tendency of TiO_2 to catalyse photo-oxidation. Organic surface treatments may be used to enhance the ease of dispersion into the application medium.

Zinc sulfide (ZnS, CI Pigment White 7) and antimony(III) oxide (Sb_2O_3, CI Pigment White 11) are white hiding pigments that find some specialist applications but their lower refractive indices mean that they are less efficient than TiO_2 in producing opacity. White lead (basic lead carbonate, CI Pigment White 1), formerly the traditional white hiding pigment, has become virtually obsolete on the grounds of both inferior technical performance and toxicity.

Non-hiding white pigments, sometimes referred to as *extenders* or *fillers*, are low cost products used in large quantities particularly by the plastics industry. They are white powders of relatively low refractive index and thus they are capable of playing only a minor role in providing opacity. They are, however, used in various other ways. For example, they may be used to modify the flow properties of paints and inks and to modify the mechanical properties and lower the cost of plastics. Commonly-used non-hiding white pigments include calcium carbonate ($CaCO_3$), barium sulfate ($BaSO_4$), talc (hydrated magnesium silicate), china clay (hydrated aluminium silicate) and silica.

9.2.2 Coloured Oxides and Oxide-hydroxides

By far the most important coloured inorganic pigments are the iron oxides, which provide colours ranging from yellow and red to brown and black. They are used extensively in paints, plastics and in building materials such as cement and concrete. Both natural and synthetic iron oxide pigments are used commercially. Oxides of iron are major constituents of some of the most abundant minerals in the earth's crust. Natural iron oxide pigments are manufactured from deposits of suitable purity by a milling process. The natural iron oxides, examples of which include yellow ochre, red hematite and burnt sienna, are cost-effective materials that meet many of the demands of the colour industry. Micaceous iron oxide is a natural pigment used in metal-protective coatings. Its flake-like particles laminate in the paint film, forming a reflective surface which reduces radiation degradation of the film, and provides a barrier to moisture as an aid to corrosion protection.

Synthetic iron oxide pigments offer the advantages over their natural counterparts of chemical purity and improved control of physical form. Several different structural types are encountered. Red iron oxides (CI Pigments Red 101 and 102) consist principally of anhydrous iron(III) oxide (Fe_2O_3) in its α-crystal modification. Yellow iron oxide pigments (CI Pigments Yellow 42 and 43), although often formulated as hydrated iron(III) oxides, are better represented as iron(III) oxide-hydroxides, $FeO(OH)$. The principal constituent of black iron oxide pigments is a non-stoichiometric mixed $Fe(II)/Fe(III)$ oxide. It is usually formulated as Fe_3O_4, however, as the two oxidation states are generally present in approximately equal proportions. Brown pigments may be derived from the mixed $Fe(II)/Fe(III)$ oxide or from mixtures containing Fe_2O_3 and $FeO(OH)$. Iron oxide pigments are characterised, in general, by excellent durability, high opacity, low toxicity and low cost. However, the yellow pigments show somewhat lower heat stability because of their tendency to lose water at elevated temperatures, in the process turning redder due to the formation of

Fe_2O_3. The colour of iron oxide pigments has been attributed principally to light absorption as a result of ligand–metal charge transfer, although probably influenced also by the presence of crystal field d–d transitions. The main deficiency of iron oxide pigments is that the colours lack brightness and intensity.

The most important synthetic routes to iron oxide pigments involve either thermal decomposition or aqueous precipitation processes. A method of major importance for the manufacture of α-Fe_2O_3, for example, involves the thermal decomposition in air of $FeSO_4.7H_2O$ (copperas) at temperatures between 500 and 750°C. The principal method of manufacture of the yellow α-$FeO(OH)$ involves the oxidative hydrolysis of Fe(II) solutions, for example in the process represented by equation (9.1).

$$4FeSO_4 + 6H_2O + O_2 \rightarrow 4FeO(OH) + 4H_2SO_4 \qquad (9.1)$$

The reaction is sustained by the addition of iron metal which reacts with the sulfuric acid formed, regenerating Fe(II) in solution. To ensure that the desired crystal form precipitates, a seed of α-$FeO(OH)$ is added. However, with appropriate choice of conditions, for example of pH and temperature and by ensuring the presence of appropriate nucleating particles, the precipitation process may be adapted to prepare either the orange–brown γ-$FeO(OH)$, the red α-Fe_2O_3 or the black Fe_3O_4.

The only other 'simple' oxide pigment of major significance is chromium(III) oxide, Cr_2O_3, CI Pigment Green 17. This is a tinctorially weak, dull green pigment but it shows outstanding durability, including thermal stability to 1000°C. The pigment is normally prepared by treatment of chromates or dichromates with reducing agents such as sulfur or carbon.

The mixed phase oxides, or mixed metal oxides, are a group of inorganic pigments that were developed originally for use in ceramics but which have subsequently found widespread application in plastics because of their outstanding heat stability and weathering characteristics combined with moderate colour strength and brightness. Structurally, the pigments may be considered to be formed from stable oxide host lattices, *e.g.*, rutile (TiO_2), spinel ($MgAl_2O_4$) and inverse spinel, into which are incorporated transition metal ions, *e.g.*, Cr^{3+}, Mn^{2+}, Fe^{3+}, Co^{2+}, Ni^{2+}. This provides a range of colours in which the excellent durability characteristics of the host crystal structures are retained. An important commercial example of a mixed oxide pigment based on the spinel lattice is cobalt aluminate blue (CI Pigment Blue 28), which is usually represented as $CoAl_2O_4$, although in practice it is found to contain slightly less cobalt than this formula would indicate. While the successful formation of a mixed metal oxide requires that the 'foreign' cation must have a suitable ionic radius to be incorporated into its lattice position, a similar valency to that of the metal ion replaced is not essential. For example, a metal ion of lower valency may be incorporated into the lattice provided that an element in a higher oxidation state is incorporated at the same time in the amount required to maintain statistical electrical neutrality. As an example, nickel antimony titanium yellow (CI Pigment Yellow 53), an important member of the series, is derived from the rutile TiO_2 structure by partial replacement of Ti^{4+} ions with Ni^{2+} ions, at the same time incorporating antimony(v) atoms such that the Ni/Sb ratio is 1:2.

Mixed metal oxide pigments are manufactured by high temperature (800–1000 °C) solid state reactions of the individual oxide components in the appropriate quantities. The preparation of nickel antimony titanium yellow, for example, involves reaction of TiO_2, NiO and Sb_2O_3 carried out in the presence of oxygen or other suitable oxidising agent to effect the necessary oxidation of Sb(III) to Sb(v) in the crystal lattice structure.

9.2.3 Cadmium Sulfides, Lead Chromates and Related Pigments

Cadmium sulfides and sulfoselenides provide a range of moderately intense colours ranging from yellow through orange and red to maroon. They are of particular importance in the coloration of thermoplastics, especially in engineering polymers that are processed at high temperatures, because of their outstanding heat stability. Cadmium sulfide, CdS (CI Pigment Yellow 37), is dimorphic, existing in α- and β-forms. The α-form is more stable and is yellow. The colour range of the commercial pigments is extended by the formation of solid solutions. When cadmium ions are partially replaced in the lattice by zinc, greenish-yellow products result, whilst replacement of sulfur by selenium gives rise progressively to the orange, red and maroon sulfoselenides (CI Pigment Orange 20) and (CI Pigment Red 108), depending on the degree of replacement. In all of these products, however, the α-CdS structure is retained. The colour of cadmium pigments has been explained on the basis of semiconductor properties and is attributed to electronic transitions between a valence band and a conductance band.[17,18] The gap between the valence and conductance bands in ZnS is 3.58 eV (347 nm), requiring light in the UV region for electronic excitation, so that the material is colourless.

CdS, with a band gap of 2.4 eV (517 nm), absorbs radiation with an energy 2.4 eV or greater, *i.e.*, the entire blue region of the spectrum, so that the reflected colour is yellow. CdSe, with a band gap of 1.8 eV, absorbs all visible light except for the low energy red region and so appears red. The colours of the solid solutions may then be explained on the basis of band gaps intermediate in energy between those of the pure species involved. This type of electronic transition gives rise to steep reflectance curves, accounting for the relatively bright colours of the pigments.

There has been considerable interest in recent years in nanoparticles based on the cadmium sulfides and sulfoselenides.[19] These materials, referred to as *quantum dots*, display unique optical and electrical properties that are quite different from the properties exhibited at pigmentary particle size, and are attributed to their semiconductor behaviour. The most apparent of these properties is their intense fluorescence, the emission wavelength of which may be tuned based on particle size. Quantum dots have potential for applications in medicine, displays, lasers and solar energy conversion.

Lead chromate pigments provide a range of colours, from greenish-yellow through orange to yellowish-red.[20] They offer good fastness properties, a high brightness of colour for inorganic pigments, and high opacity, at relatively low cost. Historically, lead chromate pigments were found to exhibit a tendency to darken, either on exposure to light (due to lead chromite formation) or in areas of high industrial atmospheric pollution (due to lead sulfide formation). These early problems have been addressed by the use of surface treatment with inorganic oxides, notably silica, and the current range of pigments now offers excellent durability. The variation in shade of lead chromate pigments is achieved by the formation of solid solutions. In this respect they resemble the cadmium sulfides, although structurally the lead chromate pigments present a more complex situation by exhibiting polymorphism. The mid-shade yellow products are essentially pure $PbCrO_4$ (CI Pigment Yellow 34) in its most stable monoclinic crystal form. Incorporation of sulfate ions into the lattice while retaining the monoclinic crystal form gives rise to the somewhat greener lemon chromes. The greenest shades (primrose chromes) consist similarly of solid solutions of $PbCrO_4$ and $PbSO_4$, stabilised chemically in an orthorhombic crystal form. Incorporation of molybdate anions into the lattice gives rise to the orange and light red molybdate chromes (CI Pigment Red 104). Molybdate chromes usually also contain small amounts of sulfate ions, which play a role in promoting the formation of the appropriate crystal form.

Cadmium sulfides are prepared by aqueous precipitation processes using suitable water-soluble sources of cadmium and sulfide ions. The zinc-containing pigments are formed when appropriate quantities of soluble zinc salts are incorporated into the process, while the sulfoselenides are prepared by dissolving elemental selenium in the sulfide solution prior to the precipitation. Since the pigments usually precipitate from solution in the less stable β-form, an essential final step in their manufacture is a controlled calcination at 600°C, which effects the conversion into the desired α-form. Lead chromate pigments are manufactured by mixing aqueous solutions of lead nitrate and sodium chromate or sodium dichromate. The mixed phase pigments result when appropriate quantities of sodium sulfate or molybdate are incorporated into the preparation.

The use of cadmium sulfide and lead chromate pigments is limited to a considerable extent on the grounds of the potential toxicity that is associated with the presence of the heavy metals cadmium, lead and chromium(VI). Their use is restricted by voluntary codes of practice reinforced by legislation in certain cases. In the European Community, for example, a Directive restricts the use of the cadmium sulfides in applications where they are not seen to be essential. The Directive relevant to lead chromates no longer permits their use by consumers and they are labelled as 'only for industrial purposes'. Their use is prohibited in coatings for toys and other consumer paints, graphic instruments and plastics for consumer goods, including food contact applications, and there are restrictions placed on their use in paints for new vehicles and packaging. However, at present, completely satisfactory substitutes for cadmium pigments are not available for use in certain high temperature plastics applications, especially in terms of thermal and chemical stability, while lead chromates remain by far the most cost-effective durable yellow and orange pigments. It is argued, particularly by the manufacturers of these products, that as a result of their extreme insolubility they do not present a major health hazard. Nevertheless, it seems likely that the trend towards their replacement by more acceptable inorganic and organic pigments will continue in an increasing range of applications.

Two types of inorganic pigments of more recent origin, bismuth vanadates and cerium sulfides, were introduced primarily as potential replacements for the so-called 'heavy metal'-containing products.[12] Bismuth vanadates, designated as CI Pigment Yellow 184, which can contain variable amounts of bismuth molybdate, are brilliant greenish-yellow to reddish-yellow pigments with high opacity and good durability.[21] They are used primarily to provide bright deep yellow shades in industrial and automotive paints and have been growing steadily in importance in the decades since their industrial introduction. The pigments are manufactured by an aqueous precipitation reaction at a controlled pH involving bismuth nitrate, sodium vanadate and sodium molybdate. The hydrated products that result are then calcined at temperatures in the range 300–700°C to remove water and develop the appropriate crystalline form. The pigments may also contain inorganic stabilisers, commonly silicates, for improved thermal stability and to reduce photochromism, a colour change when exposed to light. An exploration of lanthanide chemistry led to the development of pigments based on cerium sulfide, products that are still in their industrial infancy, providing orange through to red and burgundy shades with excellent durability and reported as being non-toxic. The colours, however, are tinctorially rather weak.

9.2.4   Ultramarines

Of this small group of pigments, ultramarine blue (CI Pigment Blue 29) is the best known and by far the most important, although violet and pink pigments are also produced. Ultramarine blue offers excellent fastness to light and heat at moderate cost. Although capable of providing brilliant reddish-blue colours in application, ultramarine blue suffers from poor tinctorial strength. As an example, the pigment has less than one-tenth of the colour intensity of copper phthalocyanine, the most important organic blue pigment. A further deficiency of the pigment is rather poor resistance towards acids. Ultramarine blue pigments have a complex sodium aluminosilicate zeolitic structure. In essence, the structure consists of an open three-dimensional framework of AlO_4 and SiO_4 tetrahedra and within this framework there are numerous cavities in which are found small sulfur-containing anions together with sodium cations maintaining the overall electrical neutrality. It has been conclusively demonstrated that the radical anion S_3^- is the species responsible for the blue colour (λ_{max} 600 nm) of ultramarine blue pigments. It is an interesting and somewhat surprising observation that products of such high durability result when species such as S_3^-, which are otherwise highly unstable, are trapped within the ultramarine lattice.[22,23]

Formerly derived from the natural mineral lapis lazuli, ultramarine blue pigments have, for more than a century, been manufactured synthetically. The materials used in the manufacture of ultramarines are china clay (a hydrated aluminosilicate), sodium carbonate, silica, sulfur and a carbonaceous reducing material such as coal tar pitch.

For the manufacture of the blue pigments, the traditional method involves heating the blend of ingredients to a temperature of 750–800 °C over a period of from 50 to 100 h, and the reaction mixture is then allowed to cool in an oxidising atmosphere over several days. Newer processes involving continuous production are reported to be generally more environmentally responsible, providing higher product quality with reduced processing time and energy consumption.

9.2.5 Prussian Blue

Prussian blue (CI Pigment Blue 27), known also variously as iron blue or Milori blue, is probably the longest established of all synthetic colorants still in use and retains moderate importance as a low cost blue pigment. On the basis of single-crystal X-ray diffraction studies, it has been concluded that Prussian blue is best represented as the hydrated iron(III) hexacyanoferrate(II), $Fe_4[Fe(CN)_6]_3 \cdot nH_2O$.[24] However, when precipitated in the fine particle size essential for its use as a pigment, significant and variable amounts of potassium or ammonium ions are incorporated into the product by surface adsorption or occlusion. In addition, the commercial products may contain indefinite amounts of water and they can exhibit variable stoichiometry and a degree of structural disorder. In the crystal structure of Prussian blue, the Fe(II) atoms are bonded exclusively to carbon atoms in FeC_6 octahedra and the Fe(III) atoms are bonded exclusively to the nitrogen atoms. Many of the pigmentary properties of Prussian blue have been explained on the basis of its crystal structure. For example, the extreme insolubility of the material has been attributed to the fact that the complex is polymeric as a result of the –Fe(II)–C–N–Fe(III)– bonding sequence. The colour of the pigment is due to metal–metal electron transfer from an Fe(II) atom to an adjacent Fe(III) atom, a phenomenon commonly encountered in mixed oxidation state transition metal compounds of this type.

The industrial production of Prussian blue is based on the reaction in aqueous solution of sodium hexacyanoferrate(II), $Na_4Fe(CN)_6$, with iron(II) sulfate, $FeSO_4 \cdot 7H_2O$ in the presence of an ammonium salt, which results initially in the formation of the colourless insoluble iron(II) hexacyanoferrate(II) (Berlin white). Prussian blue is generated by subsequent oxidation with a dichromate or chlorate.

9.2.6 Carbon Black

Carbon blacks (CI Pigments Black 6 and 7) dominate the market for black pigments, providing an outstanding range of technical properties at low cost and finding use in virtually all pigment applications, including black coatings, plastics and printing inks of all types.[25] One of the most important applications for carbon black pigments is in rubber, where as well as providing the colour, they fulfil a vital role as reinforcing agents, especially in motor vehicle tyres. Carbon blacks strongly absorb infrared and ultraviolet radiation, and can thus provide UV stabilization of plastics. Arguably, the use of black carbon pigments may be traced back to the use in prehistoric times of charcoal from burnt wood, and indeed this source provides the black in many ancient cave paintings. Nowadays, the term *carbon black* refers to a group of well-defined industrial products manufactured to provide a controlled particle size distribution which determines their optical quality. Although carbon black is virtually always classified as an inorganic pigment, there would be some justification for classifying the product amongst the high performance organic pigments. For example, the bonding in carbon black is organic in nature, while many of its properties, especially the high absorption coefficient, are arguably more closely related to those of organic than of inorganic pigments. Carbon blacks have been described as having an imperfect graphite-like structure consisting of layers of large sheets of carbon atoms in six-membered rings that are parallel but further apart than in graphite, and arranged irregularly. Carbon blacks also may contain other elements, notably oxygen, hydrogen, nitrogen and sulfur. Oxygen, in variable amounts up to 15%, and small amounts of hydrogen, are covalently bonded at the edge of the carbon layers, and nitrogen is mainly integrated as heteroatoms within the aromatic layer system. The surface composition of carbon black pigments is thus complex and can play an important role in their application. In this respect, the presence of oxygen-containing functional groups (notably carboxylic acid, ketone, anhydride, lactone and phenolic OH) is highly significant in certain pigment grades. This functionality, which is introduced by oxidative aftertreatments, confers hydrophilicity on the pigment surfaces, a feature that is of particular importance for water-based applications.

Carbon blacks are manufactured from hydrocarbon feedstocks by partial combustion or thermal decomposition in the gas phase at high temperatures. World production is today dominated by a continuous furnace black process, which involves the treatment of viscous residual oil hydrocarbons, containing a high proportion of aromatics, with a restricted amount of air at temperatures of 1400–1600 °C.

9.3 ORGANIC PIGMENTS

The synthetic organic pigment industry developed towards the end of the nineteenth century out of the synthetic textile dye industry that had become established. Many of the earliest organic pigments were prepared from water-soluble dyes rendered insoluble by precipitation on to colourless inorganic substrates such as alumina and barium sulfate. These products were commonly referred to as *lakes*. A further significant early development was the discovery and commercial introduction of a range of azo pigments, which provided the basis for the most important yellow, orange and red organic pigments currently in use. These so-called classical azo pigments offer bright intense colours although generally only moderate performance in terms of fastness properties. A critical event in the development of the organic pigment industry was the discovery, in 1928, of copper phthalocyanine. This blue pigment was the first product to offer the outstanding intensity and brightness of colour that is typical of organic pigments, combined with an excellent set of fastness properties, comparable with many inorganic pigments. The discovery stimulated the quest for other chemical types of organic pigment that could emulate the properties of copper phthalocyanine in the yellow, orange, red and violet shade areas. This research activity gained further impetus from the emergence of the automotive paint market and the growth of the plastics and synthetic fibres industries, applications that demanded high levels of technical performance. Another factor which has influenced this development is the increasing environmental concern over the use of heavy metal containing pigments such as the lead chromates and cadmium sulfides, and the perceived requirement for 'safer' alternatives. The range of high performance organic pigments which has emerged includes the quinacridones, isoindolines, dioxazines, perylenes, perinones and diketopyrrolopyrroles, together with several improved performance azo pigments.

Organic pigments generally provide higher intensity and brightness of colour than inorganic pigments. These colours are due to the $\pi-\pi$ * electronic transitions associated with extensively conjugated aromatic systems (Chapter 2). Organic pigments are unable to provide the degree of opacity that is typical of inorganic pigments, because of the lower refractive index associated with organic crystals. However, the combination of high colour strength and brightness with high transparency means that organic pigments are especially well suited to printing ink applications. The range of commercial organic pigments exhibit a rather variable set of fastness properties that are

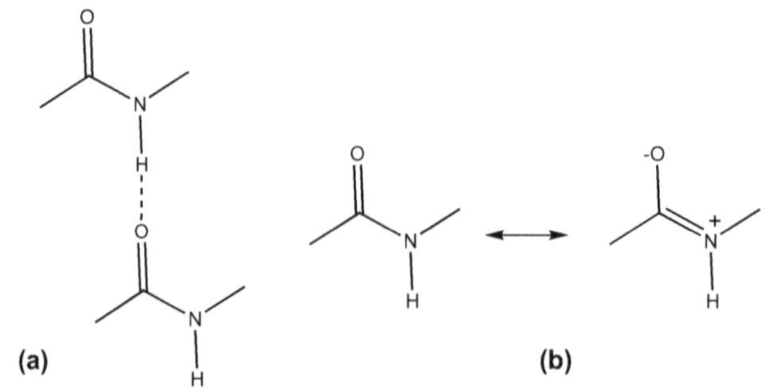

Figure 9.1   Origin of (a) intermolecular hydrogen bonding and (b) dipolar interactions due to the amide group in organic pigments.

dependent both on the molecular structure and on the nature of the
intermolecular association in the crystalline solid state. Since organic
molecules commonly exhibit some tendency to dissolve in organic
solvents, organic pigment molecules incorporate structural features
that are designed to enhance the solvent resistance. For example, an
increase in the molecular size of the pigment generally improves
solvent resistance. In addition, the amide (–NHCO–) group features
prominently in the chemical structures of organic pigments, because
its presence enhances fastness not only to solvents, but also to light
and heat, as a result of its ability to participate in strong dipolar
interactions and in hydrogen bonding, both intramolecular and
intermolecular, as illustrated in Figure 9.1. A related structural feature
is found in many heterocyclic pigments (*e.g.*, quinacridones and
diketopyrrolopyrroles) that contain both N–H and C=O groups in ring
systems, and this arrangement gives rise to a similar improvement in
fastness properties as a result of strengthening the intermolecular
association throughout the crystal structure. The incorporation,
where appropriate, of halogen substituents and of metal ions, particularly of the alkaline earths and transition elements, can also have
a beneficial effect on fastness properties. The following sections
provide an overview of the more important chemical types of commercial organic pigments, together with some discussion of the
structural features that determine their suitability for particular
applications.[1,3,5]

9.3.1   Azo Pigments

Azo pigments, both numerically and in terms of tonnage produced,
dominate the yellow, orange and red shade areas in the range of

commercial organic pigments (Chapter 3). The chemical structures of some important classical azo pigments are shown in Figure 9.2. The structures are illustrated in the ketohydrazone form since structural studies carried out on a wide range of industrial azo pigments have, in each case, demonstrated that the pigments exist exclusively in this form in the solid state.[11] Many other colour chemistry texts have followed the commonly-used convention to illustrate them in the azo tautomeric form. Simple classical monoazo pigments such as Hansa Yellow G (**9.1**) (CI Pigment Yellow 1)[26] and Toluidine Red (**9.2**) (CI Pigment Red 3)[27] are products that provide bright colours and good lightfastness, but rather poor solvent resistance. The good lightfastness of these molecules is attributed to the extensive intra-molecular hydrogen-bonding in six-membered rings. The inferior fastness to organic solvents is due to their small molecular size and the fact that the intermolecular interactions in the crystal structures involve essentially only van der Waals forces. The use of these pig-ments is largely restricted to decorative paints. The pigments resist dissolving in the solvents used in these paints (either water or ali-phatic hydrocarbons) at the low temperatures involved in their ap-plications, but they have a tendency to dissolve in more powerful solvents, especially if higher temperatures are involved, a feature that restricts their use in many industrial paint, printing ink and plastics applications.

The most important yellow azo pigments, particularly for printing inks but also for certain coatings applications are the dis-azoacetoacetanilides (diarylide yellows), which include CI Pigment Yellow 12, **9.3a**, CI Pigment Yellow 13, **9.3b**, and the more durable CI Pigment Yellow 83, **9.3c**. These pigments have been shown to exist in bis-ketohydrazone forms, structurally analogous to Hansa Yellow G, **9.1** (CI Pigment Yellow 1).[28–30] They exhibit higher colour strength and transparency than the corresponding monoazo pigments, prop-erties that are particularly suitable for printing ink applications, and improved solvent fastness which is attributable to the larger mo-lecular size. The pyrazolone oranges, such as CI Pigment Orange 34, **9.4**, are similar both in chemical structure and in properties to the diarylide yellows and are used in a similar range of applications. The naphthol reds, of which CI Pigment Red 170, **9.5**, is an important industrial example, are structurally related to Toluidine Red (**9.2**, CI Pigment Red 3). Compound **9.5** shows superior solvent resistance as a result of its larger molecular size and due to the presence of the amide groups, which provide strong intermolecular forces of attraction in the crystal lattice structure. Consequently, it is suitable for use in a

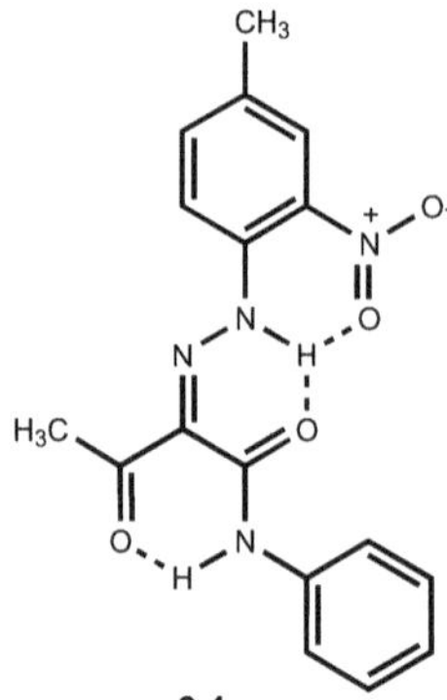

9.1

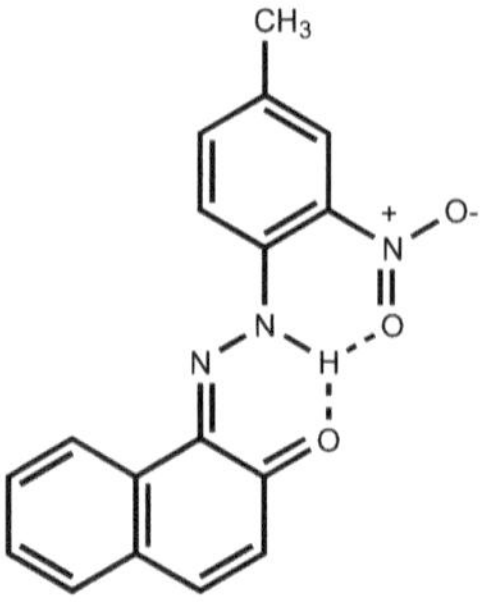

9.2

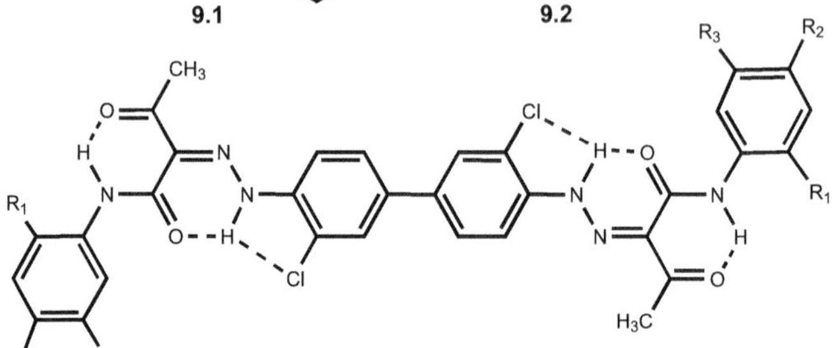

9.3a: R$_1$ = R$_2$ = R$_3$ = H
9.3b: R$_1$ = R$_2$ = CH$_3$; R$_3$ = H
9.3c: R$_1$ = R$_3$ = OCH$_3$; R$_2$ = Cl

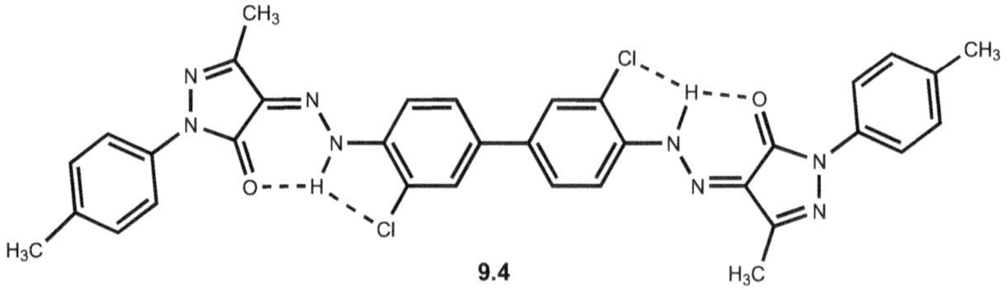

9.4

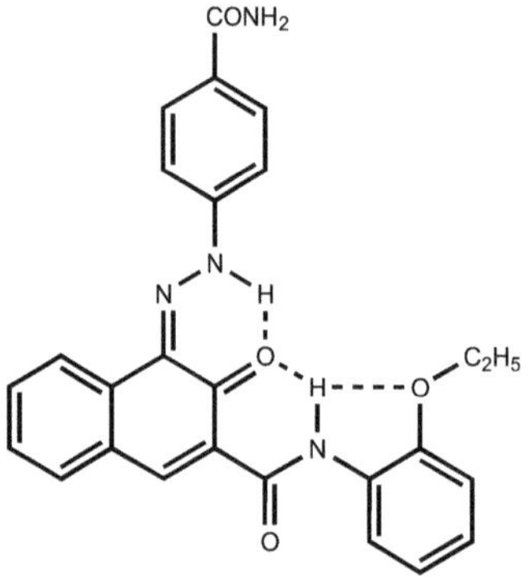

9.5

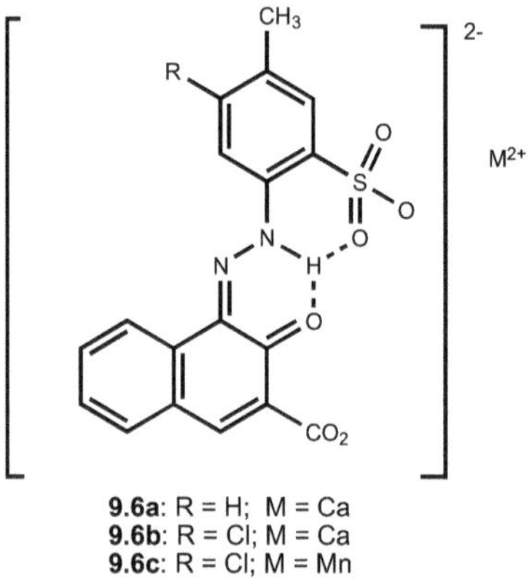

9.6a: R = H; M = Ca
9.6b: R = Cl; M = Ca
9.6c: R = Cl; M = Mn

Figure 9.2 Chemical structures of some important classical azo pigments.

wider range of paint and plastics applications. The most important classical red azo pigments are metal salts, such as compounds **9.6a–c**, which are derived from azo dyes containing $SO_3^-Na^+$ or $-CO_2^-Na^+$ groups by replacement of the Na^+ ions with divalent metal ions, notably Ca^{2+}, Sr^{2+}, Ba^{2+} and Mn^{2+}.[31] They are products of high colour strength and brightness, high transparency and good solvent resistance. They are especially important for printing ink applications. CI Pigment Red 57:1, **9.6a**, is the pigment principally used to provide the magenta inks for multicolour printing. These products are complex structurally.[32,33] The metal atoms bond to oxygen atoms derived from carboxylate, keto and sulfonate groups in the ligands and to water molecules with some bridging between the ligands, which leads to a polymeric ladder structure. Metal salt pigments have evolved from products referred to as *lakes*, which were essentially anionic azo dyestuffs precipitated on to inorganic substrates such as alumina and barium sulfate.

Azo pigments are manufactured using the two-stage process of diazotization followed by azo coupling as detailed in Chapter 3. To ensure that the pigments are obtained in high yield and purity, it is essential to maintain careful control of experimental conditions in order to minimise the formation of side products that might adversely affect the colouristic properties of the pigments. When azo coupling is complete, the aqueous suspension is generally heated to the boil for a period of time, often in the presence of surface-active agents, to refine the pigment particles and develop crystallinity. Most commercial diarylide yellow and orange pigments, including compounds **9.3a–c** and **9.4** are derived synthetically from 3,3′-dichlorobenzidine (DCB), a suspected carcinogen. The sequence leading to CI Pigment Yellow 12, **9.3a**, is presented in Chapter 3 as an example of the synthesis of a symmetrical disazo colorant. Although the pigments themselves are essentially non-toxic, evidence has been presented that they may cleave thermally at temperatures above 200 °C to give a monoazo compound and that prolonged heating above 240 °C causes further decomposition leading to release of DCB.[34] As a result of these observations, the use of these pigments is no longer recommended for applications where high temperatures are likely to be encountered, for example in thermoplastics. Metal salt azo pigments, such as compounds **9.6a–c**, are prepared by traditional diazotisation/coupling processes to form the sodium salt of an azo dyestuff which may show some solubility in water. This species is then treated with a solution of an appropriate salt of the divalent metal, which displaces the sodium to form the insoluble pigment.

9.3.2 Copper Phthalocyanines

Copper phthalocyanines provide by far the most important of all blue and green pigments. The chemistry of the phthalocyanines has been discussed in some depth in Chapter 5, in terms of the synthesis and structures, so that only a brief account is presented here. Copper phthalocyanine, **9.7** (CuPc, CI Pigment Blue 15), is arguably the single most important organic pigment. Copper phthalocyanine finds widespread use in most pigment applications because of its brilliant blue colour and its excellent resistance towards light, heat, solvents, acids and alkalis. In addition, despite its structural complexity, copper phthalocyanine is a relatively inexpensive pigment as it is manufactured in high yield from low cost starting materials (Chapter 5). Copper phthalocyanine exhibits polymorphism, the most important crystal phases being the α- and β-forms (several other forms have been reported). Both the α- and β-forms are of commercial importance. The greenish-blue β-form of the pigment is almost always the pigment of choice for cyan printing inks, and it is suitable also for use in most paint and plastics applications. The manufacture of copper phthalocyanine blue pigments, including the methods used to provide the different polymorphs and to ensure that they are prepared in a particle size form suitable for application, has been described in Chapter 5. The most important green organic pigments are the halogenated copper phthalocyanines, CI Pigment Green 7, in which the 16 ring hydrogen atoms of the CuPc molecule are replaced virtually completely by chlorine, and CI Pigment Green 36, a designation that incorporates a range of bromo- and bromochloro-copper phthalocyanines. The hue of these pigments becomes progressively yellower with increasing bromine substitution. The phthalocyanine greens exhibit the same outstanding colouristic and technical performance as the blue pigments from which they are derived and find equally widespread use in the coloration of paints, printing inks and plastics.

9.7

9.3.3 High performance Organic Pigments

Copper phthalocyanines, as discussed in the previous section, although generally regarded as classical organic pigments, exhibit outstanding technical performance and so could equally well be described as high performance organic pigments. This section contains a survey of a range of the organic pigments, encompassing a wide variety of structural types, which have been developed in an attempt to match the properties of copper phthalocyanines in the yellow, orange, red and violet shades. They include two groups of azo pigments, a series of carbonyl pigments of various types, and dioxazines. High performance organic pigments are particularly suited to applications that require bright, intense colours and at the same time place stringent demands on the technical performance of pigments, such as the coatings applied to car bodies, referred to as automotive paints. They provide excellent durability, combined with good colour properties but they tend to be rather expensive.[12]

There are two classes of high performance azo pigments: disazo condensation pigments and benzimidazolone azo pigments. The chemical structures of representative examples of these products are illustrated in Figure 9.3. Disazo condensation pigments constitute a range of durable yellow, red, violet and brown products, with structures such as compound **9.8** (CI Pigment Red 166). These pigments derive their name, and also their relatively high cost, from the rather elaborate synthetic procedures involved in their manufacture, which involves a condensation reaction (Scheme 3.8, Chapter 3). An important group of high performance azo pigments contain the benzimidazolone group, exemplified by CI Pigment Red 183, **9.9**, which range in shades from yellow to bluish-red and brown and exhibit excellent fastness properties. Their high stability to light and heat and

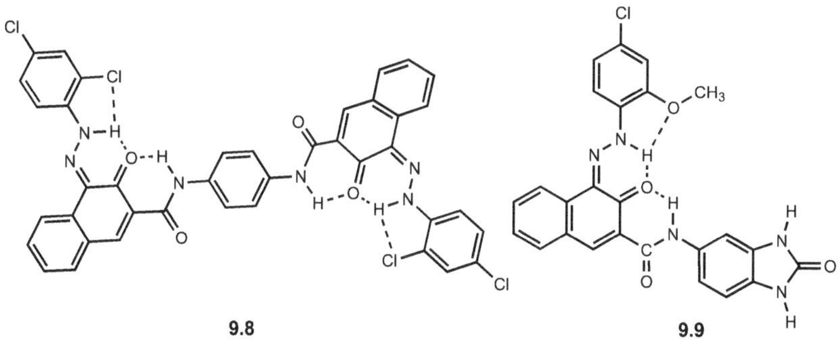

Figure 9.3 Chemical structures of some high performance azo pigments.

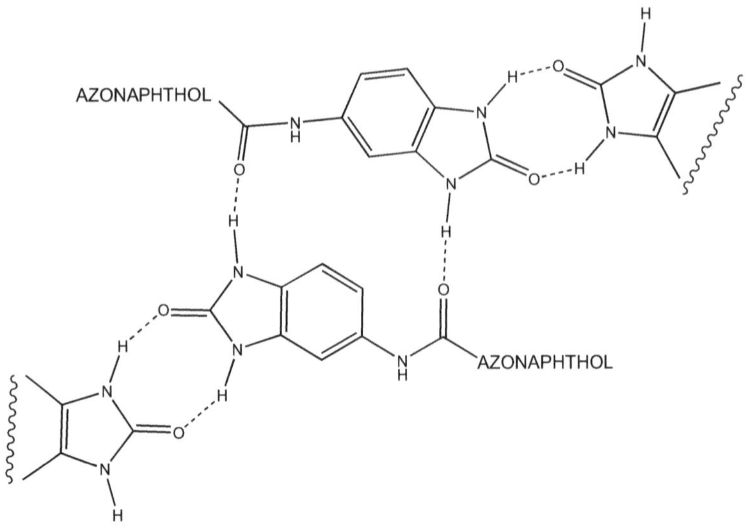

Figure 9.4   Intermolecular association in the crystal structure of a benzimidazolone azo pigment.

their insolubility is attributed to extensive intermolecular association as a result of hydrogen-bonding and dipolar forces in the crystal structure, as illustrated in Figure 9.4.[35]

Carbonyl pigments of various types may also be classed as high performance products. These include some anthraquinones, quinacridones, perylenes, perinones, isoindolines and diketopyrrolopyrroles. The chemistry of these groups of colorant has been discussed previously in Chapter 4, and so is not considered further in this section. Some representative examples of the chemical structures of important high performance carbonyl pigments are illustrated in Figure 9.5.

Several vat dyes developed originally for textile applications are suitable, after conversion into an appropriate pigmentary physical form, for use in many paint and plastics applications.[36] Examples of these so-called vat pigments include the anthraquinones, Indanthrone Blue, **9.10** (CI Pigment Blue 60), and Flavanthrone Yellow, **9.11** (CI Pigment Yellow 24), and the perinone, **9.12** (CI Pigment Orange 43). Other high performance carbonyl pigments include the quinacridone, **9.13** (CI Pigment Violet 19), diketopyrrolopyrrole (DPP) pigments, such as CI Pigment Red 254, **9.14**, perylenes, for example CI Pigment Red 179, **9.15**, and isoindolines, such as CI Pigment Yellow 139, **9.16**. The excellent lightfastness, solvent resistance and thermal stability of carbonyl pigments may be explained in many cases by

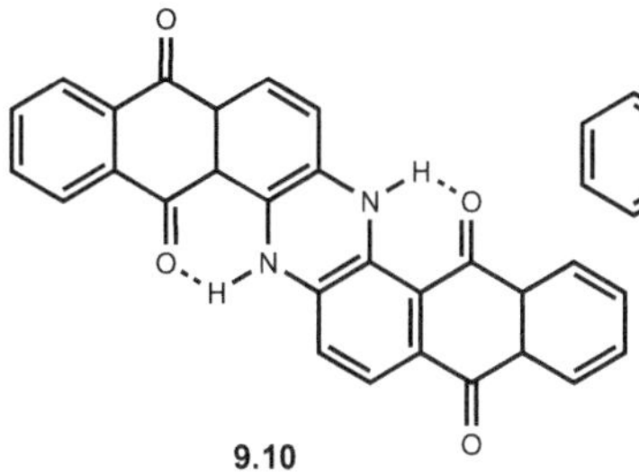

Figure 9.5 Chemical structures of a range of high performance carbonyl pigments.

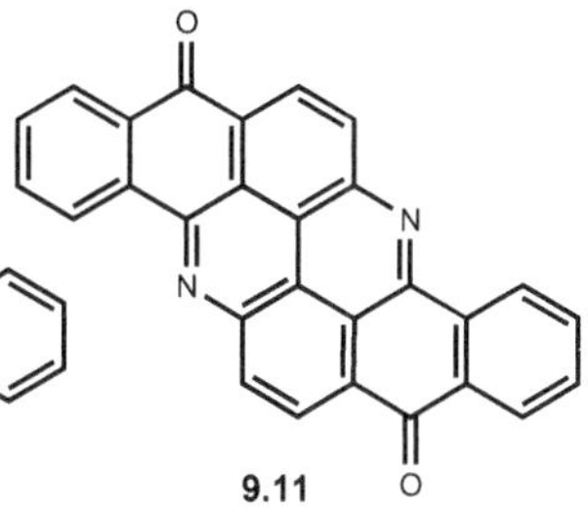

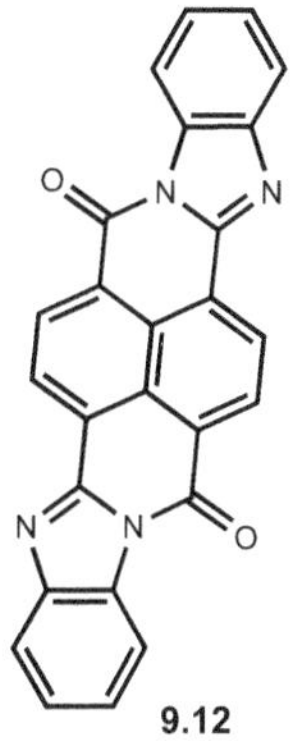

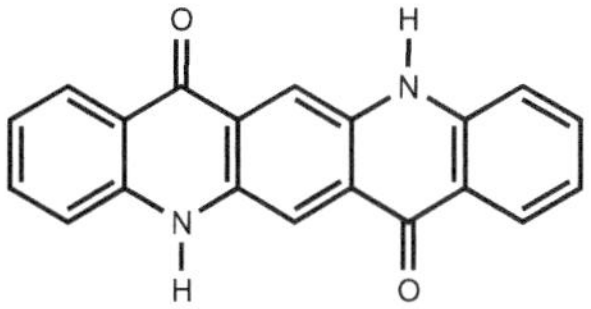

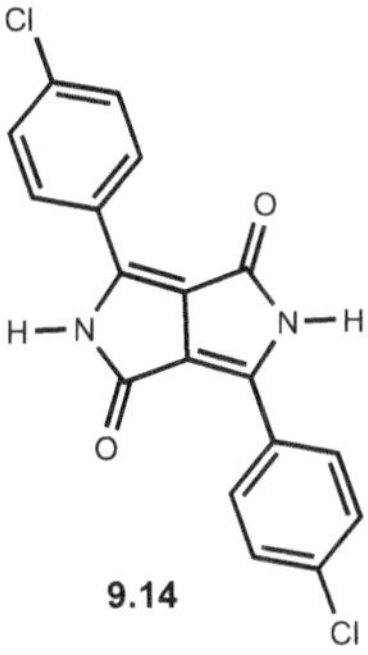

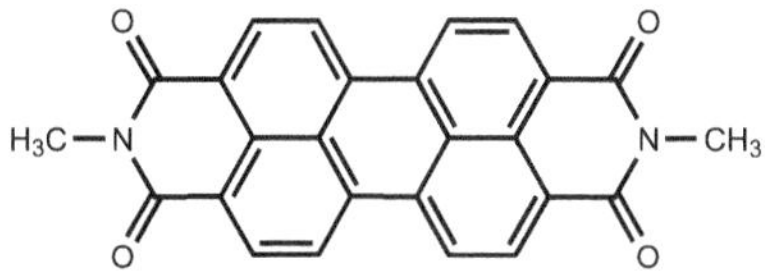

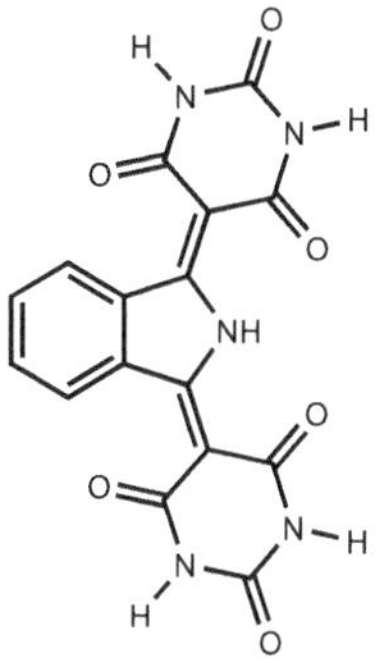

intermolecular association in the solid state as a result of a combination of hydrogen bonding and dipolar forces, similar to that illustrated for the benzimidazolone azo pigments in Figure 9.4. A diagrammatic illustration of the intermolecular hydrogen bonding that exists in the crystal lattice arrangement of all of the polymorphic forms of quinacridone, **9.13**, is given in Figure 4.6, Chapter 4.[37]

Other chemical types of high performance organic pigments are exemplified by the tetrachloroisoindolinone, **9.17** (CI Pigment Yellow 110), and the dioxazine, **9.18** (Carbazole Violet, CI Pigment Violet 23). There has been a considerable amount of research carried out in an attempt to exploit the potential of metal complex chemistry to provide high performance pigments, particularly of yellow and red shades, to complement the colour range of the copper phthalocyanines, which cannot be extended outside blues and greens. Several azo, azomethine and dioxime transition metal complex pigments have been obtained that show excellent lightfastness, solvent resistance and thermal stability. However, the products have achieved limited commercial success largely because the enhancement of fastness properties of the organic ligand that results from complex formation is almost inevitably accompanied by a reduction in the brightness of the colour. This effect may be explained by a broadening of the absorption band as a result of overlap of the band due to $\pi-\pi^*$ transitions of the ligand with those due to transition metal d–d transitions or ligand–metal charge transfer transitions.

9.17

9.18

9.4 MOLECULAR AND CRYSTAL MODELLING OF PIGMENTS

In dye chemistry, the relationship between the molecular structure of a dye and its technical performance, including colour and fastness properties, is now well-established. As discussed at length in Chapter 2, computational methods have emerged as indispensible tools in the development of our fundamental understanding of the properties of coloured molecules and in the design of new products. Molecular modelling techniques, which include a range of methods based on quantum mechanics and molecular mechanics, allow the molecular and physical properties of a particular dye to be predicted by calculation with some confidence, and without the need to resort to synthesis. The same principles apply to the molecular structures of organic pigments. As an example, the correlation between the oscillator strength values, calculated using the PPP molecular orbital approach, and the experimental molar extinction coefficient values obtained for solutions of a series of monoazo and disazo pigments is illustrated in Figure 2.17 (Chapter 2). However, structure–property relationships are much more complex in the case of pigments because they are applied as discrete crystalline solid particles, rather than as individual molecules as is often the case with dyes. Thus, the colouristic and technical performance of pigments is dependent not only on the molecular structure but also on the crystal structure arrangement, the nature of the pigment particles, particularly their size and shape distribution, and their surface characteristics. In recent decades, the application of single-crystal X-ray diffraction studies, the significant advances in structure determination from X-ray powder diffraction patterns, and advances in a variety of solid state spectroscopic techniques, including NMR and electron diffraction, have provided detailed information on the solid state structures of most of the inorganic and organic pigments in current use. As a result, our understanding of the effect on the properties of pigments of the way that molecules and ions pack in the crystal lattice structure, and the nature of the intermolecular interactions, has advanced significantly. In addition, the information has facilitated the development of computer-aided methods for the prediction by calculation of aspects of the performance of pigments in the crystalline solid state. The principles of *crystal design* in the context of pigments may be applied to assist in the development of new products for improved performance based on an understanding of their structural properties. Crystal design incorporates aspects of crystal engineering (the design of the bulk structure within the particles), control of morphology (the design of particle size and shape) and the engineering of the particle surfaces. The methods used in crystal design rely on a range of computational tools that have seen recent significant advances, due in no small measure to the dramatic increase in available computing power.[38]

Crystal engineering may be defined as an understanding of intermolecular interactions in the context of crystal packing and the utilisation of such understanding in the design of new solid materials with desired chemical and physical properties.[39] Crystal structure determination is the key feature that provides the information required for the crystal engineering. The molecules in organic pigment crystal structures and the ions in many inorganic pigment crystal structures are almost invariably close-packed, a feature that leads to desirable application properties, such as high insolubility and thermal stability. In particular, the crystal structures of organic pigments that give high performance contain specific molecular features that maximize the intermolecular interactions within the close-packed structures. Various intermolecular forces operate, including van der Waals and dipolar forces and π–π interactions. However, in this respect, hydrogen bonding commonly assumes special importance.[40] The networks of intermolecular bonds that are a feature of many crystal structures are illustrated schematically for linear trans quinacridone, **9.13**, in Figure 4.6 (Chapter 4) and for benzimidazolone azo pigments in Figure 9.4 in this chapter. Several other chemical types of high performance organic pigment exhibit similar hydrogen bonding patterns, including diketopyrrolopyrroles and isoindolines. Most crystal structure determinations that have been carried out on organic pigments are based on traditional X-ray crystallography applied to single crystals. However, due to their low solubility, suitable single crystals are often difficult to grow and special techniques are required, including high temperature vacuum sublimation and crystallization from high-boiling solvents with controlled cooling. In recent years, it has become increasingly feasible to determine crystal structures from X-ray powder diffraction patterns.[41] Ideally, the powder samples used should be highly crystalline in order to provide X-ray diffraction patterns of sufficiently high resolution to allow the indexing of the patterns, so that the *unit cell*, the small unit that describes the bulk arrangement of atoms or molecules within the crystal when stacked together in three-dimensional space, and the *space group*, which provides a description of the symmetry of the crystal, may be determined. This provides the first step towards a model from which the final structure may be obtained using mathematical refinement methods. The results of structural investigations carried out using this approach are progressively filling in the gaps in our knowledge of the crystal structures of pigments for which single-crystal determination has not proved successful, for example of some metal salt azo pigments.[33]

A crystal structure consists of a set of atoms or molecules arranged in a particular way, and a *lattice* that exhibits long-range order and symmetry. A property of crystals that provides a measure of the strength of the intermolecular interactions in a crystal is the *lattice energy*, which is sometimes referred to as packing energy. Lattice energy may be defined as the energy released when ions or molecules in the gas phase are condensed into a solid three-dimensional lattice structure. Since this is an exothermic process, the enthalpy change is invariably negative. Owing to the strong intermolecular forces of attraction in pigment crystals, they exhibit lattice energies with high numerical values. Calculation of the strength of these interactions may be carried out to assist understanding of crystal growth processes. In the so-called atom–atom method, the strength of an intermolecular force is approximated as a sum of the constituent atom–atom interactions. Thus, the lattice energy (E_{cryst}) of a crystal may be calculated by summing all of the interactions between a central molecule and all of the surrounding molecules in the crystal. These calculations commonly use methods based on *force fields*, utilising the information made available by crystal structure determination. The reader will find explanations of the basis of force field calculations in the section on molecular mechanics in Chapter 2 (Section 2.9.3). An important feature of the crystal engineering process is to verify that the calculated lattice energy values correspond reasonably closely with experimental sublimation enthalpies, although the limited availability of these values for pigments is a complication.

Crystal morphology, which describes the size and shape (or habit) of a crystal, is of fundamental importance in determining many of the application properties of pigments. It is a feature that is of prime importance in ensuring that pigments have optimum optical properties, such as colour strength and transparency/opacity, and flow properties in the media into which the pigment is dispersed. Morphology may also influence fastness properties. The crystal morphologies of a pigment are defined by the nature of the intermolecular forces. In general, pigments whose intermolecular interactions are defined mainly by π–π forces within stacks of molecules tend to form needles or rods parallel to the stacking direction, as typified by copper phthalocyanine, **9.7** (CI Pigment Blue 15) (see Figure 5.1 in Chapter 5). The presence of one other directional interaction usually leads to plate-like morphology. Crystal engineering methods that are capable of predicting crystal morphology thus offer considerable potential for the development of pigment products.

The morphology of a crystal is determined by the relative rates of growth of different crystal faces as the crystal forms. The surface attachment energy (E_{att}) may be defined for a particular crystal face as the energy released as a growth slice attaches to the surface of a growing crystal, and is related to the lattice energy (E_{cryst}) by the relationship:

$$E_{cryst} = E_{slice} + E_{att}$$

where E_{slice} is the slice energy, the energy released by the formation of a slice of thickness d, the spacing between planes at the crystal face. A simplified representation of this crystal growth process is illustrated schematically in Figure 9.6(a). It has been shown that the attachment energy at a face is inversely proportional to the rate of crystal growth at the face. This means that the faces with the lowest attachment energies are the slowest growing, and therefore the most important faces with regard to the morphology of the crystal. Methods based on force fields, using crystal structure data, may be used to calculate the attachment energies for the various crystal faces and thus allow prediction of crystal morphology.[42]

The properties of the surface of pigment particles are commonly controlled in industrial products by the use of additives that attach to the surfaces and modify the surface character. The use of 'tailor-made' additives to enhance specific aspects of application performance, traditionally on the basis of trial and error experimentation, has been extensively investigated and many such additives are now commonly used. The molecular structures of the additives may be

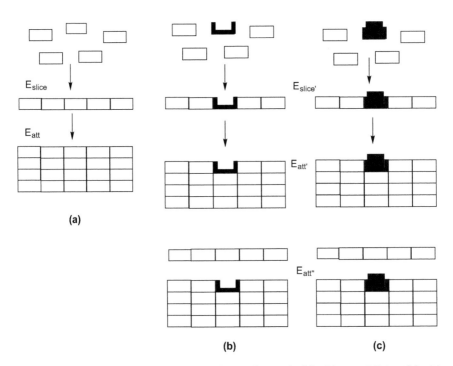

Figure 9.6 Schematic representations of crystal growth: (a) with no additive; (b) with a disruptor additive; (c) with a blocker additive.

considered as containing two specific structural features, referred to as the pigment chromophore and the functional substituent, as illustrated in Figure 9.7. The former contains structural features closely related to the chemical structure of the pigment for which it is designed, ensuring that the additive attaches strongly to the pigment surface, on the basis of molecular recognition. The functional substituent is designed to provide the desired effect at the surface in order to control the chemical character of the surface, for example hydrophilic or hydrophobic, anionic or cationic, so that compatibility with a particular application medium is enhanced and dispersion properties improved. Notable examples of such additives are a range of carboxylate, amino, sulfonate, phosphonate and amide derivatives of copper phthalocyanine that are used to modify the properties of copper phthalocyanine pigments. A further role of the additives is to influence the relative rates of growth at crystal faces by attaching to the surfaces during crystal formation, and thus modify the particle morphology. Figure 9.6 illustrates two ways in which the additive may operate, either in disruptor mode, Figure 9.6(b), where the additive is smaller than the pigment molecule, or in blocker mode, Figure 9.6(c), where it is larger than the pigment molecule. In either mode, as shown in Figure 9.6, the additive limits growth at a face by inhibiting attachment of a crystal slice. The effect of 'tailor-made' additives on crystal morphology may be predicted using crystal engineering principles on the basis of the calculation of binding energies from modified slice energies ($E_{slice'}$) and attachment energies ($E_{att'}$ and $E_{att''}$). Binding energies influence the way in which the additives adsorb selectively on specific faces of pigment crystals, and thus modify their growth rates.

The ability to predict crystal structures of molecules by modelling calculations has been an aim of crystallographers and solid-state chemists for many years, and considerable effort has been made to address the challenging issues involved. The approaches used have involved calculations of crystal packing by modelling the 3D structural arrangements such that the magnitude of the lattice energy is

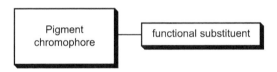

Figure 9.7 Schematic representations of an additive designed to modify the surface character of a pigment.

maximized on the basis of the strength of intermolecular interactions. However, despite strenuous efforts towards this goal applying
energy minimization techniques using force field or quantum
mechanical methods, there remains no completely successful analytical mathematical solution. The main problem is that the process
inevitably generates vast numbers of structures, which may be considered as virtual polymorphs, and this creates immense difficulties
in locating global energy minima within the numerous possible low
energy packing configurations. In general, several hundred unique
crystal packings remain after the optimization, which must be
evaluated either in terms of calculated physical properties such as
density and lattice energy or by means of empirical scoring functions.[43] A particular area in which these approaches are beginning to
offer some promise is in the prediction of new polymorphic possibilities for existing pigments.[44]

9.5 PIGMENTS FOR SPECIAL EFFECTS

There is an interesting and diverse range of pigment types which
have the ability to produce novel and unusual optical effects in
application. These products, which include metallic, pearlescent,
optically-variable and fluorescent pigments as discussed in this
section, have been experiencing significant growth in popularity and
industrial importance in recent decades, driven by factors such as
the fashion trends in automotive applications, cosmetics and other
consumer markets.

9.5.1 Metallic Pigments

The most important metallic pigment by far is aluminium flake, CI
Pigment Metal 1. Aluminium pigments are used in a wide range of
paint, printing ink and plastics applications to simulate the optical
effect that is characteristic of metallic silver and to provide a range of
coloured metallic effects when used in combination with transparent
grades of traditional coloured pigments. They are best known, however, for their use in metallic car finish paints.[45] The pigments owe
their importance to the highly reflective nature of aluminium metal,
which provides attractive bright, sparkling optical effects, and to their
stability, which owes much to the thin, tenacious oxide film on the
surface of the pigment particles. Aluminium pigments are generally
manufactured from aluminium metal by a wet ball-milling process
in the presence of a fatty acid (stearic or oleic) and a mineral oil.

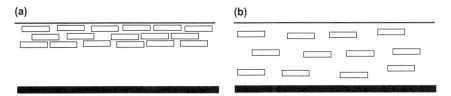

Figure 9.8 Schematic representation of the orientation of (a) leafing and (b) non-leafing aluminium pigment particles in application.

The presence of the liquid ingredients is essential to improve the efficiency of the process and to eliminate the potential explosion hazard of dry grinding.

Aluminium pigments consist of small, lamellar particles, or platelets, which facilitates their alignment in the application medium, as illustrated schematically in Figure 9.8. The shape of the particles is optimized to maximize the reflective qualities of the pigment surfaces while minimizing the scattering that takes places mostly from rough edges. The pigments are categorised according to their ability to 'leaf'. Leafing grades when incorporated into a film become oriented in a parallel overlapping fashion at or near to the surface of the film, and thus provide a continuous metallic sheath and a bright silvery finish. Non-leafing grades are distributed more evenly throughout the film producing a sparkling metallic effect. There are also a series of bronze pigments manufactured from copper or copper/zinc alloy by a dry milling process. They are used to simulate a gold or copper effect in application.

9.5.2 Pearlescent Pigments

Pearlescent pigments give rise to a white pearl lustre effect often accompanied by a coloured iridescence, similar to that observed in nature from pearls, shells, fish scales and certain minerals.[46] A wide range of particle compositions are known to produce pearl effects, although the most important commercial pearlescent pigments consist of thin platelets of mica coated with metal oxides, notably titanium dioxide but also oxides of zirconium, chromium and iron, which partly reflect and partly transmit incident light. The mica substrate acts both as a template for the pigment synthesis and as a support for the deposition of the thin oxide layers. In application, simultaneous reflection of incident light from oriented platelets creates the sense of depth that is characteristic of pearlescent lustre. The optical effects from pearlescent pigments arise from a combination of multiple transmission, refraction and reflection phenomena as the incident light interacts with successive layers of materials with different refractive indices (mica is of low refractive index, while titanium dioxide is highly refracting). In products where the layers are of an appropriate thickness, iridescent colours are produced by interference phenomena. Pearlescent pigments are used most commonly in automotive finishes, plastics and cosmetics.

9.5.3 Optically-Variable Pigments

A range of optically variable pigments have been developed, providing unusual and novelty colour-play effects that are dependent on the angle of view. These pigments show striking changes in colour and lightness depending on whether the angle of view is facing or grazing. The optical variability is based on a combination of reflection, interference and absorption effects in pigments based on multi-layered systems. These pigments show a striking angle-dependent colour-play of violet, red, gold and green hues, the nature of the effect being dependent on the chemical composition and the thickness of the layers. One type of pigment consists of a core of a plate-like micaceous iron oxide pigment, which provides the inner reflection of gold, copper or red hues, coated first with a layer of low refractive index silica, which provides interference colours depending on the film thickness, and an outer layer of selectively reflecting iron oxide. In a related type of system, the inner core is an aluminium metallic pigment. These products are suitable for use in coatings, including automotive finishes, plastics, ceramics and cosmetics.

9.5.4 Fluorescent Pigments

Daylight fluorescent pigments consist of fluorescent dyes dissolved in a transparent, colourless polymer and ground to a fine particle size.[47] In this way, they are chemically different from traditional coloured pigments. In their application, the pigments give rise to colours with a remarkable vivid brilliance as a result of the extra glow of fluorescent light. The applications of fluorescent pigments are mostly associated with their high visibility and ability to attract attention. They are used in printing inks, for example in advertising, posters, magazines and supermarket packaging, coatings applications–especially where safety is an important feature, such as fire-engines, ambulances, rescue vehicles and aircraft–and in a range of plastic products such as toys, bottles and other containers. The fluorescent dyes most commonly used are Rhodamines (Chapter 6) in the red to violet shade range, and aminonaphthalimides and coumarins (Chapter 4) in the yellow shade range. The polymers in which the dyes form a solid solution play a critical role. They must act as good solvents for the dyes as well as providing high solvent resistance and thermal stability. Traditionally, daylight fluorescent pigments have been based on a thermoset polymeric toluenesulfonamide-melamine-formaldehyde resin matrix. However, there is an increasing trend towards the use of 'non-formaldehyde' thermoplastic polymers based on polyurethane, polyamide or polyester resins. A significant issue with many fluorescent pigments is the quenching of the fluorescence as the concentration of the dye in the resin is increased. This is particularly significant with the Rhodamines, which exhibit considerable concentration quenching above about 1%. Yellow dyes, in contrast, may be used up to 5–10% in certain matrices before an excessive dulling effect, characteristic of this type of quenching, occurs. In contrast, mixtures of certain dyes may produce more brilliant fluorescence than a single dye, as a result of transfer of energy from one dye to another. In the same way certain fluorescent brightening agents (FBAs) may act as sensitisers by enhancing the fluorescence yield from yellow dyes. The lightfastness of daylight fluorescent pigments is generally inferior to that of traditional high performance organic pigments, and this limits their use in exterior applications. Nevertheless, careful selection of the resin matrix, the use of light stabilising additives, or the use of UV-absorbing overlayers can produce fluorescent colours with reasonable light stability.

REFERENCES

1. W. Herbst and K. Hunger, *Industrial Organic Pigments. Production, Properties, Applications*, Wiley-VCH Verlag GmbH, Weinheim, 3rd edn, 2004.
2. G. Buxbaum and G. Pfaff (ed.), *Industrial Inorganic Pigments, Production, Properties, Applications*, Wiley-VCH Verlag GmbH, Weinheim, 3rd edn, 2005.
3. R. M. Christie, *The Organic and Inorganic Chemistry of Pigments*, Surface Coatings Reviews, Oil and Colour Chemists' Association, London, 2002.
4. R. M. Christie, *Pigments: Structures and Synthetic Procedures*, Surface Coatings Reviews, Oil and Colour Chemists' Association, London, 1993.

5. J. Lenoir, Organic pigments, in *The Chemistry of Synthetic Dyes*, ed. K. Venkataraman, Academic Press, London, 1971, vol. 5, ch. 6.
6. P. A. Lewis (ed.), *Pigment Handbook*, John Wiley and Sons, Inc., New York, 2nd edn, 1988.
7. D. Paterson, in *Colorants and Auxiliaries: Organic Chemistry and Application Properties*, ed. J. Shore, Society of Dyers and Colourists, Bradford, 1990, vol. 1, ch. 2.
8. H. Zollinger, *Color Chemistry: Syntheses, Properties and Applications of Organic Dyes and Pigments*, Wiley-VCH Verlag GmbH, Weinheim, 3rd edn, 2003, ch. 12.
9. W. S. Czajkowski, in *Modern Colorants: Synthesis and Structure*, Blackie Academic and Professional, London, 1995, vol. 3, ch. 3.
10. J. T. Guthrie and L. Lin, *Physical-chemical Aspects of Pigment Applications,* Surface Coatings Reviews, Oil and Colour Chemists Association, London, 1994.
11. K. Hunger, *Rev. Prog. Color.*, 1999, **29**, 71.
12. H. M. Smith (ed.), *High Performance Pigments*, Wiley-VCH Verlag GmbH, Weinheim, 2002.
13. H. Skelton, *Rev. Prog. Color.*, 1999, **29**, 43.
14. W. J. Ferguson, *Plastic Rubber Int.*, 1983, **8**, 32.
15. J. H. Braun, A. Baidins and R. E. Marganski, *Prog. Org. Coat.*, 1992, **20**, 105.
16. J. H. Braun, *J. Coat. Technol.*, 1997, **69**, 59.
17. M. V. Orna, *J. Chem. Educ.*, 1978, **55**, 478.
18. W. E. Smith, *J. Oil. Col. Chem. Assoc.*, 1985, **68**, 170.
19. N. V. Hullavarad, S. S. Hullavarad and P. C. Karulkar, *J. Nanosci. Nanotechnol.*, 2008, **8**, 3272.
20. A. C. D. Cowley, *Rev. Prog. Color*, 1986, **16**, 16.
21. R. Nayak, A. Suryanarayana and S. B. Rao, *J. Sci. Ind. Res.*, 2000, **59**, 833.
22. S. E. Tarling, P. Barnes and J. Klinowski, *Acta Crystallogr., Sect. B*, 1988, **44**, 128.
23. J. Fabian, N. Komiha, R. Linguern and P. Rosmus, *J. Mol. Struct.: Theochem*, 2006, **801**, 63.
24. H. J. Buser, D. Schwarzenbach, W. Petter and A. Ludi, *Inorg. Chem.*, 1977, **16**, 2704.
25. J.-B. Donnet (ed.), *Carbon Black Science and Technology*, CRC Press, New York, 2nd edn, 1993.
26. E. F. Paulus, *Z. Kristallogr.*, 1984, **167**, 65.
27. A. Whitaker, *Z.Kristallogr.*, 1978, **147**, 99.
28. M. J. Barrow, R. M. Christie, A. J. Lough, J. E. Monteith and P. N. Standring, *Dyes Pigments*, 2000, **45**, 153.

29. M. J. Barrow, R. M. Christie and J. E. Monteith, *Dyes Pigments*, 2002, **55**, 79.
30. M. J. Barrow, R. M. Christie and T. D. Badcock, *Dyes Pigments*, 2003, **57**, 99.
31. R. M. Christie and J. Mackay, *Color. Technol.*, 2008, **124**, 133.
32. A. R. Kennedy, C. McNair, W. E. Smith, G. Chisholm and S. J. Teat, *Angew. Chem. Int. Ed.*, 2000, **39**, 638.
33. S. L. Bekö, S. M. Hammer and M. U. Schmidt, *Angew Chem. Int. Ed.*, 2012, **51**, 4735.
34. R. Az, B. Dewald and D. Schnaitmann, *Dyes Pigments*, 1991, **15**, 1.
35. K. Hunger, E. F. Paulus and D. Weber, *Farbe Lack*, 1982, **88**, 453.
36. D. Thetford and A. P. Chorlton, *Dyes Pigments*, 2004, **61**, 49.
37. E. F. Paulus, F. J. J. Leusen and M. U. Schmidt, *Cyst. Eng. Commun.*, 2007, **9**, 131.
38. P. Erk, Crystal design of high performance pigments, in *High Performance Pigments*, ed. H. M. Smith, Wiley-VCH Verlag GmbH, Weinheim, 2002, ch. 8.
39. G. R. Desiraju, J. J. Vittal and A. Ramanan, *Crystal Engineering*, World Scientific, Singapore, 2011.
40. D. Thetford, J. Cherryman, A. P. Chorlton and R. Docherty, *Dyes Pigments*, 2004, **63**, 259.
41. R. Cerny and V. Favre-Nicolin, *Z. Kristallogr.*, 2007, **222**, 105.
42. G. Clydesdale, K. J. Roberts and R. Docherty, *J. Crystal Growth*, 1996, **166**, 78.
43. D. W. M. Hoffmann and T. Lengauer, *Acta Crystallogr., Sect. A*, 1997, **53**, 225.
44. N. Panina, R. van de Ven, P. Verwer, H. Meekes, E. Vlieg and G. Deroover, *Dyes Pigments*, 2008, **79**, 183.
45. I. Wheeler, *Metallic Pigments in Polymers*, Rapra Technology Ltd., Shrewsbury, UK, 1999.
46. G. Pfaff, Special effect pigments, in *High Performance Pigments*, ed. H. M. Smith, Wiley-VCH Verlag GmbH, Weinheim, 2002, ch. 7.
47. R. M. Christie, *Rev. Prog. Color.*, 1993, **23**, 1.

CHAPTER 10

Colour in Cosmetics, with Special Emphasis on Hair Coloration

With contributions from Olivier X. J. Morel[†]

10.1 INTRODUCTION

Cosmetics may be broadly defined as materials that are used to enhance the appearance or appeal of the human body. This definition covers a vast range of substances including facial and eye make-up, hair colouring and conditioning products, skin-care creams, nail varnishes, coloured contact lenses, fragrances and perfumes, deodorants, cleansing and sanitising products, and a host of other products.[1-3] Many manufacturers distinguish between cosmetics that are used for decorative purposes and those whose purpose is broadly in the area of care, such as cleansing and skin-care products, including those used for protection against the harmful effects of the sun.[4] Cosmetics may also be described by the physical composition of the product. For example, they may be formulated as liquids, lotions, creams or gels (as emulsions or dispersions), sticks (as in lipsticks) or powders (either in loose or compressed form), or as aerosol sprays. Colour is of prime importance in the subset of cosmetics generally

[†]Xennia Technology, Monroe House, Works Road, Letchworth, Hertfordshire, SG6 1LN, UK
E-mail: omorel@xennia.com

Colour Chemistry, 2nd edition
By Robert M Christie
© R M Christie 2015
Published by the Royal Society of Chemistry, www.rsc.org

referred to as *make-up*, which are coloured products applied to the human body, mainly to the facial area, to enhance the appearance and attractiveness of the users, who are mostly female. This chapter deals with the principles involved in the colorants used in these decorative cosmetic products, followed by a more detailed discussion of the chemistry that has been developed specifically for use in the coloration of human hair.

The manufacture of cosmetics is a multibillion dollar global industry, currently dominated by a relatively small number of multinational corporations in the USA, Europe (mainly in Germany, France, Italy and UK) and Japan. This is an industry sector that has not thus far experienced the level of competition from emerging economies, such China and India, which has led to the substantial transfer of the manufacture and application of traditional textile dyes and pigments to those parts of the world. This feature may be attributed, to a certain extent, to the cultural association of products branded as cosmetics with the fashion and design industry as led by Europe and the USA, often linked with specific Western brand names.

The formulation of cosmetic products is strictly controlled by legislation, which varies from country to country. However, most countries have some formal legislation that either restricts or prohibits certain ingredients and a regulatory agency that controls what can be incorporated. In the USA, the regulating body is the Food and Drug Administration (FDA). The main directive in the European Union affecting the manufacture, labelling and supply of cosmetics and personal care products is the Cosmetics Directive 76/768/EEC, applying also to Iceland, Norway and Switzerland. This Directive was extended in 2009 to include the use of nanoparticles and stricter rules on animal testing of cosmetics. The Canadian health authorities operate a Cosmetic Ingredient Hotlist, a regularly-updated list of substances that are restricted or prohibited in cosmetics.

10.2 COLORANTS FOR DECORATIVE COSMETICS

Cosmetics have been used since ancient times to impart colour to the human body. The purpose is to enhance appearance, for example by hiding skin imperfections or by intensifying or highlighting the eyes, lips and nails.[5] The technology used in the manufacture and use of coloured decorative cosmetics is broadly similar to that used in the coatings and printing ink industries, mainly involving dispersions of pigments as the colorants of choice (Chapter 9). There are, however, certain additional features required in cosmetic formulations, for example that they must be non-toxic and non-irritant in use, and there is commonly also a requirement to provide an appealing look, feel and odour. The colorants used are subject to specific government regulations, mostly modelled on those specified by the US FDA, the European Commission or the Japanese Ministry of Health and Welfare. Dyes and pigments are approved by the FDA for use in food, drugs and cosmetics (FD&C) or drugs and cosmetics (but not food) (D&C). In addition to specifying those colorants permitted for use in cosmetics, the authorities require batch certification in terms of composition and purity.

The familiar categories of make-up include foundation, blusher, mascara, eyeliner, eye shadow, lipstick and nail varnishes.[1] The function of foundation is to impart a smooth, appealing finish to the skin, masking minor imperfections and levelling out skin tones. The white pigment, titanium dioxide is most commonly used to provide the required coverage, on the basis of its high opacity (Chapter 9). Colour is added to foundation products mainly using combinations of yellow, red and black iron oxides. Lipsticks are dispersions of colorants in a base consisting of a blend of oils, fats and waxes, moulded as sticks. The colour of a commercial lipstick is critical as it is often its main selling feature. Usually, the colour is based on reds, although commercial products may range from orange–yellow through pure reds to purples and browns. Iron oxides may be used for some russet and brown lipstick shades. However, organic pigments provide the bright, clean orange, red and violet hues that are the mainstay of modern products. The pigments used are mainly *lakes*, which are insoluble pigments prepared by the precipitation of a water-soluble dye with a metal ion on to a substrate.[6] The dyes used mainly belong to the azo (Chapter 3) or triarylmethine (Chapter 6) classes. As an example, the calcium lake of D&C Red 7, **10.1**, equivalent to CI Pigment Red 57:1, is one of the most important lipstick pigments. Some lipsticks may also contain a small proportion of a dye. An example is eosin, D&C Red 21, a brominated fluorescein that is effectively an acid–base indicator. The dye is applied in its insoluble orange acid form, **10.2**, which changes colour and stains the lips a purple–red colour as it is converted into the alkaline form, **10.2a**, by neutralisation within the lip tissue (Scheme 10.1). This chemistry forms the basis of colour-changing 'mood lipsticks', the popularity of which varies with fashion trends. The aluminium lake formed from the alkaline form, **10.2a**, is used as a cosmetics pigment.

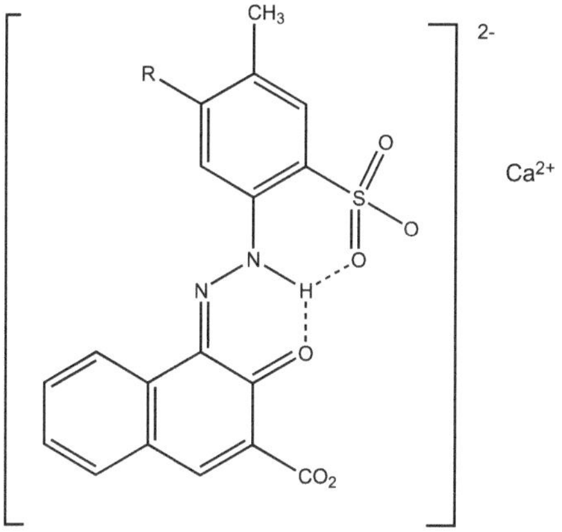

10.1

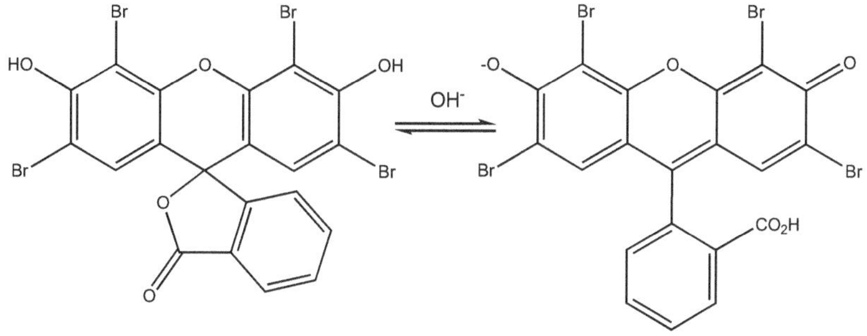

10.2 **10.2a**

Scheme 10.1 Eosin, an acid/base indicator used in mood lipsticks.

The colorants permitted for cosmetic application in the area of the eye are rather more restricted. This is especially true in the USA, where only inorganic pigments, such as titanium dioxide and the iron oxides, and a few selected organic pigments, for example lakes based on the triarylmethine dye FD&C Blue 1, **10.3**, are permitted. For many years, black pigments for eye preparations were restricted to black iron oxide in the USA, although carbon black was permitted in other geographical areas. However, in 2004, the FDA approved one form of carbon black, a high purity furnace black, D&C Black 2, for use in eye cosmetic products. Effect pigments, notably pearlescent and metallic pigments (Chapter 9), have experienced rapid recent growth in popularity for a range of cosmetic applications, in view of the eye-catching glitter and sparkle effects that they provide.

10.3

10.3   HAIR COLORATION

Human hair has always possessed powerful, symbolic and evocative properties. Its coloration is one of the oldest acts of human adornment, reflecting our common dissatisfaction with the natural colour. It is used to conceal the greying process as we grow older, as a means to express individuality, as a fashion statement or to project an image. As a result of an increasingly ageing, and thus greying, population, demand for hair colouring products has been increasing rapidly, a trend that seems likely to continue into the future. Indeed, the demand for hair dyes may well accelerate due to growing individual expectations in developing global economies. Human hair is a protein fibre with certain physical and chemical similarities compared with protein-derived textile fibres, such as wool, as discussed in Chapter 7. However, hair coloration has certain unique features that distinguish it from textile dyeing, and thus completely different dye types and dyeing processes are required. As a process carried out directly on the human head, hair cannot be dyed above *ca.* 40 °C, with a dyeing time generally not exceeding *ca.* 40 min, and using a low ratio of dye formulation to hair fibre, a parameter that is referred to as *liquor ratio* in textile dyeing. The colour on the hair is required to be stable to agencies such as air, light, friction, perspiration and chlorinated water, and to other hair treatments. The colouring process is required to cause minimum damage to the hair and to avoid staining the scalp. Critically, the process must be toxicologically safe to the individual concerned and to the hair-dresser who applies the formulation.

Until synthetic dyes appeared around the mid-nineteenth century (Chapter 1), natural materials were used to colour hair. The best-known natural hair colouring product is *henna*, obtained from the leaves of *Lawsonia inermis*.[7] The colouring qualities of henna are due to lawsone, **10.4**, a naphthoquinone derivative. The antiquity of the use of henna was established by analysis of the hair of Rameses II, which revealed the presence of lawsone, and, around 5000 years later, hair coloration with henna is still practised, especially in regions of Asia. With this exception, modern hair coloration now uses the extensive range of synthetic hair dyes that have been developed.[8–13] These synthetic products may be categorised either according to the chemistry involved or to the degree of permanence of the colour on the hair. In chemical terms, they may be classed as involving either *oxidative* or *non-oxidative* processes. Oxidative dyeing products, constituting *ca.* 70% of the hair colorant market, are dominated by *permanent* hair dyes, which owe their popularity to the long lasting effect, ease of application and versatility in terms of the range of achievable colours. Human hair grows by *ca.* 0.3 mm each day, the growth cycle lasting about 3 years, at which point the hair falls out. Thus, the designation as 'permanent' requires qualification because further treatment is required periodically to cover new growth. Non-oxidative hair dye products are categorised mainly as either *semi-permanent* or *temporary*. In addition to the colouring agent and ingredients used to assist application, hair dyeing products, especially of the permanent type, often contain a bleaching agent (hydrogen peroxide) which removes some of the natural hair colour.

10.4

Human hair is naturally coloured, the colour commonly reflecting the individual's geographic or ethnic origin. Natural hair colours occupy a small segment of CIELAB colour space, corresponding to dominant absorption wavelengths in the range 586–606 nm, while the lightness (L^*) varies over a wide range, from 1.8% to 90%. Natural hair owes its colour to the pigment *melanin*, which occurs as granules *ca.* 1 µm long and 0.3 µm in diameter.[14] Melanin is formed in pigment-producing cells referred to as melanocytes. A melanocyte produces essentially two types of melanin: *eumelanin* and the less prevalent *pheomelanin*. The pigments are complex, irregular polymers whose structures remain incompletely characterised. Generally, hair contains a mixture of the two. The more eumelanin is present, the darker the hair. The variety of natural hair colours is determined not only by the concentrations of the two pigments, but also the size and shape of the granules, their distribution patterns and their crystal structures.

The biosynthetic routes to the two melanin types are closely related. The amino acid tyrosine, **10.5**, the common starting material, is initially hydroxylated to form 3,4-dihydroxyphenylalanine, **10.6** (DOPA), which is oxidised to dopaquinone, **10.7**, in enzyme-mediated reactions (Scheme 10.2). Thereafter, the pathways diverge.[15]

Scheme 10.3 illustrates the route that ultimately generates eumelanin. Dopaquinone, **10.7**, is converted, *via* leucodopachrome, **10.8**, dopachrome, **10.9**, 5,6-dihydroxyindole-2-carboxylic acid, **10.10** (DHICA), and 5,6-dihydroxyindole, **10.11** (DHI), into the reactive intermediate indolequinone, **10.12**, from which an oxidative polymerisation leads to eumelanin.[16] Scheme 10.4 illustrates the route towards pheomelanin, whereby dopaquinone, **10.7**, reacts with the amino acid cysteine, **10.13**, to give cysteinyl DOPA derivatives **10.14a/ b**, which are converted *via* intermediates **10.15a/b** and **10.16a/b** into dihydrobenzothiazines **10.17a/b**.[17] Details beyond this stage are uncertain.

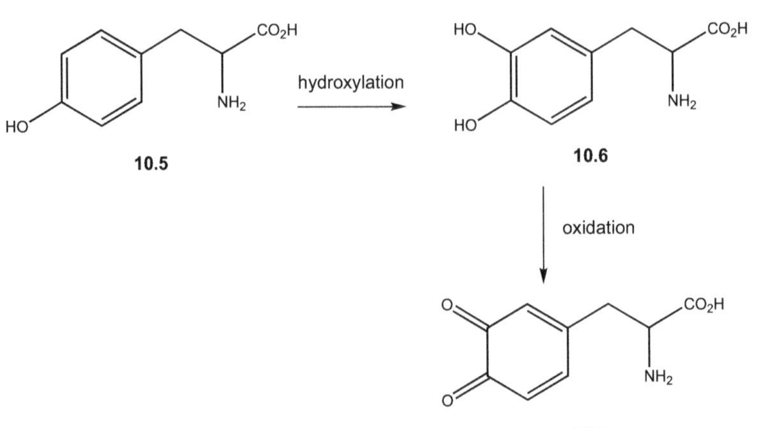

Scheme 10.2 Oxidation of tyrosine, **10.5**, to dopaquinone, **10.7**, *via* DOPA, **10.6**.

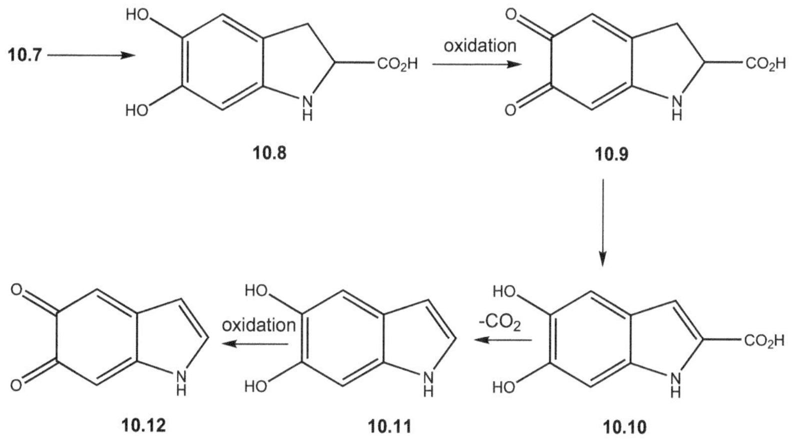

Scheme 10.3 Biosynthesis of precursors to eumelanin.

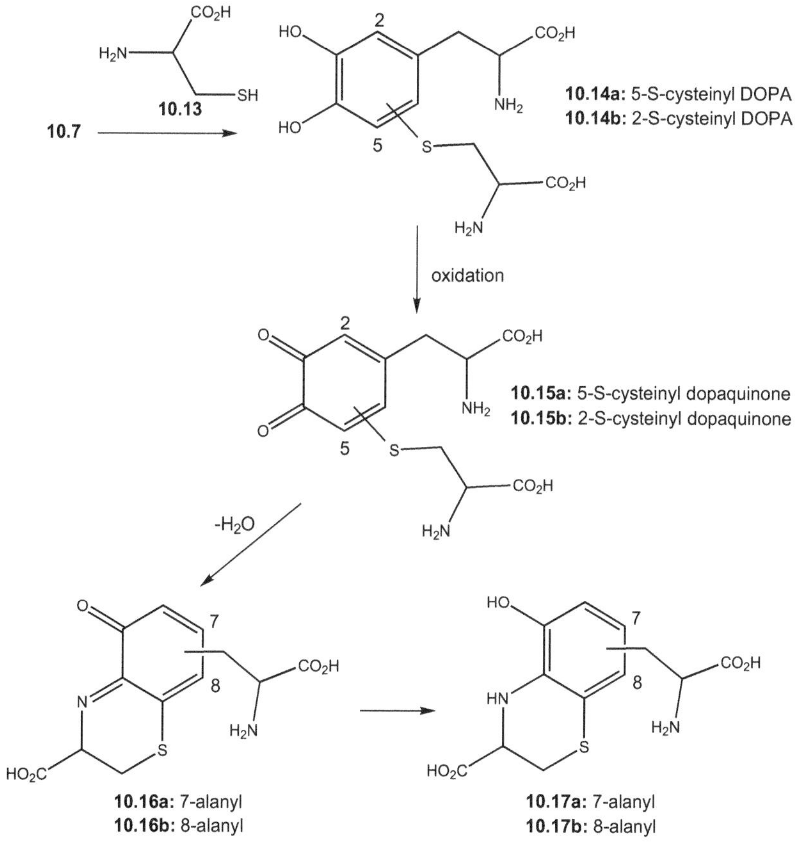

Scheme 10.4 Biosynthesis of precursors to pheomelanin.

The molecular size and shape of a dye are critical factors governing its ability to penetrate into the hair fibre. Computer-aided molecular modelling studies aimed at predicting the ability of a dye molecule to penetrate into the fibre have used particular molecular size descriptors devised to take some account of molecular shape, for example S_{LD}, defined as 'the longest diagonal line of the smallest shadow of the projected figure',[18] and L_D, 'the longest dimension of the smallest cross-section of the optimum parallelepiped enclosing the molecule'. A size limit of *ca.* 9.5Å has been proposed for non-ionic dyes, with slightly larger limits for ionic dyes.[19] The diffusion of dye molecules into hair is facilitated by fibre swelling, which is in turn influenced by the dyeing conditions and the use of swelling agents in hair dye formulations. Hair dyeing is commonly conducted in the pH range 9–10. At alkaline pH, hair acquires a negative charge and significant swelling is observed, due mainly to ionisation of amino acid residues containing carboxylic acid groups. Most traditional hair dyes are non-ionic, with a few anionic dyes. However, there has been recent interest in the development of cationic hair dyes, which have stronger affinity for the negatively-charged hair fibres.

Concerns about the human safety profile of ingredients in hair coloration products, especially certain precursors used in oxidative hair colouring, have been raised over the years. In view of their extensive use in regular direct contact with humans, the nature of the chemical structures of some ingredients, especially aromatic amines, and a few signals in the epidemiological literature, the human safety profile of hair dye components has been studied intensively, and some original ingredients are now prohibited.[20] In certain individuals, hair coloration may cause skin irritation or allergic reactions at various levels of severity. To prevent or limit allergic reactions, a patch test is commonly conducted on an area of skin before a particular product is used. International epidemiological studies on the link between hair coloration and more serious human conditions have been reported.[21,22] These studies indicate that hair coloration may play a role in the risk of certain lymphomas. The increased risk is described as moderate and much more significant among women who had used the dyes before 1980, when the compositions would have been more likely to include potentially carcinogenic components. In parallel, the industry continues to provide reassurances over the formulation ingredients in current use. Nevertheless, a change to the situation cannot be excluded in the future as stricter legislation, regulations and controls on the use of chemicals emerge. In Europe, the safety of hair dyes is controlled by the Cosmetic

Directive. In the USA, the legal responsibility rests with the FDA. In Japan, the Ministry of Health, Labour and Welfare regulates the safety of hair dyes, considering them as 'quasi-drugs', subject to approval based on evidence of their safety, and there are similar regulations in other Far Eastern countries.

10.3.1 Oxidative Hair Coloration

The chemistry of permanent hair dyeing is based on the 150 year old observation by Hofmann that *p*-phenylenediamine, when exposed to oxidizing agents, produced brown colours on certain substrates. In view of advances in chemistry and also concerns around the safety profile of some precursors, it is perhaps surprising that this chemistry still forms the basis of the most common hair coloration method.[23] The process requires three main components. The first is an *o*- or *p*-substituted (hydroxy or amino) aromatic amine, referred to as the *primary intermediate*, *oxidation base*, or *developer*. Primary intermediates include *p*-phenylenediamine, *p*-aminophenol and their derivatives. The second component, the *coupler*, is commonly an aromatic compound, either a benzene derivative with electron-donating groups arranged *m*- to each other, including *m*-phenylenediamines and resorcinol, or certain naphthols. Couplers alone do not produce significant oxidation colours but rather modify the colour generated when used together with the primary intermediates and an oxidising agent. Couplers are commonly classified into three groups, according to the colour obtained when used with the primary intermediates: yellow–green (mainly resorcinol and its derivatives), red (mainly phenols and naphthols) and blue (mainly *m*-phenylenediamines). The third component is the oxidising agent, almost exclusively hydrogen peroxide. The oxidant serves two main purposes: to oxidize the primary intermediates and to lighten the natural hair colour by bleaching. The oxidation is promoted by alkaline conditions, most commonly using ammonia, which also assists swelling of the hair fibre, thus facilitating dye penetration. Alternatives to ammonia, such as monoethanolamine, are also used but generally provide lower bleaching efficiency. Permanent hair colouring preparations are generally marketed as a two component kit, mixed immediately before use. The first component contains the mixture of dye precursors, often around 5–6 species, in the combinations required to produce the desired colour, together with ammonia. The second component is a stabilised hydrogen peroxide solution. The mixture is applied initially near to the hair roots for a period to allow exposure to undyed new growth, followed by application to the rest of the hair.

The dye precursors are small molecules that penetrate readily and deeply into the hair fibre. The first stage of colour formation involves oxidation of the primary intermediate with alkaline hydrogen peroxide to form reactive electrophilic species. In the case of *p*-phenylenediamine, **10.18**, this species is *p*-benzoquinonediimine, **10.20**, in equilibrium with its conjugate acid, **10.20a**, while *p*-benzoquinonimine, **10.21**, is formed from *p*-aminophenol, **10.19** (Scheme 10.5). These species undergo an electrophilic substitution reaction with the coupler, **10.22**, preferentially at a position *para* to an electron-releasing amino or hydroxyl group, forming leuco compounds, **10.23**, which oxidize to form dinuclear indo dyes, **10.24** (Scheme 10.6). In the case where R=H, reaction of the dinuclear species with a further molecule of the reactive species may lead to formation of trinuclear indo dyes, **10.25**. The presence of a blocking group (*e.g.* R=CH$_3$) in a *p*-position prevents this reaction. The coloured molecules formed are much larger than the precursors, and so become permanently entrapped within the hair fibre.

In formulated products containing mixtures of primary intermediates and couplers, the colour developed depends on competing reactions between the various precursors inside the fibre, influenced by such factors as concentrations, diffusion rates, redox potentials and pH. Countless numbers of alternative dye precursors have been examined over the years for their ability to fulfil the requirements of oxidative hair dyeing, addressing issues associated with toxicology, colouristic performance and fastness properties. For example, investigation of heterocyclic analogues has led to the introduction of primary intermediates based on derivatives of dihydropyrazolone **10.26**. They undergo oxidation to generate the quinonediimine

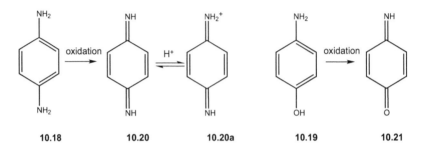

Scheme 10.5 Generation of the reactive species from oxidation of *p*-phenylenediamine, **10.18**, and *p*-aminophenol, **10.19**.

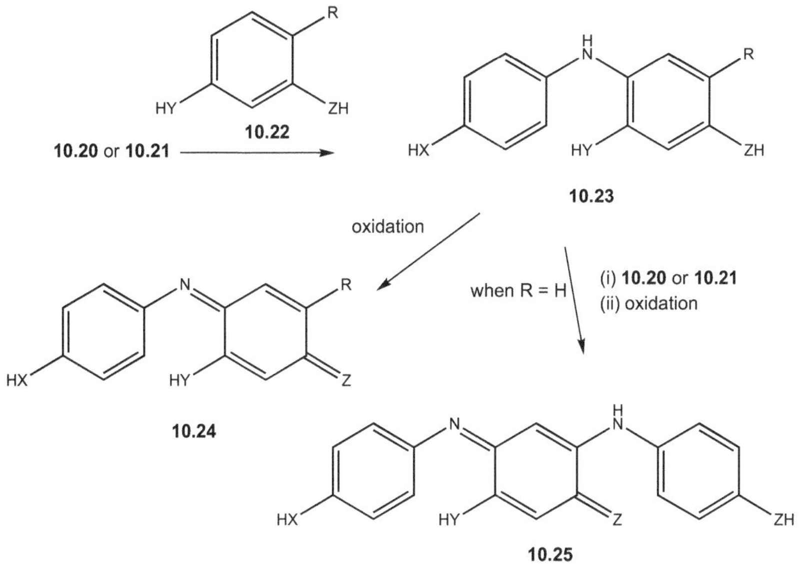

Scheme 10.6 Pathway proposed for the formation of dinuclear, **10.24**, and trinuclear indo dyes, **10.25**, in the oxidative coupling involved in permanent hair dyeing. X, Y and Z may be independently O or NH.

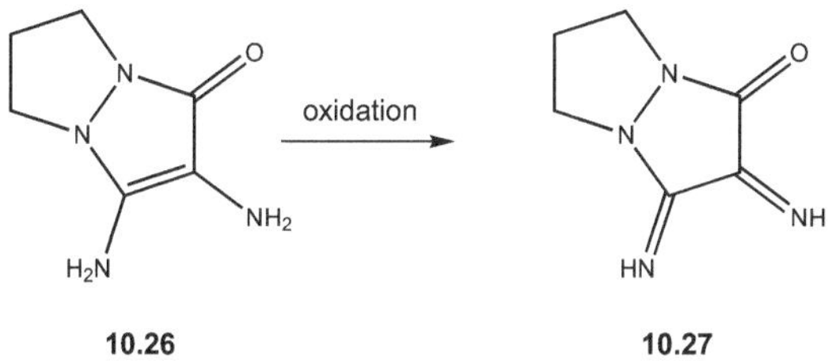

Scheme 10.7 Oxidation of dihydropyrazolone **10.26** to the quinonediimine analogue **10.27**.

analogue, **10.27**, as the reactive species (Scheme 10.7). However, while the research intensity has produced a few individual successes, most of the compounds in current commercial dye formulations remain those that have been used for decades.

Several processes have been investigated in attempts to provide alternatives to traditional oxidative hair dyeing aimed broadly at safe dyeing systems that give intense coloration and good fastness properties, while avoiding hair damage from the use of oxidizing agents and strongly alkaline conditions.[23] Similar to the oxidative process, these methods mostly also involve chemical reactions that generate colour *in situ*. For example, processes based on the synthesis of insoluble azo, methine or azomethine colorants within the hair fibre have been proposed. Attempts to exploit successful textile dyeing processes include the use of fibre-reactive dyes that form a covalent bond between the dye and the hair fibres in the manner of textile fibres (Chapter 8) and solubilized vat dyes, water-soluble sulfate esters of leuco vat dyes, which, after application to hair, are subjected to acid hydrolysis and oxidation to generate the insoluble vat dye (Chapter 7). Processes generating indigo in the hair have been proposed either from naturally-occurring or specifically-designed synthetic indigo precursors.[24] However, none of these processes have reached a successful commercial outcome, probably due to various technical issues and the prohibitive cost of the requirement to assess the toxicological profile of the agents used in order to achieve approval and registration as hair dyes.

Processes that attempt to synthetically mimic melanin formation inside the hair fibre have attracted attention because of perceived market advantages, notably the potential to provide colours with natural appearance and the ability to designate the product as 'natural'. However, the challenges are immense since natural hair colour is determined not only by the melanin structure but also by the granule sizes and their distribution throughout the hair. The experimental approaches have been aimed primarily at synthetic imitation of the biosynthetic pathways leading to the natural pigments, focussed mostly on eumelanin, by oxidation of the various intermediates as identified in Schemes 10.1–10.3. An early process reported the use of tyrosine, **10.5**, or DOPA, **10.6**, with a tyrosinase to give a light brown hair colour, with repeated dyeing necessary to build up colour. Subsequently, improved processes were described using DOPA in combination with compounds such as hydroquinones, diamines or aminophenols and with hydrogen peroxide oxidation replacing the tyrosinase, and also using DHI, **10.11**, the closest isolable intermediate in melanin biosynthesis. Comparable approaches towards synthetic pheomelanin generation have been restricted by difficulties in synthesizing the intermediates. It is, however, likely that the synthetic melanins are formed dispersed throughout the fibre, rather than deposited in granules as in the natural process, so that the nuances in colour are not imitated.

10.3.2 Non-oxidative Hair Dyeing

Non-oxidative hair dyeing processes are often referred to as *direct* dyeing, since the dyes are applied directly and the molecules remain chemically unchanged within the hair after application, thus providing a contrast with the oxidative dyeing processes. These products yield non-permanent colour effects on hair in that they resist a few washings, while slowly and evenly fading. The dyes are classified according to the duration of the colour on the hair as either *semi-permanent* or *temporary*.

Semi-permanent products provide hair colour lasting roughly 4–6 washings. They offer the advantages of ease of use, no requirement for mixing, and a degree of reversibility since an undesirable result can be removed by repeated shampooing. Semi-permanent dyeing relies on the diffusion of small, coloured non-ionic molecules into the hair and the principal interactions between the dye and fibre molecules are relatively weak dipolar and van der Waals' forces. The N-hydroxyethyl substituent (occasionally also N-hydroxypropyl) is widely encountered in the structures of all chemical classes of semi-permanent hair dyes, some relevant examples of which are illustrated in Figure 10.1. The role of these groups is to promote water-solubility while also ensuring an appropriate water/lipid partition coefficient, a property that has an important impact on dye uptake by the hair. Nitro dyes (Chapter 6) constitute by far the most important chemical class of semi-permanent hair dyes, providing colours ranging through yellow, orange and red–brown to violet.[25] Simple nitroanilines

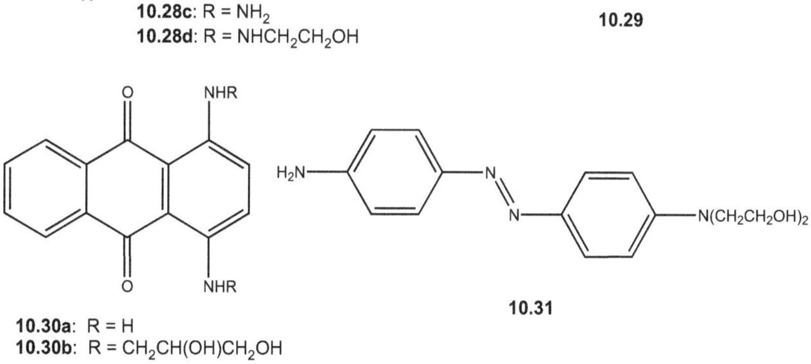

Figure 10.1 Structures of some semi-permanent hair dyes.

provide a few yellow semi-permanent hair dyes, for example dye **10.28a**. Nitroaminophenols and some related ether derivatives provide most of the yellow to orange dyes, as exemplified by the orange dye **10.28b**. Nitrophenylenediamine dyes offer a wide colour range, providing dominant products from orange through to violet. The colour of these dyes depends on the electron-donating power of the amino substituent. For example, dye **10.28c** is red and dye **10.28d** violet. Nitrodiphenylamines provide a few yellow dyes, including compound **10.29**. Anthraquinone dyes are important to complete the semi-permanent hair colour palette by contributing violet and blue colours, since nitro dyes cannot so readily provide compounds absorbing at long wavelengths. Appropriate dyes are often selected from the range used for dyeing synthetic textile fibres, for example dye **10.30a** (CI Disperse Violet 1), although a few have been specifically designed for the application, for example the blue dye **10.30b**. A few azo dyes, including the yellow–orange compound **10.31**, are also used as semi-permanent hair dyes.

Semi-permanent dyes are usually applied to natural, unbleached hair after shampooing. The formulations use mixtures of dyes, blended to the desired shade. It is common to use several dyes of similar colour but with different molecular sizes to provide even coloration. Larger dye molecules tend to be retained by the more damaged tip of the hair, but do not penetrate so readily into the roots, while smaller molecules penetrate the entire hair fibre but wash out of the more porous tip.

Temporary hair coloration generally lasts only from one shampoo to the next. Originally, temporary hair dyes were aimed mainly at women with greying hair and were referred to as *colour rinses*. A modern use of temporary hair dyeing is to produce a colour for a single event, hence referred to as 'party' colours. They may also be used after permanent or semi-permanent hair dyeing, or after bleaching, to provide an immediate shade correction. The dyes colour less intensely than permanent or semi-permanent hair dyes, but they are simple to use and involve minimal commitment or health risk. Temporary hair colouring products are formulated exclusively with dyes that have been approved by the US FDA. A typical representative example is the triarylmethine dye FD&C Blue 1, **10.3**. Temporary hair colour formulation incorporates some apparently contradictory constraints. The colour should be removed easily by shampooing, but should be colour-fast towards rain and perspiration.

While it seems likely that the permanent oxidative process in an optimised form and the current range of semi-permanent dyes will remain dominant for hair coloration into the foreseeable future, there remains scope for research towards new technologies for hair coloration. The rapidly developing science of genetics and an emerging understanding of the molecular basis of hair pigmentation may be key elements in the development of systems encouraging natural, biotechnological or semi-synthetic hair re-pigmentation. While there has been some preliminary success in this area, delivering colouring agents to hair follicles, further development will be required before the technology is ready for the consumer. A growing understanding of the principles of the greying process, and the possibility of its inhibition or prevention, may well accelerate this development.[26] The explosive growth in nanotechnologies is certain to continue to attract the interest of hair colour chemists, for example to exploit the potential of photonics to make use of materials that manipulate light to create colour without using traditional colorants, as one way of addressing potential environmental and toxicological concerns.

REFERENCES

1. R. G. Harry and M. M. Rieger, Color cosmetics, in *Harry's Cosmeticology*, 8th edn, Chemical Publishing Co., New York, 2000, ch. 26.
2. D. F. Williams (ed.), *Chemistry and Technology of the Cosmetics and Toiletries Industry*, Springer, New York/Heidelberg, 1996.
3. S. Houlton, *Chem. Br.*, 1998, **34**(11), 33.
4. Z. D. Draelos, Photoprotection in colored cosmetics, in *Clinical Guide to Sunscreens and Photoprotection*, Informa Healthcare, London, 2008, ch. 14.
5. J. A. Graham and A. J. Jouhar, *Int. J. Cosmetic Sci.*, 1980, **2**, 77.
6. R. M. Christie and J. Mackay, *Color. Technol.*, 2008, **124**, 133.
7. C. K. Dweck, *Int. J. Cosmetic Sci.*, 2002, **24**, 287.
8. J. F. Corbett, *Rev. Prog. Color.*, 1985, **15**, 52.
9. K. C. Brown, Hair colouring, *Cosmet. Sci. Technol. Ser.*, 1997, **17**, 191.
10. J. F. Corbett, *Hair Colorants - Chemistry and Toxicology*; Micelle Press, Weymouth, UK, 1998.
11. J. F. Corbett, Synthetic dyes for human hair, in *Colorants for Non-Textile Applications*, ed. H. S. Freeman and A. T. Peters, Elsevier Science B. V., Amsterdam, 2000, p. 456.
12. J. S. Anderson, *J. Soc. Dyers Colourists*, 2000, **116**, 193.
13. R. M. Christie and O. J. X. Morel, The coloration of human hair, in *The Coloration of Wool and other Keratin Fibres*, ed. D. M. Lewis

and J. A. Rippon, John Wiley & Sons, Inc., Chichester, UK, 2013, ch. 11.

14. P. A. Riley, The nature of melanins, in *The Physiology and Pathophysiology of Skin*, ed. A. Jarret, Academic Press, London, 1974, vol. 3, p. 1102.
15. I. Castanet and J. P. Ortonne, Hair melanin and hair color, in *Formation and Structure of Human Hair*, ed. P. Jollès, H. Zahn and H. Höcker, Birkhäuser Verlag, Basel, 1997, p. 209.
16. S. Ito, *Biochim. Biophys. Acta*, 1986, **883**, 155.
17. G. Prota, *Melanins and Melanogenesis*, Academic Press, San Diego, 1992.
18. M. Sakai, S. Nagase, T. Okada, N. Satoh and K. Tsujii, *Bull. Chem. Soc. Jpn.*, 2000, **73**, 2169.
19. O. J. X. Morel, R. M. Christie, A. Greaves and K. M. Morgan, *Color. Technol.*, 2008, **124**, 301.
20. J. F. Corbett, *Dyes Pigments*, 1999, **41**, 127.
21. S. de Sanjosé, Y. Benavente, A. Nieters, L. Foretova, M. Maynadiè, P. L. Cocco, A. Staines, M. Vornanen, P. Boffetta, N. Becker, T. Alvaro and P. Brennan, *Am. J. Epidemiol.*, 2006, **164**, 47.
22. Y. Zhang, S. de Sanjosé, P. M. Bracci, L. M. Morton, R. Wang, P. Brennan, P. Hartge, P. Boffetta, N. Becker, M. Maynadiè, L. Foretova, P. Cocco, A. Staines, T. Holford, E. A. Holly, A. Nieters, Y. Benavente, L. Bernstein, S. H. Zahm and T. Zheng, *Am. J. Epidemiol.*, 2008, **167**, 1321.
23. O. J. X. Morel and R. M. Christie, *Chem. Rev.*, 2011, **111**, 2537–2561.
24. J.-I. Setsune, H. Wakemoti, T. Matsueda, T. Matsuura, H. Tajima and T. Kitao, *J. Chem. Soc., Perkin Trans. I*, 1984, 2305.
25. R. Raue and J. F. Corbett, Nitro and nitroso dyes, in *Ullmann's Encyclopedia of Industrial Chemistry*, Wiley-VCH Verlag GmbH, Weinheim, 2002.
26. J. M. Wood, H. Decker, H. Hartmann, B. Chavan, H. Rokos, J. D. Spencer, S. Hasse, M. J. Thornton, M. Shalbaf, R. Paus and K. U. Schallreuter, *FASEB J.*, 2009, **23**, 2065.

CHAPTER 11

Functional or 'High Technology' Dyes and Pigments

11.1 INTRODUCTION

The chemistry of the most important dyes and pigments used in the coloration of traditional substrates, including textiles, paints, printing inks, plastics and cosmetics, has been dealt with extensively in the preceding chapters of this book. It is likely that this range of well-established products will remain for the foreseeable future as the most important materials manufactured for the purpose of providing colour. In recent decades, new products for such conventional applications have appeared with much reduced frequency. The traditional colour manufacturing industry has been consolidating its product range and its research effort has been concentrated more on process and product improvement, and addressing a range of environmental issues (Chapter 12). However, in the same period, there have been exciting developments in organic colour chemistry as a result of the opportunities presented by the emergence of a range of new applications that place significantly different demands on dyes and pigments. Indeed, this remains broadly the most active current area of colour chemistry research and development, both academic and industrial, linked to a variety of potential applications. These colorants have commonly been termed *functional* because the applications in question often require the dyes or pigments to perform certain functions beyond the simple provision of colour.[1-5]

Colour Chemistry, 2nd edition
By Robert M Christie
© R M Christie 2015
Published by the Royal Society of Chemistry, www.rsc.org

Alternatively, they have been referred to as *high-technology* colorants, because they are designed for use in applications derived from advances in fields to which this particular term commonly refers. The term *functional π-electron systems* has also been used to describe these materials, drawing attention to the fact that it is commonly the extensive π-electron system that is the characteristic structural feature of the molecules giving rise to particular physical properties that are exploited in their application. Consequently, the fact that the molecules are coloured may become almost irrelevant. Applications of functional dyes and pigments continue to appear and to evolve at such a pace that any text dealing with them is likely to date more quickly than almost any other area of colour chemistry. Currently, they encompass a range of electronic applications including flat screen displays, solar energy conversion, lasers and optical data storage, reprographic techniques, such as electrophotography and inkjet printing, and various biomedical uses. There is also intense interest in *chromic materials*, which are chemical species that respond to various external stimuli by exhibiting reversible colour change. While the range of functional colorants is unlikely to rival the traditional dyes and pigments in terms of the quantities required, they are potentially attractive to manufacturers due to the possibility of high added value. In this chapter, an overview of the principles of some of the more important of these applications is presented, necessarily selective in approach because of the diversity of the topics, together with a discussion in each selected case of the chemistry of the colorants that may be used. For some of these applications, traditional dyes and pigments may be suitable, although often the colorants may require special purification procedures, a feature that is not a common requirement for conventional applications. For others, new colorants tailored to the needs of the particular application have been designed and synthesised.

It may be argued that certain colorants found in nature may be considered as functional dyes. In animals, they play important functions, for example as sources of attraction and in the defence mechanisms that are associated with camouflage. Plant pigments, especially chlorophyll, play a vital role in the light-harvesting process involved in photosynthesis. Naturally-occurring colorants may also serve important biochemical functions. For example, they may provide protection for cells and organisms against the harmful effects of exposure to light, and they are also important in biological energy transfer processes. In addition, evidence has emerged for the potential therapeutic benefits of some natural dyes in our diet, including the possibility that they may provide protection against cancer on the basis, at least in part, of their antioxidant activity, by a mechanism that involves the ability to quench singlet oxygen.

11.2 ELECTRONIC APPLICATIONS OF DYES AND PIGMENTS

The rapid advances in electronics technology that have been a dominant feature of the last few decades have had an immense impact on our lives. The numerous examples of these developments include the growth in the ownership, sophistication and diversity of personal computers, mobile telephones, digital cameras, the variety of electronically-controlled goods that have become commonplace in the home and in the workplace, including kitchen appliances, televisions, hi-fi and recording equipment, and the numerous electronic devices that are increasingly used in cars. The list is virtually endless and there is little evidence to suggest that the rate of development of new electronic products will slow down in years to come. It may be a cliché, but we are indeed living in a mature electronic age. Examples of applications in electronics, in which dyes and pigments play an important functional role, especially important in those areas encompassed by the term *optoelectronics*, include display technologies, lasers, solar energy conversion and optical data storage.

11.2.1   Colour in Electronic Displays

Electronic multicolour displays have become essential, indispensible components in the devices that we use in our everyday lives. They are commonplace in an extensive range of applications, including television and computer monitor screens, mobile telephones, car dashboards and aircraft cockpit instrumentation. For many years, electronic display systems capable of generating full colour images were more or less restricted to devices based on cathode ray tube (CRT) technology. However, we are currently presented with a proliferation of approaches towards development of display technologies, and new systems continue to appear frequently. In the case of television sets, the evolution from traditional monochromatic technology towards full colour displays occurred mainly in the 1960s and early 1970s. Subsequently, the dramatic advances in computing technology have increased the availability and popularity of computer-controlled devices, leading to much more widespread use of colour displays. In addition, technical advances have provided the level of processing power that is necessary to generate colour images and to provide the means to digitally control and manipulate the colours. In recent years, the rapidly-changing technologies used in electronic colour displays, not only in current commercial use but also those that are under development, have been well-documented.[6–10] In this section, an overview of selected important display technologies is presented, with particular emphasis placed on the colour science involved. Almost invariably, the multicolour images are produced using the principles of additive colour mixing (Chapter 2), since the observer is directly viewing the source of the light. Thus, this requires the use of colorants that provide the three additive primary colours–red, green and blue (RGB), commonly organised in matrices. Display systems may be categorised in various ways, for example according to the viewing mode (either projection or direct view) or in terms of whether the technology either intrinsically generates the light required for the display (emissive) or modulates light from a separate external or internal source (non-emissive). Recent advances in the commercial systems have been focussed on particular features, such as increasing the size of the display, enhancing the colour and image quality and reducing cost and, as consumers, we have seen considerable consequent benefits from the developments that have taken place. Indeed, this is such a rapidly developing field that it is likely that the discussion presented, in common with other contemporary reviews on the subject, may well quickly become dated.

Cathode ray tubes (CRTs) have traditionally underpinned conventional television. In these displays, an electron beam, generated from a cathode at one end of a glass vessel under vacuum, scans across the glass front of the vessel that is coated with inorganic phosphors.[11] When the electron beam impacts this surface, the phosphors emit visible light, a process known as *cathodoluminescence*. The inorganic phosphors that are used are semiconductors, commonly based on zinc sulfide or on lanthanide compounds. The impact of a high energy electron beam on these materials results in promotion of electrons from the valence band to the conduction band, leaving behind a positively charged 'hole'. When an electron and a hole recombine, visible light of a particular wavelength may be emitted. The long-established blue phosphor, which is based on a combination of ZnS with a small amount of silver, provides a strong blue glow with a maximum at 450 nm. The green phosphors use ZnS and copper, emitting around 530 nm. Originally, red CRT phosphors were based on (Zn,Cd)S:Ag, although yttrium oxide/sulfide containing some europium is now usually used, providing emission around 610 nm. The use of CRT displays has declined significantly, especially as flat-screen displays have become the norm and as newer technologies have emerged. Their main drawbacks are their bulk and the difficulties that are associated with scaling up to the large formats that are increasingly in demand.

The most important multicolour display technology in current use involves liquid crystal displays (LCDs), which overtook CRTs in popularity around the mid-2000s. Liquid crystals (LCs), commonly referred to as the fourth state of matter, are materials that are intermediate in character between the crystalline solid and liquid states.[12,13] Unlike normal isotropic liquids in which the molecules essentially adopt a randomised orientation, liquid crystals show some time-averaged positional orientation of the molecules. In this sense, they resemble solid crystalline materials, although they retain most of the properties of liquids, notably the ability to flow. They are formed most commonly from molecules with rod-like geometry, which are referred to as *calamitic*. These molecules may orientate in various ways to form different types of LC phases (*mesophases*). There are three main mesophase types: *smectic, nematic* and *chiral nematic*. In the smectic mesophase, as illustrated in Figure 11.1(a), the molecules are arranged in 'raft-like' layers with their molecular axes parallel. These layers can pass over each other as the material flows. In the

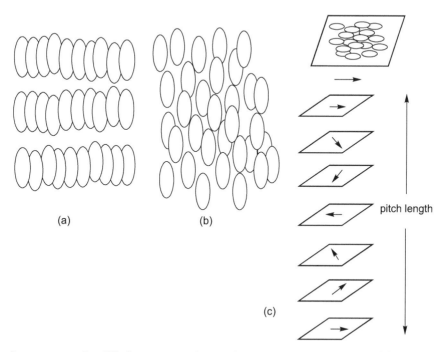

(a)   (b)   (c)

Figure 11.1 Simplified representations of molecular alignment in (a) smectic, (b) nematic and (c) chiral nematic liquid crystals.

nematic mesophase (Figure 11.1b), the molecules are also aligned parallel, but there is no separation into layers. The main feature that distinguishes a chiral nematic (alternatively referred to as *cholesteric*) from a nematic phase is, as the name implies, that the molecular structure is chiral, *i.e.*, not superimposable on its mirror image. The chirality causes the molecules in the phase to adopt a twisted structure, resulting in a screw-like, helical arrangement. The structure of the chiral nematic phase, illustrated in a somewhat idealised way in Figure 11.1(c), may be envisaged as composed of nematic LC layers. The structure of this phase is discussed further in Section 11.5.2, in the context of its ability to exhibit thermochromism. In recent decades, liquid crystals have played an increasingly important part in our lives, their most familiar application being in displays.[14]

LCDs have superseded those based on CRT technology in view of various advantages that they offer, including visual appeal, low power consumption and their suitability for manufacture on a wide range of scales, from the displays on miniaturised electronic devices through to large screen displays. LCDs are non-emissive. They operate by utilising the ability to control ambient light in order to provide contrast, *i.e.*, areas of dark and light, within the display. This is achieved as a result of a change in orientation of the liquid crystal molecules within certain sections of the display as an electric field is applied. In most LCDs, the liquid crystal is held in a cell between two closely spaced glass plates. Transparent electrodes, constructed using indium tin oxide (ITO) coated on the inside of the glass and in contact with the LC material, are used to apply a voltage that causes the LC molecules to switch direction, thus providing control of the molecular orientation. The presence of polarising filters on each side of the cell gives rise to light and dark areas within the display. The light source may involve either ambient light with a reflective mirror or alternatively fluorescent backlighting behind the cell. The most important LCD technologies are based in twisted nematic (TN) LCs and rely on the ability of the molecules to twist the plane of polarised light through 90°. The simplest LCDs, such as those used in calculators or digital watches operate on the basis of direct driving of segmented displays. Large area multicolour displays utilise the simultaneous addressing of arrays of pixels, for example using a matrix arrangement of thin-film transistors.

To produce the multicolour effect, microcolour filters are incorporated into the display panel, with a construction based on the RGB primary colours, whereby one-third of the pixels are red, one-third green and one-third blue.[15] Pigments, especially from the high performance organic category (Chapter 9), are generally preferred to dyes for these applications because of their superior durability. The pigments for colour filter applications require good thermal stability and lightfastness, and appropriate spectral characteristics. Suitable red pigments, include the anthraquinone **11.1** (CI Pigment Red 177), perylene **11.2** (CI Pigment Red 179) and quinacridone **11.3** (CI Pigment Red 122) (Figure 11.2). The blue pigment used is the ubiquitous copper phthalocyanine, **11.4** (CI Pigment Blue 14), while green is provided by copper phthalocyanine derivatives, notably the poly-halogenated derivative, CI Pigment Green 36. The microfilters may be manufactured by a variety of processes, although commonly involving deposition of pigments of the three colours from a fine dispersion in a process solvent or a photopolymerisable resin to form a matrix in an appropriate design.

Through the 1970s and 1980s, there was a considerable research effort aimed at the development of LCDs based on so-called *guest/host* (GH) systems.[16] These displays involved the use of *dichroic* dyes dissolved in the LC host, commonly of the chiral nematic type. In the operation of these devices, the dye molecules align with the LC molecules and change direction with the application of an electric field, an effect that causes a change in colour intensity, as illustrated simplistically in Figure 11.3. It was found that sufficient contrast could be provided by the use of certain specifically-designed dyes, a

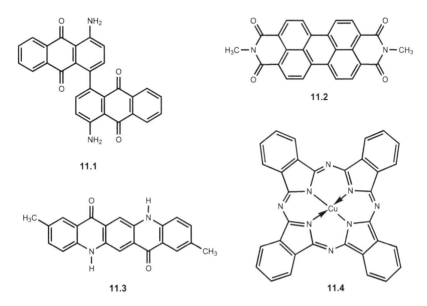

Figure 11.2 Organic pigments used in colour filters for flat-screen displays.

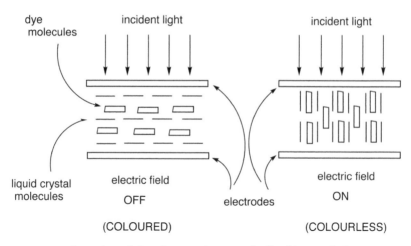

Figure 11.3 Orientation of dyes in on/off states of a liquid crystal display.

feature which meant that the use of a polariser to enhance contrast became unnecessary. The use of these systems is currently restricted to monochromatic displays, although the development of three layer multicolour displays based on yellow, magenta and cyan dyes has been proposed,[17] as also has their potential for use in erasable electronic paper. The chemical principles underlying the dyes used is presented here as an interesting classical example in the development of functional dyes during the late twentieth century.

To explain the change in colour intensity with the change in orientation of the dye molecules with respect to the direction of incident light, as indicated in Figure 11.3, it is important to recognise that electronic excitation caused by light absorption is accompanied by a change in the polarity of the dye molecule. The electronic transition that occurs involves charge transfer and may be ascribed a *transition dipole moment*, which has not only magnitude but also direction. The transition moment in aminoazobenzene dye **11.5**, for example, is directed from the electron-releasing amino group through to the electron-withdrawing nitro group, as illustrated in Figure 11.4 (see Chapter 2 for a justification of this principle based on the application of the valence-bond approach to colour and constitution). It is of particular note that, in the case of dye **11.5**, the transition moment is more or less aligned with the long axis of the molecule. For light absorption to occur, the electric vector of the incident light, which is perpendicular to the direction of propagation of light as illustrated in Figure 11.4, must oscillate in the same direction as the transition moment of the dye. If the transition moment of the dye is

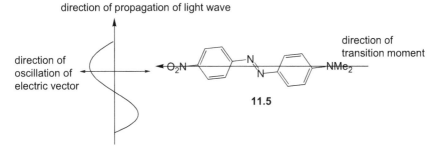

Figure 11.4 Orientation of an aminoazobenzene dye, **11.5**, for maximum light absorption in a liquid crystal display.

perpendicular to the direction of the light source, *i.e.*, parallel to the electric vector, then maximum colour intensity will be achieved. This situation exists when the electric field is in the 'off' position as illustrated in Figure 11.3. If the molecule then rotates through 90°, for example when the electric field is switched on as illustrated in Figure 11.3, then a change to minimum colour intensity will be observed.

The rod-like organic molecules that give rise to LC behaviour are commonly relatively hydrophobic aromatic systems, although with a degree of polarity that is located mainly in the end-substituents. For application in LCDs, the dyes require good solubility in and compatibility with the organic LC host, and thus the dye molecules are invariably non-ionic, often with substituents that enhance hydrophobic characteristics and thus solubility in the host. The dyes are also required to be of high purity and to show good lightfastness. Most important of all, from the point of view of providing high contrast, is the ability of the dye molecule to align, and switch, with the liquid crystal host. This feature may be quantified in terms of the *order parameter*, S, a quantity that may be measured spectroscopically ($= 0$ for non-alignment and $= 1$ for perfect alignment). The target value for a good LCD dye is $S \cong 0.8$. Simple LCDs of this type are required to produce black/white contrast. Thus, to ensure that absorption occurs throughout the visible spectrum, a combination of yellow, red and blue dyes is required. Azo dyes, such as compound **11.5**, are capable of providing only reasonable order parameters and also generally suffer from inadequate light stability. The dyes that most effectively satisfy the requirements for guest–host LCDs belong to the carbonyl chemical class. Anthraquinone dyes, such as the yellow dye **11.6**, the red **11.7** and the blue **11.8** (illustrated in Figure 11.5), are especially suitable for these applications. Intuitively, a rod-like molecular shape might be expected to provide the required

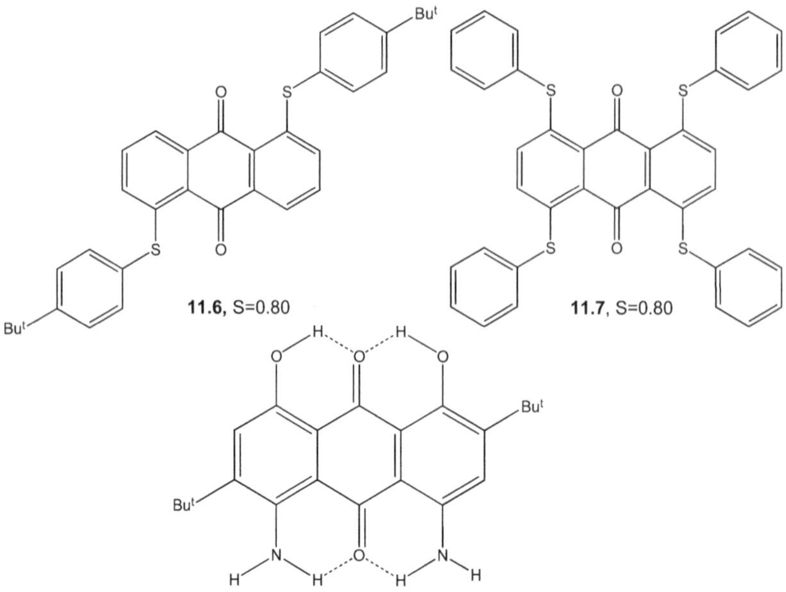

11.6, S=0.80   **11.7**, S=0.80

11.8, S=0.73

Figure 11.5   Anthraquinone dyes used in liquid crystal displays (*S* is the order parameter).

orientational behaviour for LCD dyes. This feature is apparent in the structure of blue dye **11.8** while, additionally, the extensive intra-molecular hydrogen-bonding in this dye promotes good lightfastness. However, this particular molecular shape is not so apparent in the sulfur-containing anthraquinone dyes **11.6** and **11.7**, yet both provide suitably high order parameters. These examples demonstrate that the nature of the interactions between the dye and LC molecules in these displays are probably more complex than those that are presented in this simplistic approach.

In recent decades, there has been growing interest in the electrical properties of organic materials. In the context of colour chemistry, special interest has developed in materials that are capable of par-ticipating in the conversion of electrical energy into emitted light, a phenomenon referred to as *electroluminescence*.[18] A particular con-sequence of the intense research activity that has taken place in this area is the development of organic light-emitting diodes (OLEDs), devices that offer immense potential as sources of illumination (either of white or coloured light) and as emissive flat screen displays.[19–21] In contrast to LCDs, displays based on OLED technology offer the advantage that they are self-luminous. The displays may

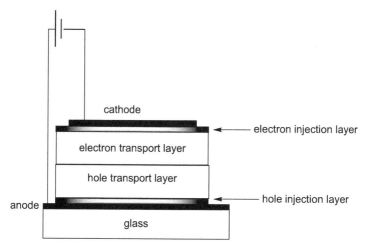

Figure 11.6   Construction of a simple organic light-emitting diode (OLED).

consequently be thinner and lighter in construction than LCDs, and they can achieve higher visual contrast. There are two principal types of OLED, based either on small organic molecules or on polymers. Frequently, the term OLED is used to describe the former type, although the more precise acronym SMOLED is often used, while the latter type may also be referred to as a polymer light-emitting diode (PLED). OLEDs are multilayer electronic devices that can vary in the complexity of construction. A relatively simple construction is illustrated in Figure 11.6, although in practice additional layers are generally used to provide improved device efficiency. The device illustrated consists essentially of a sandwich constructed on a glass substrate, coated with conductive indium tin oxide (ITO), which forms a transparent anode. Thin layers containing the organic components that are required for its operation are coated separately and successively on to this substrate to provide both conductive hole-transporting and electron-transporting layers. A calcium/aluminium cathode layer completes the construction. The organic molecules used in the devices are essentially semiconducting materials that contain extensive delocalised π-electron systems. In the operation of the device, a voltage is applied across the OLED such that the anode becomes positively-charged with respect to the cathode. Electric current flows through the device as electrons (negatively-charged) are injected into the lowest unoccupied molecular orbitals (LUMOs) in the organic layer at the cathode, and are withdrawn from highest occupied molecular orbitals (HOMOs) in the organic layer at the anode. The latter process may alternatively be described as injection of positively-charged 'holes'

at the anode. Electrons and holes come together as a result of electrostatic attraction, and they combine to form an *exciton*. The decay of
this excited state is accompanied by emission of visible light, which is
believed to occur closer to the electron-transporting layer since holes
are generally more mobile than electrons.

The hole-transporting layer contains highly electron-rich aromatic
compounds, such as the triphenylamine derivative, *m*-MDTA, **11.9**,
whose role is to provide the required charge transport properties
(Figure 11.7). Fluorescent dyes with high quantum yields are used to

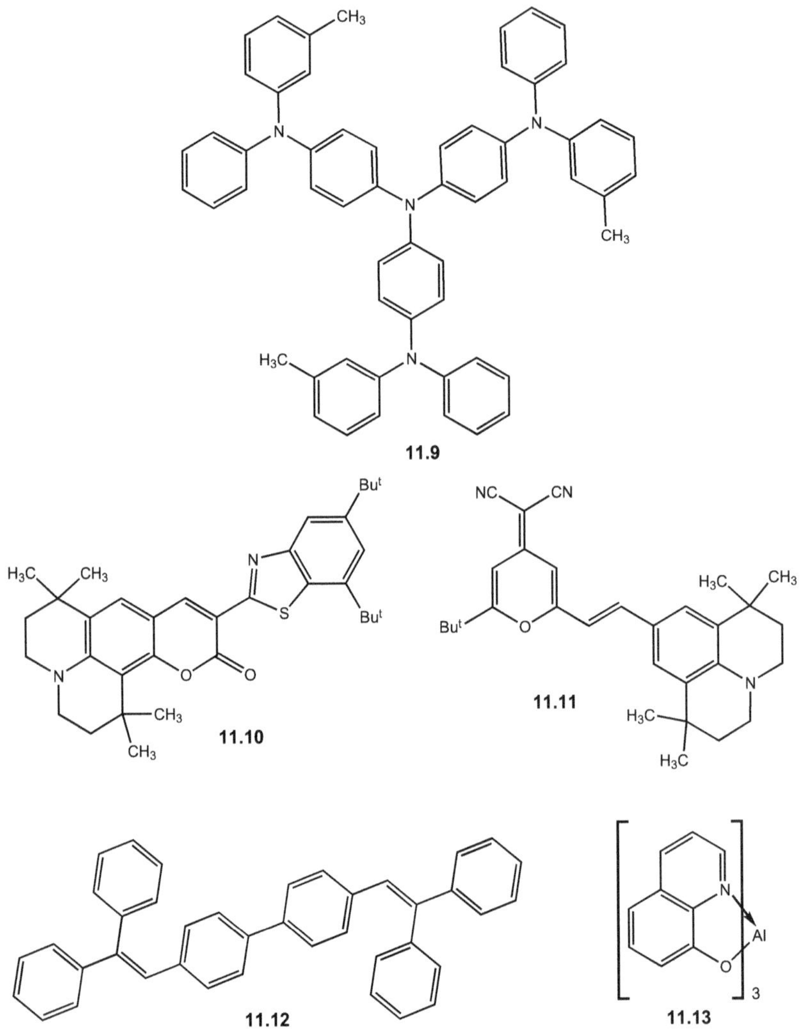

Figure 11.7   Some organic materials used in OLED displays.

promote light emission at different wavelengths, including coumarins such as dye **11.10** (green) and the pyran derivative **11.11**, DCJTB (red), while blue emitters are generally polycyclic hydrocarbons such as the distyrylarylene **11.12**, DPVBI.[22] The aluminium complex of 8-hydroxyquinoline, **11.13**, is a useful multi-purpose component in OLED devices, as a green light emitter, an electron transport material and a host for red-emitting dyes.

The second approach to the production of OLEDs emerged from the discovery that poly(*p*-phenylenevinylene), **11.14** (PPV), was capable of producing a greenish glow when a thin film of the polymer was subjected to a high voltage.[23] Since this discovery, a large number of light-emitting polymers have been prepared and investigated. Mostly these are based on polymers of the PPV type, containing a variety of substituents, either as homopolymers or copolymers. However, light-emitting materials based on other highly-conjugated polymers, such as polyfluorenes and polythiophenes, have also been investigated in the search for materials with improved efficiency of emission over a range of wavelengths, and with enhanced stability and ease of processing.[24] The principles of the electroluminescence from the polymers are based on semiconductor properties that are essentially analogous to those involved in the small molecule systems, with the π-orbitals arising from their extensively conjugated structure forming the valence band and the π*-orbitals the conduction band. These materials offer significant potential for future developments in flat screen display technology, not least because of the potential for the relative simplicity and ease of device construction.

11.14

Flat-screen plasma displays represent another commonly-encountered commercial technology. These devices are constructed from panels consisting of millions of micro-cells sandwiched between two glass panels. Pulses of light from each pixel element are created by a plasma discharge, obtained by applying a voltage to a noble gas environment, generally a mixture of xenon and neon, contained in each cell in the display matrix. The plasma emits high energy UV radiation (140–190 nm) that induces emission of visible light from a matrix of RGB-emitting inorganic phosphors.

The discussion presented in this section has centred on the principles of the most important colour display technologies, and is by no means exhaustive. There are numerous other technologies, currently either available or under development. Examples include electrochromic, field emission and thin-film electroluminescent displays, the commercial viability of which will be determined in the years to come.

11.2.2 Laser Dyes

Lasers are devices that provide an intense beam of highly focussed radiation.[25] The term *laser* is an acronym, referring to *l*ight *a*mplification by *s*timulated *e*mission of *r*adiation. A laser beam is generated in a part of the device referred to as the *gain medium*, which may be a solid, liquid or gas. Historically, laser technology has utilised various inorganic materials to produce the required emission. Several different types of inorganic laser were developed to emit either in the ultraviolet, visible or infrared region of the electromagnetic spectrum. These conventional inorganic lasers are commonly low cost, robust devices. However, they have the disadvantage that they emit at only a few selected wavelengths, and in very narrow bands. An example of a commonly-encountered inorganic laser is the gallium-arsenic diode laser, which emits in the near-infrared region at around 780 nm. Dye lasers, in contrast, emit over a broad band of wavelengths, thus offering the advantage of tunability through a wide wavelength range, and on this basis they have become well-established in laser technology.[26–28] In addition, by simply changing the dye, emission may be achieved in different spectral regions. Conventional dye lasers are based on specific fluorescent dyes dissolved in an appropriate solvent, and this solution is used as the liquid gain medium. Dye lasers have been used in a wide range of applications including communications technology, microsurgery, spectroscopy, photochemistry, studies of reaction kinetics, isotope separation in nuclear fuel enrichment and microanalysis.

The principles of the function of the fluorescent dye in a dye laser may be explained with reference to the mechanism of fluorescence as described in Chapter 2 (see Figure 2.6). The gain region of the laser contains the dye solution and is surrounded by a system of mirrors. This part of the device, referred to as the resonant cavity, provides the amplification and controls the beam direction. As a result of absorption of a quantum of light, the dye molecule is promoted from its ground state, S_0, to its first excited state, S_1^*. Lasing occurs when incident radiation interacts with the dye molecule in its excited state, thereby causing (stimulating) the molecule to decay to the ground state by emission of radiation (fluorescence). For lasing to be achieved, a situation has to be brought about in which the dye molecules exist predominantly in the excited state. This so-called *population inversion* is obtained by pumping the system with an appropriate source of energy, either continuously or as pulses, using a powerful inorganic laser. In contrast to spontaneous fluorescence emission, the stimulated emission of radiation from dye lasers is strictly coherent (same phase and polarisation) and is of high intensity.

There are a number of general requirements for laser dyes.[29] Strong absorption at the excitation wavelength is clearly required but there should be minimal absorption at the lasing wavelength, so that there is as little overlap as possible between absorption and emission spectra. Additional important requirements are a high quantum yield (0.5–1.0), a short fluorescence lifetime (5–10 ns), low absorption in the first excited state at the pumping and lasing wavelengths, low probability of intersystem crossing to the triplet state and good photochemical stability. Laser dyes are required to exhibit high purity, since traces of impurities can seriously reduce fluorescence intensity as a result of quenching, thereby causing a significant reduction in lasing efficiency. By appropriate dye selection it is possible to produce coherent light of any wavelength from 320 to 1200 nm. Examples of laser dyes are given in Figure 11.8. For shorter wavelengths (up to *ca.* 470 nm), aromatic hydrocarbons, such as anthracenes and polyphenyls, and fluorescent brightening agent-type materials such as stilbenes, oxazoles, coumarins and carbostyrils are most commonly used. Coumarins, such as the benzothiazolyl derivative **11.15a** (Coumarin 6) and the benzimidazolyl derivative **11.15b** (Coumarin 7), are suitable for use in the 470–550 nm range, while xanthene derivatives, such as Rhodamine 6G, **11.16**, are of prime importance in the 510–700 nm region of the visible spectrum. Rhodamine 700, **11.17**, exhibits highly efficient laser action in the range 700–800 nm, providing an excellent example of molecular design for this particular application. The 'double butterfly' structure causes the molecule to become rigid and thus increases the efficiency of fluorescent emission, while the trifluoromethyl ($-CF_3$) group enhances photostability. For dye lasers operating at longer wavelengths into the near-infrared region, oxazines and polymethine dyes are most suitable. For example, polymethine dye **11.18** provides a lasing maximum at 950 nm. More recently, difluoroboron complexes of dipyrromethenes (pyrromethenes or BODIPY complexes), exemplified by compound **11.19** (PM 567), have developed special importance for laser applications in view of their high photostability and their potential to provide enhanced gain.

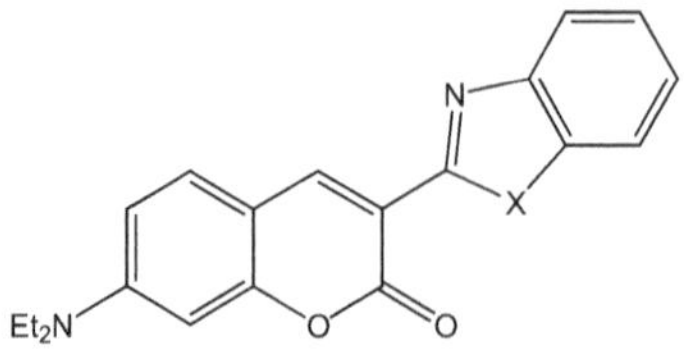

11.15
a: X = S; **b**: X= NH

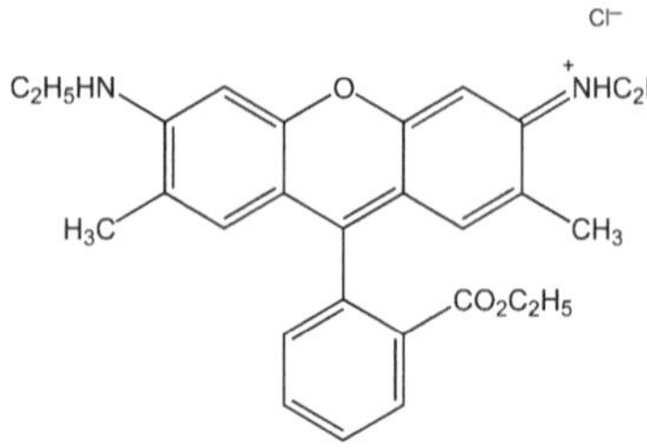

11.16

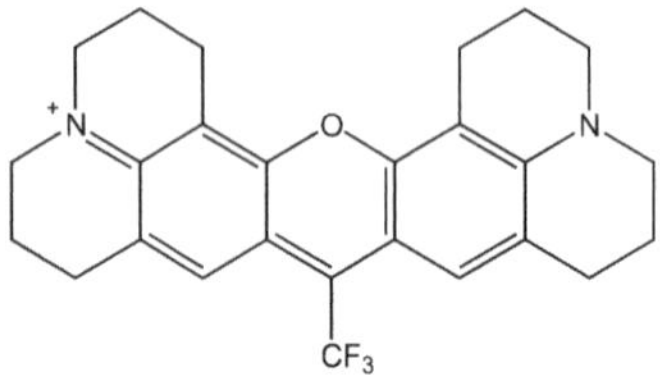

11.17

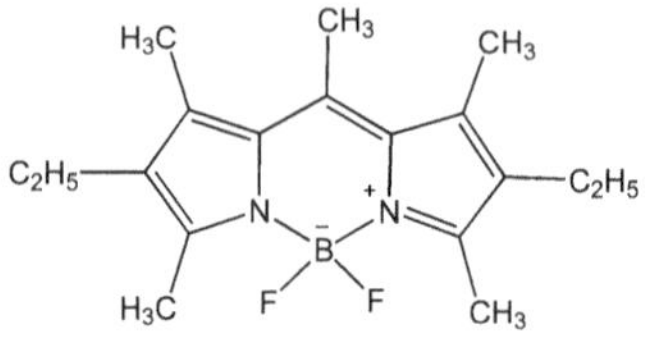

11.19

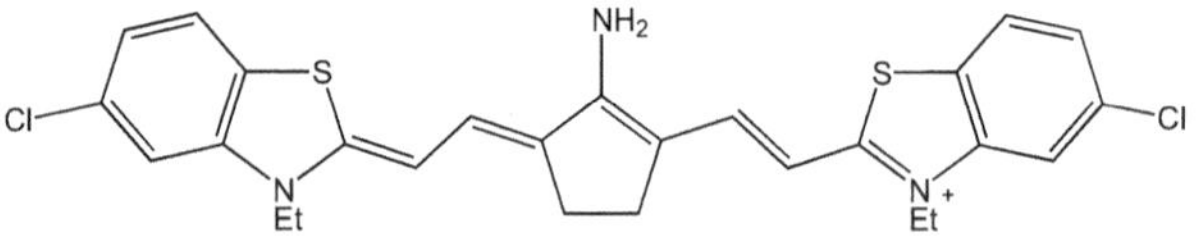

11.18

Figure 11.8 Examples of laser dyes.

There are certain inherent technical disadvantages of liquid-phase lasers, particularly in terms of the need to circulate the dye solution in order to ensure that dye molecules in the triplet state and also products of degradation of the dyes are removed. Recent important advances in solid-state and semiconducting lasers based on inorganic materials have meant that devices based on these technologies have gradually displaced liquid-based dye lasers from their previously pre-eminent position. However, there have been important recent advances aimed at the development of tunable solid-state dye lasers. The research into this laser technology has been mainly associated with the development of organic polymeric materials that act as hosts for the fluorescent dyes, with the aim to improve the operational performance of the gain medium and thus enhance the lifetime of practical devices.[30]

11.2.3 Dyes in Solar Energy Conversion

One of the main technical challenges facing mankind is to ensure the security of our future energy supply, especially at a time when global demand is increasing. Currently, it may be argued that we have become over-dependent on energy generation from the combustion of fossil fuels, which not only contributes towards air pollution and potential climate change based on carbon dioxide emission, but is also based on shrinking, finite and non-renewable resources. At the same time, there is significant opposition in some quarters to power generation using nuclear fission based on concerns over its safety, which has been fuelled by certain dramatic events, such as those that occurred at Chernobyl and Fukushima. Therefore, the search for economically-viable alternatives based on renewable energy sources, including wind, tidal, wave and solar power, has assumed a high priority, politically and technologically, and is arguably essential for a globally sustainable society.[31] There has thus been intensifying effort over many years aimed at the development of the means to convert solar energy into electrical energy, a process that utilises a powerful, plentiful, non-diminishing energy source and suffers little from global environmental problems. In a sense, such systems mimic the natural harvesting of solar energy by plants and bacteria as part of the process of photosynthesis that is responsible for their growth.

The most important photovoltaic solar cell technology is currently based on inorganic materials, notably silicon, which can be in single-crystal, polycrystalline or amorphous form, making use of its semiconducting properties.[32–34] Other inorganic semiconducting materials, for example GaAs, InP and GaSb, may be used to produce highly efficient solar cells, although they are expensive. Conventional cells contain a sandwich of n-type and p-type doped silicon layers. Interaction with light produces a free electron and a positive hole. At the junction between the layers, the electron moves towards the n-type doped silicon layer and the hole towards the p-type, providing current flow through an external circuit. The efficiency of inorganic photovoltaic cells may reach up to 24% in the most expensive types, although it is more commonly around 15%, and significantly less in cheaper commercial cells.

Dye-sensitised solar cells (DSCs), which utilise specific dyes for the purpose of solar energy conversion, have been under intense investigation, as alternatives to traditional silicon photocell technology.[35–37] Indeed, this technology currently represents one of the most active areas of research in functional dye chemistry and technology. One type of construction for such a photocell is shown in Figure 11.9. The electrodes for the photocell are constructed from coated conductive glass. At one electrode, which serves as the current collector, there is a semiconducting layer of an inorganic substrate, usually nanosized, mesoporous titanium dioxide in its anatase polymorphic form, on to which the dye is adsorbed. Another important component

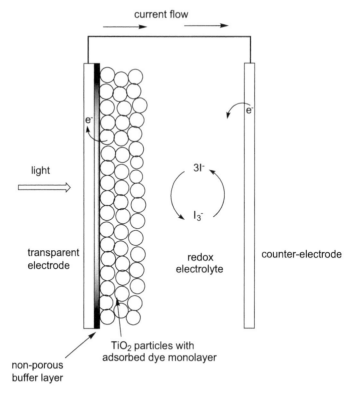

Figure 11.9   Design of a dye-sensitised photocell.

of the photocell is an electrolyte, containing a redox couple dissolved in an appropriate organic solvent in the intra-electrode space. The electrolyte is generally an iodine/tetraalkylammonium iodide mixture.

The mechanism that has been proposed for the operation of this type of photocell is illustrated in Figure 11.10, in which the dye acts as a sensitiser for the photoelectrochemical process. Light is absorbed by a monolayer of the dye (D), which is anchored on the semi-conducting TiO_2 surface. Consequently, the dye is raised to its first excited state (D*). At the electrode, the excited state D* releases an electron into the conduction band of the TiO_2, at the same time forming an oxidised species, D^+. At the counter-electrode, an electron is transferred to the electrolyte and, mediated by the I_2/I^- ($I_3^-/3I^-$) redox equilibrium, the original state of the dye (D) is regenerated by reduction. The electrical circuit is completed by electron flow through the external load. For efficient cell operation, the rate of electron injection from the dye excited state (D*) must be faster than its decay back to the ground state. In addition, the rate of reduction of oxidised

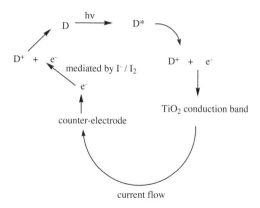

Figure 11.10 Proposed mechanism for the operation of a dye-sensitised photocell.

sensitiser (D^+) by the electron donor within the electrolyte must be higher than its reverse reaction with injected electrons, and also higher than the rate of reaction of the injected electrons with the electron acceptor within the electrolyte.

An obvious feature required by a dye for application in DSCs is the ability to absorb solar energy strongly. The dye should be firmly attached to the surface of the semiconductor and inject electrons into its conduction band with a high quantum yield. Its redox potential should be sufficiently high that it can be regenerated rapidly from the oxidised species. Finally, extremely high stability is of vital importance in this particular application because of the need to withstand exposure to light and also countless redox conversions over a period of many years. Of the numerous types of dye that have been investigated, the most successful have been based on octahedral ruthenium(II) complexes of 2,2'-bipyridyl derivatives. Of particular current importance is dye **11.20** (N719), which has been reported to be capable of providing a conversion efficiency in DSCs greater than 11%. The role of the carboxylate groups in dyes of this type is to enhance the attachment of the dye to the TiO_2 surface. In addition, a wide range of metal-free organic dyes has been investigated in view of their higher extinction coefficients and potentially lower cost, particularly because the noble metal ruthenium is a limited resource and is expensive. An example of such a dye specifically-designed for DSC applications is dye **11.21**. Dye molecules of this type are described as having a donor–conjugated bridge–acceptor architecture. They may be considered as composed of three parts: (a) an electron-donating group (in this case the tertiary amino group) to enhance electron injection into the conduction band of the semiconductor; (b) a conjugated bridge, a polyene sometimes containing a heterocyclic ring system, to tune the light absorption characteristics; and (c) terminal cyanoacrylic acid functionality acting as the electron accepting group and to provide fixation of the dye on the surface of the semiconductor film electrode. DSC technology now appears to offer considerable industrial potential as a means of solar energy conversion, especially as technological developments are providing efficiencies that are beginning to approach those given by traditional silicon photocells.

11.20

11.21

Another way in which dyes may be useful in solar energy conversion is in fluorescent solar collectors (FSCs), alternatively referred to as spectral converters or solar concentrators. These devices, first introduced in the 1970s,[38,39] consist of thin sheets of clear plastic, generally poly(methyl methacrylate), doped with fluorescent dyes. They utilise the ability of fluorescent dyes to absorb certain wavelengths of incident sunlight and to re-emit the energy at higher wavelengths. The emitted radiation is trapped by internal reflection at faces and edges and is channelled towards the edge of the sheet where the photovoltaic solar cell is located, as illustrated in Figure 11.11. Multiple layers of polymer containing dyes with different absorption and emission characteristics may be used to maximise the solar energy collection throughout the spectrum. The requirements of dyes for this application are broadly similar to those for laser dyes. These include a high fluorescence quantum yield, compatibility with the plastic material, high lightfastness in the polymer matrix, little overlap of the absorption and emission spectra (generally associated with a large Stokes shift) and high purity. Dyes from the coumarin and perylene classes (see Chapter 4 for a discussion of the structure and properties of carbonyl fluorescent dyes) have most commonly been employed for these applications. The attraction of FSCs is their potential to assist in addressing the high cost of solar energy conversion, as compared with energy generation from other resources, which continues to limits its exploitation on a large scale. FSCs combine the strong light absorption characteristics of dyes with the conversion efficiency of traditional solar cells, and significantly reduce the amount of semiconductor material required. However, their development has been restricted by the limited photostability of the dyes currently available and by low concentration efficiency within the devices.

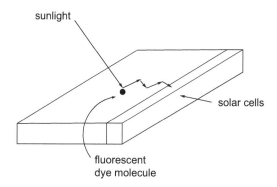

Figure 11.11   Design of a fluorescent dye-based solar collector.

Nevertheless, interest in these systems remains high for the future if these limiting issues can be addressed, for example by advances in materials science, nano-optics and device design.[40,41]

11.2.4 Dyes in Optical Data Storage

Optical data storage refers to systems that store data digitally, usually on a small circular disc, and where the system is addressed using a highly-focussed beam of laser light.[42,43] The data are stored 'bit by bit' as depressions or 'pits' and regions without depressions or 'lands'. The first commercially successful optical data storage system was the audio compact disc (CD), introduced in the early 1980s, which offered certain advantages over traditional vinyl records, including robustness, ease of handling together with good sound quality.[44] The introduction of the standardised pre-recorded CD-audio, and its rapid popular acceptance, provided the stimulus for the further technological developments of optical storage systems that have taken place at a remarkable pace. Within a few years, CD-ROM (read only memory) was introduced for the digital storage of computer application software and data. Discs in the CD-audio and CD-ROM format are constructed from polycarbonate, already containing the data as pits and lands, and coated with a reflecting layer of aluminium. They are addressed by a laser with a wavelength of 780 nm, and provide a memory capacity of around 700 MB. The demand for higher storage capacity led to the subsequent introduction of the DVD (digital versatile disc), which uses a laser wavelength of 650 nm, thus enabling a smaller pit structure and consequently a higher capacity (17 GB) in its standardised format. The successor to the DVD is the BluRay disc, which uses an even shorter laser wavelength (405 nm) and provides significantly higher storage capacity. Soon after the introduction of pre-recorded CDs, demand developed for storage systems that allowed information to be recorded by the individual. The first product developed was referred to as the WORM (*w*rite *o*nce *r*ead *m*any) system. This system was based on a dye layer with a strong absorption at the laser wavelength (780 nm). The absorbed laser radiation is converted into heat which destroys the dye. The differences in reflection that result can be read by a laser of lower intensity. However, the system was not commercially successful, mainly because it required special recording and playback apparatus, which was expensive. In 1989, the recordable CD (CD-R) was introduced, which utilises reflection at a metal surface, in a similar way to CD, while also incorporating a weakly absorbing dye layer. During recording, the heat that results when the dye layer absorbs the laser radiation destroys the dye molecules, melts the polymer and deforms the metallic layer, thus forming a pit. The commercial success of this system was mainly due to its compatibility with standard CD-audio and CD-ROM drives. Subsequently, analogous technologies leading to recordable DVD-R and BluRay-R have been devised. Rewritable systems (CD-RW, DVD/RW) were also developed, in these cases using materials that undergo crystalline-amorphous phase change induced by local heating for the purpose of information storage.[44]

From the early 1980s, intense research activity developed into compounds that absorb in the near-infrared (NIR) region of the electromagnetic spectrum (*i.e.*, beyond 750 nm), a spectral region that had previously been regarded mainly as a scientific curiosity rather than offering practical application significance.[45,46] Interest in this region emerged particularly as a result of the development of inexpensive and robust inorganic semiconductor lasers and the potential that these devices offered for the development of optoelectronic devices, most notably for optical data storage applications. An important example was the gallium-arsenide diode laser emitting in the 780–830 nm range, as used for example in CD-R systems. Such compounds are commonly referred to as near-infrared absorbing dyes, even though, in the strict sense, they are not dyes since their absorption is outside the visible region. Nevertheless, these compounds were commonly designed and synthesised by an extension of conventional dye chemistry so that it is convenient to consider them in this context. In addition, the organic materials absorb NIR radiation as a result of electronic transitions (generally $\pi \rightarrow \pi^*$), in the same way as materials that absorb in the UV and visible region. Important examples of NIR absorbing dyes include a range of phthalocyanine derivatives (Chapter 5) and cyanine dyes together with structurally-related systems such as croconium dyes (Chapter 6). However, dyes with NIR absorption characteristics have been synthesised based on most chromophoric systems, including azo and carbonyl dyes (Chapters 3 and 4, respectively) by the introduction of appropriate substitution patterns that produce the required bathochromicity. A range of other applications has emerged for NIR absorbing dyes, including security printing, for example in banknotes and 'invisible' but laser-readable bar-coding for the identification of branded products, infrared photography and laser filters. Appropriately designed compounds also have potential for use in photodynamic therapy, a cancer treatment that is described in Section 11.4.2.

Dyes for optical recording applications require absorption characteristics that are appropriate for the system in question, including
high molar extinction coefficients, decomposition temperatures
above the glass transition temperature of the polycarbonate used to
construct the disc, good solubility in the solvent used to spin-coat the
discs and stability to environmental and processing conditions. Selected examples are given in Figure 11.12. The CD-R system uses dyes
with a λ_{max} value in the range 690–710 nm and with a small but finite
absorbance at the laser wavelength (780 nm). Examples of appropriate
dyes include the cyanine **11.22a** and phthalocyanine **11.23**. DVD-R
uses dyes absorbing at the lower laser wavelength (650 nm), for example the cyanine **11.22b** and the nickel azo dye complex **11.24**. An
example of a dye designed for BluRay-R applications (405 nm) is the
methine dye **11.25**.

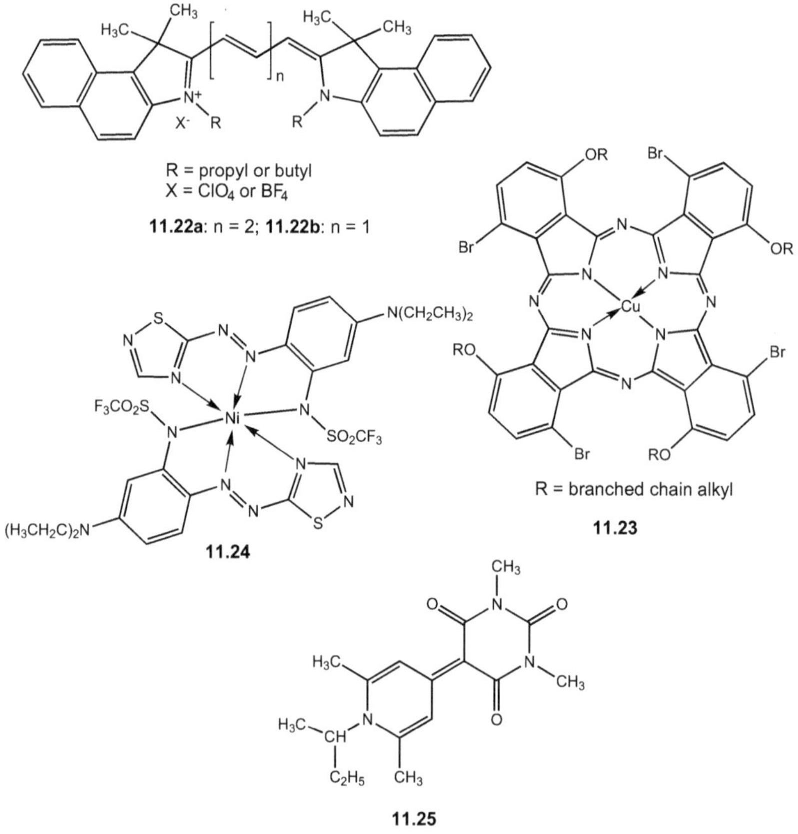

Figure 11.12   Examples of dyes for optical data storage.

11.3 REPROGRAPHICS APPLICATIONS OF DYES AND PIGMENTS

The term *reprography* was first used around the 1960s to encompass the new imaging techniques for document reproduction, including xerography, electrofax and thermography, which were emerging at the time. The term is still used to refer to those newer imaging processes that may be distinguished from conventional printing techniques, such as lithography, flexography, screen and gravure printing, and traditional film photography. These processes have experienced remarkable growth mainly because of the rapid advances in computing technology and the consequent growth in demand for small office and home (SOHO) printing of high-quality, using relatively inexpensive printers. The most important of these techniques currently are electrophotography and inkjet printing, both of which offer the advantage of printing on to plain paper.

11.3.1 Electrophotography

The term *electrophotography* encompasses the familiar techniques of photocopying (also commonly called xerography) and laser printing. In both of these printing systems, the ink is a toner, which is generally a powder consisting mainly of pigment, charge control agent, and a low melting binder. Toner printing systems use optical or electrical methods to form an electrostatic latent image to which the toner is attracted and subsequently transferred to the substrate. Most photocopiers and laser printers operate on a similar basis, so that it is convenient to discuss them together, but there are also some key differences.

The photocopying process, as illustrated in Figure 11.13, may essentially be separated into six stages.[47] One of the key aspects of photocopying is a photoconducting material, which, as the name implies, is a conductor of electricity in the presence of light but an insulator in the dark. In the first stage of the process, the dark photoconducting surface of a drum is given a uniform electrostatic charge. In the imaging step, the document to be copied is illuminated with white light, commonly by sweeping a linear array of photodiodes across the paper. In colour photocopying, there are separate rows of photodiodes, each row covered with either a red, green or blue filter. Where there is no image, light is reflected on to the photoconducting surface and this causes the charge to be dissipated. Where there is an image, light is absorbed and does not reach the photoconductor. In this way, a latent electrostatic image is formed on the drum. The drum is then exposed to toner particles that have been given an opposite charge to that on the drum. The toner particles are thus

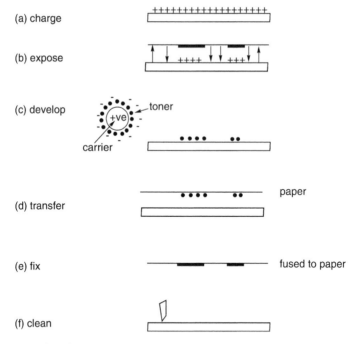

(a) charge

(b) expose

(c) develop

(d) transfer

(e) fix

(f) clean

Figure 11.13   The photocopying process.

attracted to the image areas of the photoconducting surface. The next step involves transfer of the image from the drum to the substrate, usually plain paper, by applying to the back of the paper an electrical potential of opposite charge to that of the toner particles. The next step fixes the image to the paper. This is achieved by a heat treatment, which melts the toner resin and fuses it to the paper. Finally, the photoconducting surface is cleaned to make it ready for the next copying cycle.

In a laser printer, the first step also involves giving the photoconducting drum a uniform electrostatic charge. However, in the second step a laser is used to write on the charged drum the information stored in the memory of the printer. In contrast to the photocopying process, therefore, the charge is dissipated in the image areas. In this case, the toner carries a charge that is the same as that on the drum, so that it is repelled from the non-image areas on to the uncharged image areas. From this stage onwards, the principles of the laser printing process are similar to those of the photocopying process.

Organic colorants of various types are key components in the operation of photocopiers and laser printers. The photoconducting drum contains two surface layers. The first is a thin charge-generating layer, which is on top of a thicker charge-transport layer. The charge-generating layer contains a charge-generating material that is usually a highly-purified pigment, commonly of the perylene, polycyclic anthraquinone or phthalocyanine type, with carefully controlled crystallinity and particle size. Light interacts with the pigment to form an ion-pair complex and at the same time releasing an electron that passes to earth. A positive hole is thus generated that migrates to the interface with the charge-transport layer. This layer contains highly electron-rich compounds, such as polyarylamines, not usually coloured materials, which facilitate transfer of the positive holes to the outer surface.

The toners used in photocopying and laser printers are powders, which consist mainly of pigments, charge control agents and resin, a relatively low melting (60–70 °C) polymer. The most extensively used pigment is carbon black since most printing remains monochrome. However, the production of multicoloured prints by this method, using pigments of the three subtractive primary colours, yellow, magenta and cyan, is increasing in popularity. The yellows are commonly provided by disazo pigments, the magenta by quinacridones and the cyan by copper phthalocyanine (see Chapter 9 for a description of the chemical structures of these types of pigments). Charge control agents are materials that, as the name implies, assist in the control of the electrostatic charge applied during the printing process. These are ionic materials (either anionic or cationic), which may be coloured or non-coloured. For example, 2:1 chromium complexes of azo dyes, exemplified by compound **11.26**, which is structurally similar to those used as anionic premetallised dyes for application to wool (Chapter 7), are commonly used as the charge control agents in black toners.

11.26

11.3.2   Inkjet Printing

Digital inkjet printing has experienced tremendous growth in recent years.[48,49] The technique provides a non-impact means of generating images, which involves directing small droplets of ink in rapid succession, under computer control, on to the substrate. There are several types of inkjet printing methods. Two of the principal types are the continuous jet method and the impulse or 'drop on demand' method, although only the latter is considered here as it has assumed the highest industrial importance. In this method, as illustrated in Figure 11.14, pressure on the ink is applied by firing a droplet when it is needed to form part of the image. The pressure may be generated by either thermal or mechanical means. The former mechanism is involved in bubble-jet printers, in which the print-head is heated at the nozzle to a temperature above 300 °C in rapid pulses, causing the formation of a bubble as the ink solvent boils, exerting the pressure that fires an ink droplet. The collapse of the bubble causes ink to refill the nozzle ready for the next pulse. In the latter mechanism, a pressure wave is generated by a piezoelectric crystal, which deforms when activated by an electrical impulse. An array of nozzles is used to generate the image, with the print-head as close as possible to the substrate surface for accuracy of delivery.

Inkjet printing requires the use of inks that meet stringent physical, chemical and environmental criteria. Inkjet inks are very low viscosity fluids as is required by the non-impact method of ink delivery. They have a relatively simple composition, consisting of a solvent, almost invariably water, and a colorant together with other additives for specific purposes, for example water-miscible organic solvents in aqueous inks to assist dye solubility and to minimise crystallisation as ink dries at the nozzle, viscosity modifiers and surfactants for control of surface tension. When first introduced, the colorants used were

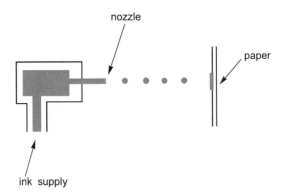

Figure 11.14   'Drop-on-demand' inkjet printing.

mainly dyes because pigments showed a tendency to block the nozzles and to settle from the ink on storage. Initially, the dyes used in inkjet printing were selected from the range of conventional water-soluble products used in textile applications, notably from the acid, direct, or reactive dye application classes, or from food dyes, this last group offering the particular advantage of clearly-established non-toxicity. The earliest drop-on-demand printers were designed for monochrome printing and thus only required a black ink. An example of such a 'first generation' inkjet dye is CI Food Black 2 (**11.27** in Figure 11.15). These early inkjet dyes performed reasonably in many respects, but the principal problem was inadequate water-fastness that could, for example, lead to smudging of the print when handled with moist fingers or when splashed accidentally. In the 'second generation' inkjet dyes, this particular feature was improved by the design of specific dyes that are soluble at the slightly alkaline pH (7.5–10) of the ink, but are rendered insoluble by the weakly acidic pH conditions (pH 4.5–6.5) on the paper substrate. One of the simplest ways in which this was achieved involved the incorporation of carboxylic acid groups, which are ionised (as $-CO_2^-$) at the higher pH thus enhancing water-solubility, but are protonated (as $-CO_2H$) as the pH is reduced, thus reducing solubility. An example of such a dye is compound **11.28** (Figure 11.15), whose molecular structure is closely related structurally to CI Food Black 2, **11.27**, but with sulfonic acid groups replaced, to a certain extent, with carboxylic acid groups. Further improved resistance to wet treatments on paper was provided by applying the dyes as their ammonium salts, which facilitates free carboxylic acid formation through loss of ammonia.

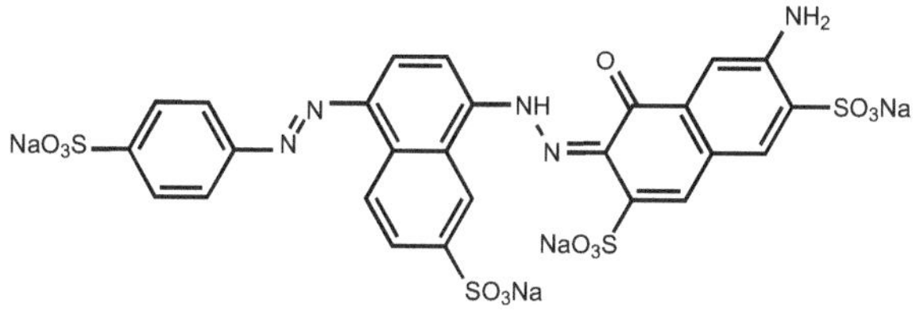

11.27

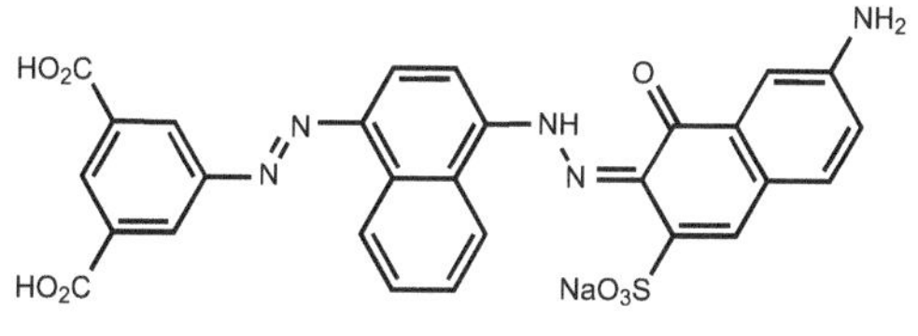

11.28

Figure 11.15 Examples of water-soluble dyes used in ink jet printing.

Most simple SOHO inkjet printers now use a CMYK four-colour system. In printers working to more demanding quality specifications, additional colours are used, for example light cyan (LC) and light magenta (LM) in six-colour printers. The dyes are required to provide high intensity and brightness of colour (chroma), good fastness to light, water and heat (especially for bubble-jet printers in which high temperatures are generated), low toxicity and environmental impact, low inorganic salt content, high water-solubility and good storage stability in the ink. In the early days of inkjet printing, the dyes were selected from the range commercially available for textile applications. However, even the best of these generally had inadequate fastness properties and did not easily provide the ideal colour for the desired trichromatic system. Consequently, considerable effort was devoted to investigate new molecules designed specifically for inkjet applications.[50,51] The most important yellow and magenta dyes in current use are derived from the azo chemical class while the cyans are based on the copper phthalocyanine system. Originally, the water-soluble sulfonated copper phthalocyanines CI Direct Blues 87 and 199 were used since they provided the desired hue with high tinctorial strength. However, these dyes demonstrated inadequate technical performance, including poor fastness to the atmospheric ozone contamination that is prevalent in the SOHO environment. It was demonstrated that this was due to the susceptibility of the carbon–nitrogen bonds in the internal macrocyclic ring of the phthalocyanine system to attack by ozone. A series of interesting molecular design concepts were adopted to address the technical deficiencies, leading for example to cyan dye **11.29**, which is reported to have high durability in inkjet applications.[52] The electron density in the ring system, and hence the susceptibility to oxidative attack, is lowered by the introduction of electron-withdrawing functionality into the system. This provides not only improved ozone resistance but also enhanced lightfastness, a feature that can also involve oxidative degradation processes. The sulfonyl groups in dye **11.29** provide the appropriate electron-withdrawing properties and also promote water solubility. A further feature of the molecular design is aimed at enhancing the molecular aggregation of the dyes in application, which improves fastness properties, for example by presenting a physical barrier to attack from ozone. In the traditional textile dyes obtained by sulfonation of copper phthalocyanine, the sulfonic acid group is introduced more or less randomly into α- and β-positions, leading to a complex mixture. To promote aggregation, the uniformity of the dye mixture and the molecular symmetry are enhanced by control of the position of substituents. In dye **11.29**, this is achieved by incorporating a single substituent into a β-position in each of the four benzene rings. Similarly creative molecular design principles have been applied to develop the yellow and magenta azo dyes that are currently used for inkjet applications.[52]

R : R^1 / R^2 = 3/1
R^1 = -SO$_2$(CH$_2$)$_3$SO$_3$M
R^2 = -SO$_2$(CH$_2$)$_3$SO$_2$NHCH(OH)CH$_3$

11.29

Many of the early difficulties encountered in using pigments as colorants for inkjet printing, especially the tendency of the pigment to settle on storage and to block the jet nozzles, have now been resolved, and pigment-based inks are now used extensively. The inks are generally aqueous dispersions of appropriate CMYK pigments in a fine nanocrystalline form, with specific surface treatments applied to the pigment particles to ensure that the dispersions remain stable for a reasonable length of time. The pigments are also required to be bound to the substrate, a feature that may be achieved by micro-encapsulation of the pigment in a binder, often used in conjunction with a receptive substrate surface, or by incorporating a resin system that is polymerised by UV irradiation after printing. Carbon black pigments are highly effective in inkjet printing and are widely used. Yellows commonly use suitable azo pigments, magenta often employs quinacridones and cyan utilises copper phthalocyanine (Chapter 9). As a broad generalisation, SOHO printing uses dye-based inks because they are capable of producing brighter colours and high storage stability in the ink, while pigment printing is preferred commercially when the hard copies require better longevity.

The use of inkjet printing as a means of producing multi-coloured designs on textiles is growing in importance commercially.[53] The dyes used for particular fabric substrates and the coloration principles are essentially the same as those involved in screen printing on textiles (Chapters 7 and 8). There are, however, significant differences between these printing techniques in terms of the liquid formulations and the way they are used, although the ingredients are broadly similar. The nature of the ink delivery process in inkjet printing means that it is

inappropriate to incorporate into the ink the auxiliary ingredients necessary to ensure image quality (thickening agent), dye penetration
(urea) and dye fixation (*e.g.*, alkali in the printing of cellulosic fibres
with reactive dyes). Instead, the fabrics are pre-treated with these ingredients. An attractive feature of inkjet printing on textiles is that the
designs produced digitally may be applied directly to the fabric using
the digital computer interface, allowing a fast and flexible response to
the consumer and a consequent lowering of the cost for short runs of
product. For longer runs, however, the technique cannot as yet compete economically with the efficiency of traditional rotary screen
printing. Industrial applications of inkjet printing technology continue
to expand beyond traditional paper and textile printing, offering considerable potential in the manufacture, for example, of wall-coverings,
printed electronic circuits, displays and solar panels.

11.3.3 Dye Diffusion Thermal Transfer (D2T2) Printing

The D2T2 printing process employs a set of polyester ribbons coated
with dyes that are transferred to a receiver sheet by the application of
heat at discrete points using a thermal head, which can reach a temperature of up to 400 °C. Consequently, the dye sublimes on to and
migrates into the substrate, which is either a polymer film or paper
coated with a polymer that is receptive to the incoming dye. Since the
D2T2 process involves dye sublimation, initial attempts were made to
use the range of commercial disperse dyes used for textiles (Chapter 7).
However, it became clear that specifically designed dyes were necessary
to meet the requirements of the application, and an extensive range
has been investigated.[54,55] The chemical structures of the dyes in the
CMY trichromatic system used in current printers are broadly

11.30

confidential for reasons of commercial sensitivity. Dyes of the methine type, typified by dye **11.30**, are reported to be suitable. The need for special paper and the consequent expense has limited the growth of the D2T2 technique, but it has an important place in the market for colour printing of high quality and durability.

11.4 BIOMEDICAL APPLICATIONS OF DYES

11.4.1 Biological Stains and Probes

The use of dyes in biological research and medicine is an immense field, which continues to grow year by year. A commonly-used and long-established technique is *biological staining*, in which the selective coloration of biological features is used to enhance their visibility under an optical microscope and thus aid characterisation.[56] The methods involve selective binding of the dye molecules to substances, tissue, cells or microorganisms and their subsequent detection by microscopic techniques. The mechanism of the binding of the dyes commonly involves ionic attraction of dye anions or cations to the opposite charges carried by structural components of tissue. There may also be contributions from weaker forces of attraction such as hydrogen bonding, dipolar and van der Waals forces. In many cases, however, the mechanism of binding is not fully established. A few stains react with biological species to form covalent bonds. The best-known example is Schiff's reagent, **11.32**, which is formed from the reaction of para-rosaniline, a triarylmethine dye, **11.31**, with sulfurous acid (or an aqueous solution of sulfur dioxide). The colourless reagent contains a primary amino group that is capable of reacting with aldehyde functionality to form coloured species, such as **11.33**, as illustrated in Scheme 11.1. Further reaction of the other primary amino group in species **11.33** with aldehyde groups may proceed. Aldehyde groups may be present or chemically generated at specific sites in biological systems, for example from polysaccharides or by partial hydrolysis of DNA.

There is often advantage in using strongly fluorescent dyes or brightening agents to colour specific components of living cells, because of their ease and sensitivity of detection, for example by fluorescence microscopy, and their ability to provide high specificity. Fluorochromes used in this way are referred to as *fluorescent probes*.[57] In terms of the dye chemistry, established dyes may be used but, more commonly, dyes are synthesised in small quantities for specific research purposes and are consequently very expensive. As a molecular design tool, quantitative structure–activity relationship (QSAR)

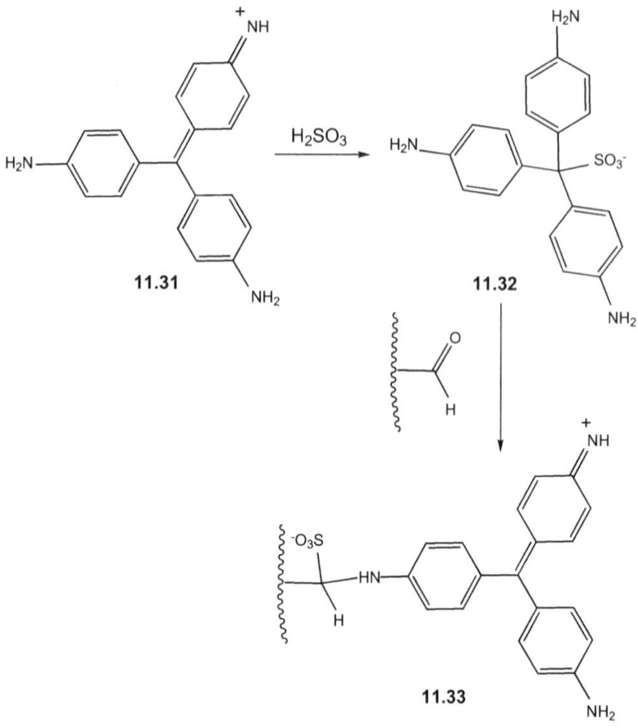

Scheme 11.1   Biological staining using Schiff's reagent, **11.32**.

methods (Chapter 2) are useful in predicting the location of a probe molecule in a cell.[58] Fluorescent probes are capable of detecting, identifying and quantifying biomolecules, such as antibodies, proteins, lipids and polysaccharides in cells. They may also be used to de-code cell components (DNA, RNA, proteins and enzymes)[59] and to probe cell structures and functions, often for diagnostic purposes. Variation in the spectral characteristics of a bound dye are indicative of changes in the conformation of biomolecules and also the nature of their environment, thus providing a useful tool in the investigation of biological activity and the mechanism of biological reactions. This technology has also been incorporated into practical devices for biomedical applications, including optical chemical sensors (*e.g.*, for oxygen, carbon dioxide, pH and metal ions), optical biosensors (*e.g.*, for enzymes, glucose, anti-bodies, and DNA)[60] and optical sensors of cell temperatures.[61]

11.4.2   Photodynamic Therapy

Photodynamic therapy (PDT) has been the subject of an immense body of research over the last 30 years or more and has emerged as

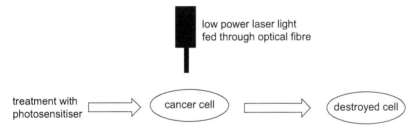

Figure 11.16 PDT (photodynamic therapy) cancer treatment.

arguably the most significant development in medical treatment that involves the use of dyes.[62–65] PDT is a treatment for certain cancers, some already established in clinical use and some undergoing trials, as well as for age-related macular degeneration (ARMD, a significant cause of blindness in the elderly), skin disorders and bacterial and viral infections. The treatment process, illustrated schematically in Figure 11.16, involves the use of a photosensitising compound (the dye) interacting with low energy light. The photosensitiser is applied to the patient, either topically or intravenously, after which some time is allowed so that it equilibrates within the body. During this time, the photosensitiser selectively accumulates in the tumour cells (or lesion). Irradiation of the cells with laser light, fed *via* an optical fibre for example, is then carried out, initiating their destruction and thus providing a potential means for destroying the tumour. For skin treatments, lamps fitted with appropriate filters may be used as the light source. An important advantage of PDT is that treatment is localised and thus may give rise to fewer side-effects than conventional cancer treatments, such as chemotherapy.

There are two main mechanisms by which light may react with the photosensitiser, ultimately causing damage to the tissue.[65] Both mechanisms involve the generation of a triplet state from which energy is transferred to nearby molecules, with the involvement of molecular oxygen in oxidative degradation processes. In both cases, laser light interacts with the photosensitiser and promotes it from its ground state (^{1}Sens) into its singlet excited state (^{1}Sens*), which then undergoes intersystem crossing to a triplet excited state (^{3}Sens*). In the *type I* mechanism, this triplet state reacts with adjacent biomolecules to form free radicals, for example by electron transfer or hydrogen abstraction. These radicals undergo further reaction involving molecular oxygen, which exists in a triplet ground state (^{3}O$_2$), to form various reactive oxygen-containing species, such as hydroperoxides, hydroxyl radicals and superoxide. In the *type II* mechanism

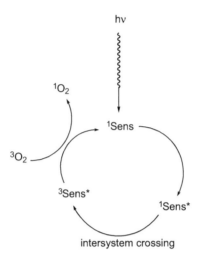

Figure 11.17   Photo-induced mechanisms involved in photodynamic therapy.

(Figure 11.17), which is believed to be prevalent in most PDT processes, the photosensitiser triplet state (^{3}Sens*) interacts directly with triplet oxygen (^{3}O$_2$) to generate singlet oxygen (^{1}O$_2$) and the photosensitiser consequently returns to its ground state (^{1}Sens). In this case, the photosensitiser is a catalyst for formation of singlet oxygen, which is the effective photodynamic agent, being highly reactive, for example towards unsaturated centres in the proteins and lipids that construct the cell membrane.

The most commonly used photosensitiser for PDT is referred to as Photofrin. To produce Photofrin, haematoporphyrin, obtained from blood, is first treated with acetic and sulfuric acids followed by an alkaline work-up to give a derivative, HpD, which is a complex mixture of porphyrin monomers and oligomers. Photofrin is obtained from HpD by removal of the non-active monomeric components. Structures such as **11.34** (Figure 11.18) have been proposed for the active components of Photofrin. Although still clinically the most widely used PDT agent, this first generation photosensitiser has several inherent deficiencies. It is a complex mixture, presenting issues of reproducibility, exhibits only modest activity and has low absorption at the red-light wavelength (630 nm) used in traditional PDT light sources. As a side-effect, it may cause sensitivity of the skin to sunlight for a prolonged period of time after the treatment. Consequently, there has been a vast amount of research carried out in the search for new generations of improved photosensitising agents. Among the requirements considered for the second generation agents were that the

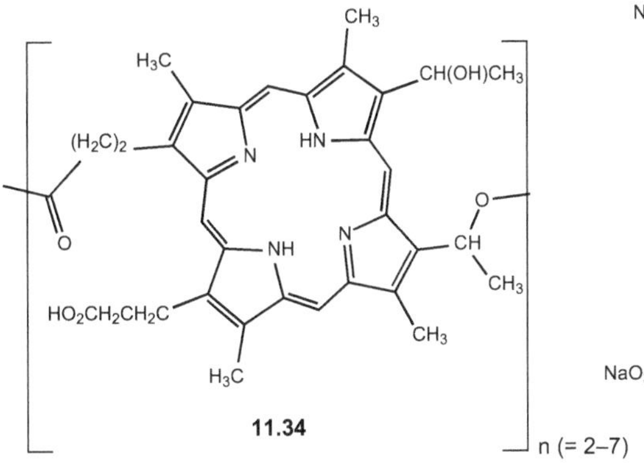

Figure 11.18 Photosensitisers used in photodynamic therapy (PDT).

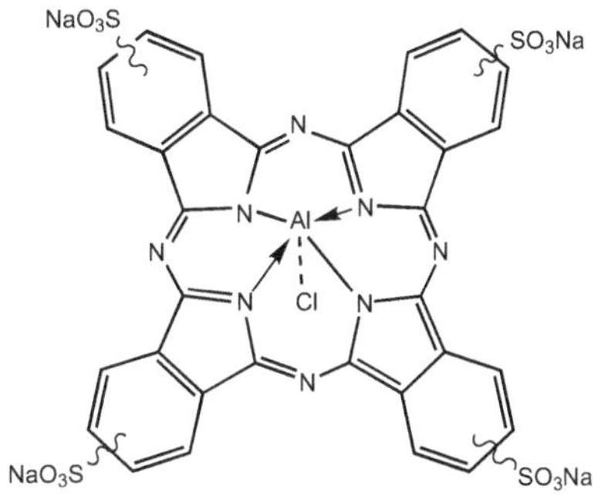

dye should absorb at higher wavelengths for improved absorption of the laser light. The dyes should provide solubility in the body fluids, improved specificity for tumour cells, low toxicity until light is applied, and should clear from the body after treatment. They should also have a triplet excited state that is sufficiently long lived to facilitate the formation of singlet oxygen. For this last purpose, it is commonly advantageous to include heavy atoms in the molecular structure of the dyes, such as metals, sulfur, bromine and iodine. A wide range of dyes has been investigated for this purpose, with some undergoing clinical trials, including a variety of modified porphyrins. Particular promise has been shown by phthalocyanine derivatives that absorb bathochromically compared with the porphyrins, for example the sulfonated aluminium(III) phthalocyanine **11.35**. Various issues are addressed in the development of so-called third generation PDT products. One of the most significant of these issues centres on the so-called near-infrared 'therapeutic' window (700–950 nm), in which light absorption by the body is minimised and its depth of penetration into tissue is increased.[66] Human tissue absorbs light below 700 nm due to skin pigmentation, haemoglobin and some polymeric biomolecules, while water absorbs above around 950 nm. There is consequently interest in investigating photosensitisers that absorb at wavelengths in this region, for example taking advantage of the absorption wavelength range that is available from the range of cyanine dyes.

11.5   CHROMIC MATERIALS

Chromic materials is a term used increasingly to refer to dyes and pigments that exhibit a distinct colour change induced by an external

Table 11.1 Colour change phenomena with the associated stimulus.

Colour change phenomenon	Stimulus
Ionochromism	Ions
Thermochromism	Heat
Photochromism	Light
Electrochromism	Electric current flow
Solvatochromism	Solvents
Vapochromism	Vapours
Mechanochromism	Mechanical action
Chronochromism	Time
Radiochromism	Ionizing radiation
Magnetochromism	Magnetic field
Biochromism	Biological sources

stimulus, for example exposure to light, heat or electric current. Traditional applications of dyes and pigments, such as those described in Chapters 7–9, are required to provide a constant, predictable and reproducible colour and, as far is technically feasible, a permanent colour in terms of exposure to external effects. Any variation in the colour, for example when exposed to a change in temperature or to light, would normally be regarded as highly undesirable, *i.e.*, a defect. However, it has been recognised in recent decades that there are potential important commercial niche applications for dyes and pigments that change colour when exposed to an external stimulus, especially when that change is controllable and reversible, providing impetus for extensive research into new materials and their potential for technical applications.[67]

Specific colour change phenomena are named using the suffix *chromism* and a prefix that describes the stimulus causing the change. Table 1 provides a list of phenomena and the stimulus involved. Ionochromism, thermochromism, photochromism, and electrochromism are the most extensively studied. The others listed are generally less well-established and have either experienced limited commercial exploitation or are essentially academic curiosities. Indeed, the list in the table is probably not exhaustive.

11.5.1 Ionochromism

Ionochromism involves a reversible colour change caused by interaction with an ionic species. Dyes exhibiting a wide range of colour changes are available, either from colourless to coloured or from one colour to another. The most common ionochromic materials, arguably the longest-established useful group of chromic materials, are pH-sensitive dyes. These dyes are sensitive to the hydrogen ion (H^+),

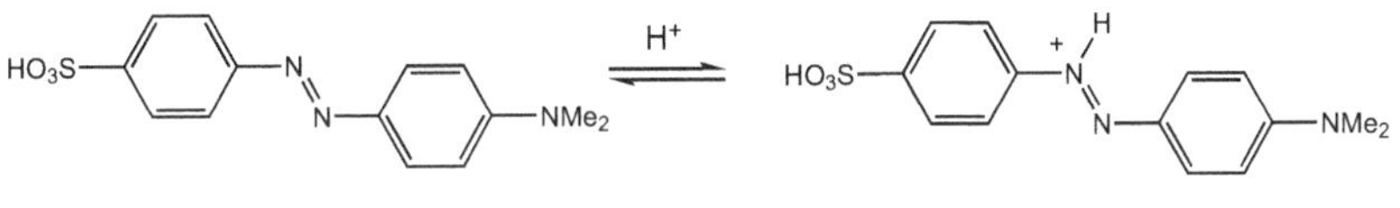

11.36 orange   **11.36a**, red

Scheme 11.2   Colour change in the protonation of methyl orange, **11.36**, an acid–base indicator.

and are referred to as *halochromic*. An immense range of coloured organic molecules show halochromic behaviour.[68] The main chemical classes of technical importance are phthalides, triarylmethines and fluorans. However, many other chromophores can undergo halochromism, including simple azo dyes, such as methyl orange, **11.36**, which changes from orange to red with the formation of the protonated species **11.36a** as the pH is lowered (Scheme 11.2). Halochromic phthalides and fluorans are discussed in the next section in the context of their use as colour formers in leuco dye thermochromic systems, in which the colour change is induced by a temperature-sensitive pH change within a composite system. Another category of ionochromism is *metallochromism*, in which the dyes change colour as they bind, often selectively, to specific metal ions. The dyes that show this property commonly consist structurally of chelating ligand functionality, notably crown ethers or cryptands, attached to chromophores.

Ionochromic materials have many important technological applications. Halochromic dyes have been used extensively over the years in analytical chemistry as reversible pH indicators in detecting the end-point of acid–base titrations and in spot papers, such as litmus. Similarly, metallochromic materials have been used in the qualitative and quantitative analysis of metal ions. The importance of these traditional applications has diminished significantly as modern instrumental analytical methodology has developed to supersede their use. However, new applications have emerged more recently, for example in absorbance-based ion-selective optical sensors, which have applications in chemical process control, medical diagnostics and environmental monitoring.[69]

11.5.2   Thermochromism

Thermochromism is the term used to describe a change of colour as a result of a temperature change.[70] Thermochromic dyes and pigments find application where this colour change indicates a temperature change, for example in plastic strip thermometers, medical thermography and non-destructive testing of engineered articles and

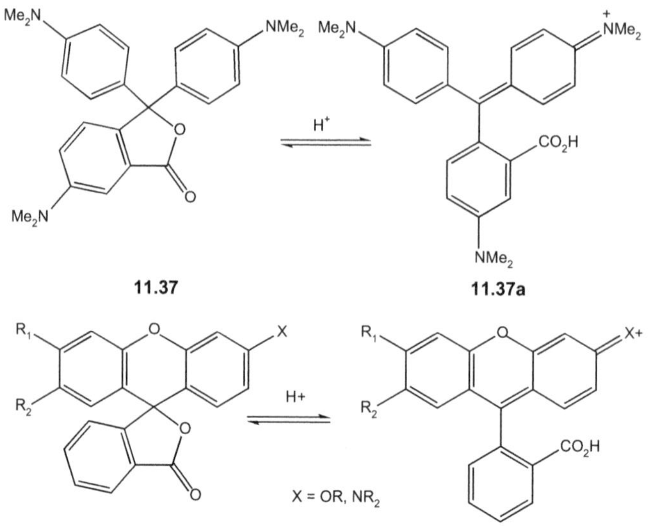

Scheme 11.3 Colour formation in the ring-opening of crystal violet lactone and related compounds.

electronic circuitry. They may also be used in thermal imaging and for a variety of decorative or novelty effects. Crystal violet lactone, **11.37**, CVL, the most important colour-forming material used in thermochromic colorants, is a non-planar molecule and is colourless. In contact with acid, the lactone ring is opened to give the violet/blue arylcarbonium ion (triarylmethine) structure, **11.37a**, as illustrated in Scheme 11.3. Similar chemistry is exhibited by xanthene derivatives (fluorans), represented by structure **11.38**, which extend the colour range of the ring-opened forms, **11.38a**, to orange, red, green and black. Colour formation reactions of this type are utilised in carbonless copy paper, based on colour formation on the copy as a result of pressure of writing or typing in the master sheet. In such systems, the underside of the master sheet contains the colour former, encased in microcapsules, which are tiny spheres with a hard polymer outer shell. Pressure on the master sheet breaks the microcapsules and brings the colour former in contact with an acidic reagent coated on the copy sheet, thus causing an irreversible colour formation reaction. This particular application involves only halochromism. The colour formers are not inherently thermochromic. However, they may be used to generate colour thermally when used in combination with other materials. In thermally-sensitive paper, the colour former and an acidic developer, usually a phenol, are dispersed as insoluble particles in a layer of film-forming material. When brought into contact with a thermal head at around 80–120 °C, the composite mixture melts as a result of the localised heating. As a result, the colour former and developer diffuse together and react to form a colour. This process is assisted by the presence of a third component, a sensitiser, such as dibenzyl terephthalate, which assists diffusion by acting as a solvent. In this case, the thermochromism is irreversible and a permanent image is formed.

The most widely-used industrial thermochromic system is the referred to as the *leuco* dye type. The term *leuco* describes a dye that can acquire two forms, one of which is colourless. This micro-encapsulated composite system relies on colour formation from the interaction of three materials: a colour former (the leuco dye), an acidic developer and a low-melting, non-volatile hydrophobic solvent, a set of ingredients similar to those used in thermally-sensitive paper as described above. However, in the leuco dye thermochromic systems, it is curious that, even though similar chemistry is involved, the composite material is coloured at low temperatures and decolourises as the temperature is raised. To explain this effect, a mechanism illustrated in simplified form in Scheme 11.4 has been proposed.[71–73] At temperatures below the melting point of the solvent, the system is heterogeneous with the colour former existing as the coloured

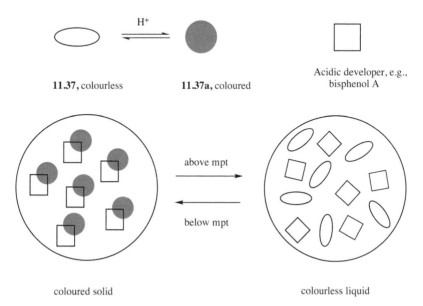

Scheme 11.4 Schematic representation of the mechanism of colour change in a microencapsulated leuco dye thermochromic system.

ring-opened protonated species (*e.g.*, **11.37a** in the case of CVL). With the anion derived from the acidic developer, commonly a phenol such as bisphenol A, this cation forms a complex which is insoluble in the solvent, and its formation is favoured by a localised polar environment. As the temperature is raised and the solvent melts, the components dissolve to form a homogeneous liquid in which the hydrophobic environment favours formation of the colourless neutral ring-closed form (**11.37** in the case of CVL) together with undissociated developer. As the mixture cools and the solvent solidifies, phase separation occurs and the coloured dye–developer complex reforms.

Thermochromism is also shown by certain liquid crystalline materials. As discussed in Section 11.2.1 in the context of displays (LCDs), the three main sub-divisions of the liquid crystalline state are *smectic*, *nematic* and *chiral nematic*. Chiral nematics are capable of showing a thermochromic effect that is quite different to that of leuco dye thermochromic types. They provide a continuously changing spectrum of colours over a range of temperatures (referred to as *colour-play*) when observed against a dark (ideally black) background. The colours arise from changes in the orientational structure of the liquid crystal with temperature and from the way that light interacts with the liquid crystals to produce coloured reflection by interference.[74] The phase adopts a screw-like, helical arrangement that, as illustrated in a somewhat idealised way in Figure 11.1(c), may be envisaged as composed of nematic liquid crystal layers. The director (representing the average molecular axis direction) of an individual layer is rotated through a small angle with respect to the director in adjacent layers until it eventually turns through 360°. The thickness between identically-oriented layers is the pitch length (p) of the helix. When the magnitude of the pitch length is similar to the wavelength of visible light, interaction of white light with the liquid crystal causes selective reflection of a wavelength band, in a manner analogous to Bragg X-ray reflection. The remaining wavelengths are transmitted, and when these are absorbed by a black background the colour associated with the reflected light is observed.

The reflected wavelength is proportional to the pitch length of the helix. Thermochromism arises because the pitch length varies with temperature as illustrated in Figure 11.19. The colour changes are especially pronounced and occur within a particularly narrow temperature band when there is a transition from one liquid crystal phase to another. For example, as a material passes through a transition temperature leading from a smectic into a chiral nematic phase, the helix forms and the rapid decrease in pitch length with increasing

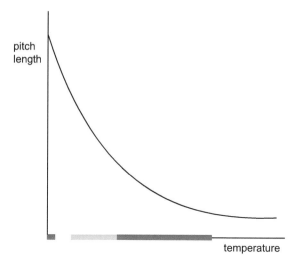

Figure 11.19 Variation of pitch length with temperature for a thermochromic chiral nematic liquid crystal.

temperature gives rise to colours changing from red initially through the spectrum to blue. As illustrated in Figure 11.19, the change in pitch length with increasing temperature is rapid initially and becomes progressively less pronounced as the isotropic phase is approached. Thus, as the temperature increases steadily, the colours corresponding to longer wavelengths (reds and yellows) are observed only fleetingly and the longer lasting visual impression is of greens and blues.

11.5.3 Photochromism

Photochromism is commonly defined as a process in which a compound undergoes a reversible change between two chemical species with different absorption spectra, *i.e.*, with different colours, on irradiation with light.[70] Most photochromic dyes convert reversibly from colourless into coloured when exposed to UV light, as illustrated in Scheme 11.5. The dyes may be categorised into two broad types. In the case of T-type dyes, the reverse reaction is thermally-driven, *i.e.*, occurs when the light source is removed, while with P-type dyes, the reverse reaction is photochemically-induced using light of a different wavelength. Photochromic materials are used most extensively in ophthalmic sun-screening applications, notably the familiar spectacles which become sunglasses when exposed to UV light, and also ski-goggles and motorcycle helmet visors. For lens applications, the

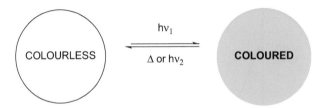

Scheme 11.5   General scheme for reversible photochromism.

dyes are required to react rapidly to UV light to produce a strong colour from a colourless inactivated state, and to fade back to the colourless state at a controlled rate, so that the kinetics of coloration and fading are key features, requiring careful matching when mixtures are used. The dyes are also required to provide more or less constant performance over many cycles of photocoloration, a feature known as fatigue resistance.

While most industrial dyes are of the T-type, there is considerable interest in P-type photochromics and their potential applications.[75]

The most important industrial T-type photochromic dyes are spirooxazines and naphthopyrans (Scheme 11.6). The importance of spirooxazines is due to their ability to impart intense photocoloration, good photo-fatigue resistance and relative ease of synthesis.[76] Spirooxazines, **11.39**, contain a spiro (sp³) carbon atom that separates the molecule into two, each half containing heterocyclic rings whose π-systems are orthogonal. The localized π-systems mean that they absorb only in the UV region. The oxazine C–O bond is ruptured on exposure to UV radiation to give a coloured photomerocyanine, **11.39a**. When the UV source is removed, the molecule reverts to the more stable colourless ring-closed form as the oxazine bridge is re-established. Merocyanine **11.39a** is one of several possible isomers of the coloured forms. The ring-opening is believed to proceed initially through metastable *cisoid* isomers that subsequently rearrange to a more stable *transoid* isomer, such as **11.39a**.[77] The simplest spirooxazines give blue photocoloration, although dyes giving red and violet through to turquoise colours in their activated forms may be produced by appropriate molecular design. Naphthopyrans, represented by general structure **11.40**, also undergo light-induced ring-opening to give a coloured form, **11.40a**.[78] By appropriate substituent pattern variation, naphthopyrans provide photochromic colours across the spectrum from yellows through oranges, reds, violets and blues. In addition to modifying colour, molecular changes may be used to fine-tune the kinetics of photocoloration and fading, and to enhance photostability.

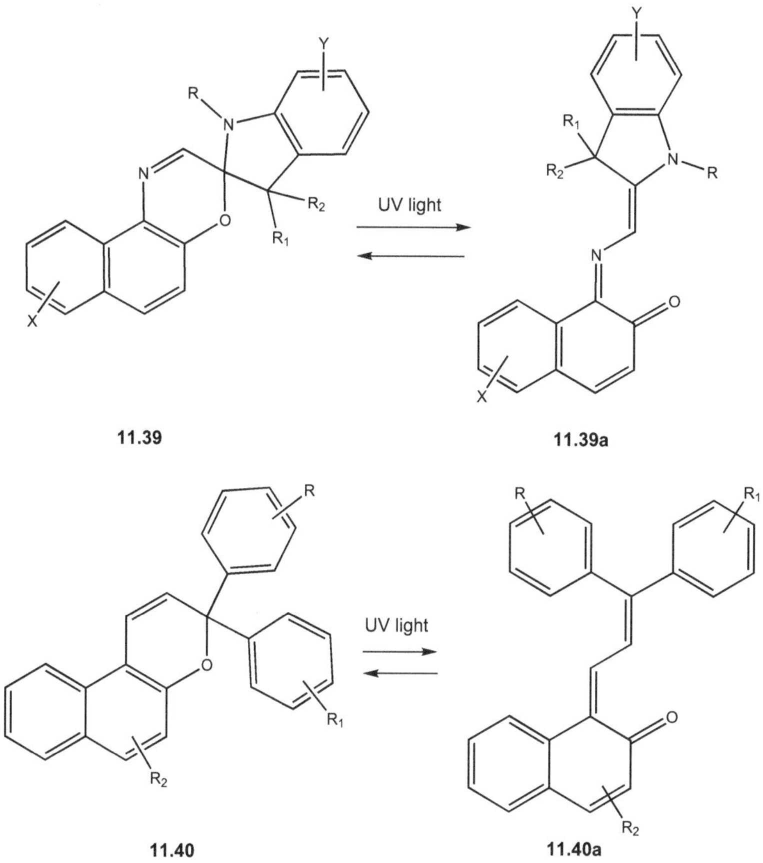

Scheme 11.6 Photochromism of T-type dyes: spirooxazine **11.39** and naphthopyran **11.40**.

The most important two groups of P-type photochromic dyes are fulgides, such as dye **11.41**, and diarylethenes, exemplified by dye **11.42**. As illustrated in Scheme 11.7, the reaction initiated by UV light is in both cases an electrocyclic cyclisation in which a colourless or weakly coloured ring-opened species converts into a coloured ring-closed species (**11.41a** or **11.42a**). This chemistry contrasts with the T-type photochromics where the photocoloration is due to ring-opening of a colourless species. A further contrasting feature is that the coloured species does not readily undergo thermal reversion, but requires the absorption of certain wavelengths of visible light to return to its original state. The P-type photochromics are considered to have considerable potential for application in devices using optical switching, including data storage.

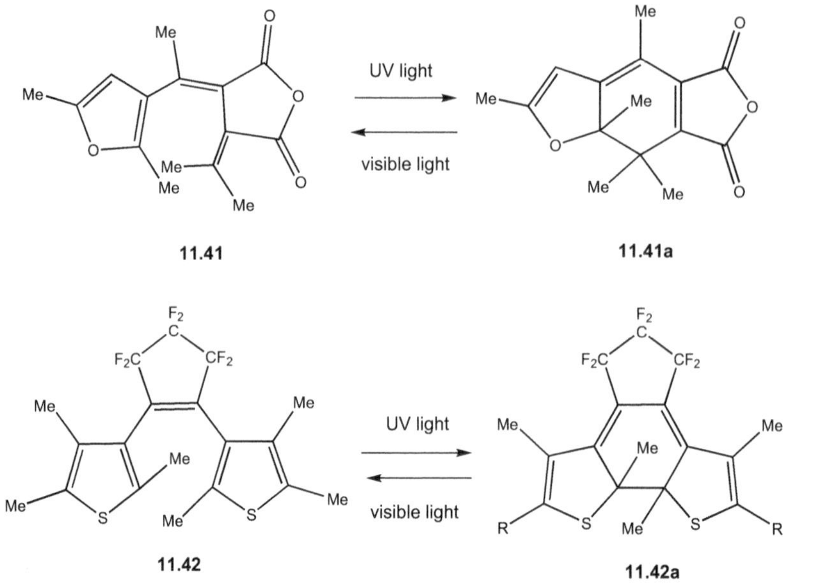

Scheme 11.7   Photochromism of P-type dyes: fulgide **11.41** and diarylethene **11.42**.

11.5.4   Electrochromism

Electrochromism involves a reversible colour change resulting from a flow of electric current. The colour change is due to electron transfer reactions at an electrode: oxidation at an anode and reduction at a cathode.[79] Many commercially important electrochromic materials are inorganic. Several transition metal oxides, the most important being tungsten oxide (WO_3), show electrochromism in the solid state. In the case of WO_3, the pure compound is pale yellow, although appearing essentially colourless in a thin film. The blue colour generated electrochemically is due to partial reduction of W(VI) to W(V) at a cathode, the depth of colour developed being proportional to the charge injected. Other solid-state electrochromes include the blue pigment Prussian blue (Chapter 9) and certain metal phthalocyanines. The best-known example of a compound that exhibits electrochromism in solution is methyl viologen, **11.43**, a bipyridylium dication, which is colourless and undergoes reduction at a cathode to give the bright blue radical cation **11.43a** as illustrated in Scheme 11.8. Other organic electrochromes include 1,4-phenylenediamines and thiazines. There have also been significant recent developments in polymeric electrochromic materials, including polyaniline and polythiophene derivatives.

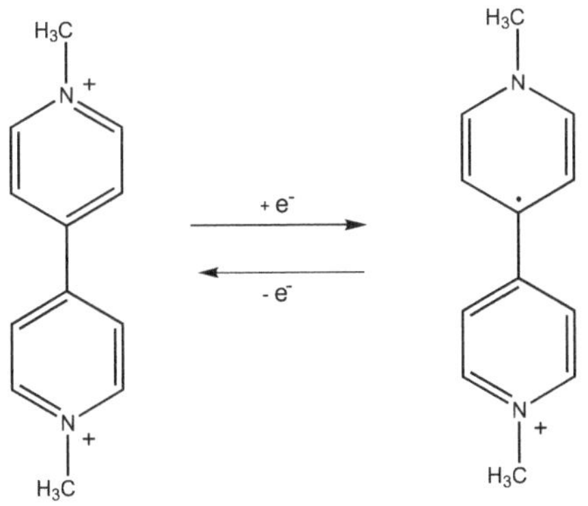

11.43   **11.43a**

Scheme 11.8 Reversible electrochromism of methyl viologen.

From the time of the earliest developments, it was envisaged that reversible electrochromic colour change might be useful in the production of multicolour flat screen displays. Although there has been some progress towards certain niche display applications, other technologies, notably liquid crystal displays (LCDs) and light emitting diodes (LEDs), as discussed earlier in this chapter, have proved much more successful commercially in achieving the full colour range over a uniform large area at relatively low cost. The most important applications of electrochromism are in electrically-switchable, anti-dazzle rear-view car mirrors and in smart windows for the control of light and temperature in buildings. The mirrors contain an indium tin oxide (ITO) coated glass surface with the conductive side facing inwards as one electrode and a reflecting metal electrode at the back. The gap between the electrodes consists of an electrolyte solution or gel containing two soluble electrochromes, one of which is oxidised at the anode and the other reduced at the cathode as electric current flows. Smart windows may be constructed from two transparent sheets of glass with conductive surfaces that act as the electrodes, on which solid state electrochromic materials are coated, and between which there is a conducting layer of a polymeric electrolyte. The windows darken electrochromically according to the level of incident sunlight. The units generally incorporate a light sensing system, and may also use integrated solar cells.

11.5.5 Miscellaneous Colour Change Phenomena

Various other chromic materials are known, as listed in Table 1. *Solvatochromism* is an extensively-studied phenomenon whereby a dye dissolved in different solvents exhibits different colours, mostly associated with solvent polarity (see Chapter 2 for further discussion of this topic). *Mechanochromism* refers to colour change in materials, usually in the solid state, when they are placed under mechanical stress. *Piezochromism,* a form of mechanochromism that occurs as a result of the application of pressure or compression, is exhibited by certain organic polymers, such as polydiacetylenes and polythiophenes, and by some palladium complexes. *Tribochromism* is a form of mechanochromism that is exhibited when certain crystalline compounds become more highly coloured as they are ground to a fine powder. Other colour change phenomena include *vapochromism* (exposure to vapours), *chronochromism* (time), *radiochromism* (ionising radiation), *magnetochromism* (magnetic fields) and *biochromism* (interaction with biological sources).

REFERENCES

1. Z. Yoshida and T. Kitao, *Chemistry of Functional Dyes*, Mita Press, Tokyo, 1989.
2. P. Gregory, *High Technology Applications of Organic Colourants*, Plenum Press, New York, 1991.
3. P. Bamfield and M. G. Hutchings, *Chromic Phenomena: Technological Applications of Colour Chemistry*, Royal Society of Chemistry, Cambridge, 2010.
4. S.-H. Kim (ed.), *Functional Dyes*, Elsevier, Amsterdam, 2006.
5. R. M. El-Shishtawy, *Int. J. Photoenergy*, 2009, article ID 434897, DOI: org/10.1155/2009/434897.
6. S. Matsumoto, *Electronic Display Devices*, John Wiley & Sons, Inc., New York, 1990.
7. R. R. Hainich and Oliver Bimber, *Displays: Fundamentals and Applications*, CRC Press, Taylor & Francis Group, Boca Raton, FL, 2011.
8. C. Sergeant, *Rev. Prog. Color.*, 2002, **32**, 58.
9. T. L. Dawson, *Rev. Prog. Color.*, 2003, **33**, 1.
10. L. D. Silverstein, *IS&T Rep.*, 2006, **21**(1), 1.
11. L. Ozawa, *Cathodoluminescence and Photoluminescence: Theories and Practical Applications*, CRC Press, Taylor & Francis Group, Boca Raton, FL, 2007.
12. P. J. Collings, *Liquid Crystals: Nature's Delicate Phase of Matter*, Princeton University Press, Princeton, New Jersey, 2002.
13. L. M. Blinov, *Structure and Properties of Liquid Crystals*, Springer, Heidelberg, 2011.

14. R. H. Chen, *Liquid Crystal Displays: Fundamental Physics and Technology*, John Wiley & Sons, Inc., Hoboken, 2011.
15. G. H. Heilmeier and L. A. Zanoni, *Appl. Phys. Lett.*, 1968, **13**, 91.
16. R. A. Sabnis, *Displays*, 1999, **20**, 119.
17. H. Iwanaga, *Materials*, 2009, **2**, 1636.
18. Z. H. Kafafi, *Organic Electroluminescence*, CRC Press, Taylor & Francis Group, Boca Raton, FL, 2005.
19. T. Tsujimura, *OLED Display Fundamentals and Applications*, John Wiley & Sons, Inc., Hoboken, 2012.
20. K. Müllen and U. Scherf (ed.), *Organic Light Emitting Devices, Synthesis, Properties and Applications*, Wiley-VCH Verlag GmbH, Weinheim, 2006.
21. Z. Li and H. Meng (ed.), *Organic Light Emitting Devices*, CRC Press, Taylor & Francis Group, Boca Raton, FL, 2007.
22. A. A. Shoustikov, Y. You and M. E. Thompson, *IEEE J. Selected Top. Quantum Electron.*, 1998, **4**(1), 1.
23. R. H. Partridge, *US Pat.* 3995299, 1976.
24. D. F. Perpichka, I. F. Perpichka, H. Meng and F. Wudl, in *Organic Light Emitting Devices*, ed. Z. Li and H. Meng, CRC Press, Taylor & Francis Group, Boca Raton, FL, 2007, pp. 45–293.
25. K. Thyagarajan and A. K. Ghatak, *Lasers: Fundamentals and Applications*, Springer, Berlin, 2nd edn, 2010.
26. F. J. Duarte and L. W. Hillman (ed.), *Dye Laser Principles: with Applications*, Elsevier, London, 2012.
27. D. H. Titteron, *Rev. Prog. Color.*, 2002, **32**, 40.
28. C. V. Shank, *Rev. Mod. Phys.*, 1975, **47**, 649.
29. G. S. Shankarling and K. J. Jarag, *Resonance*, 2010, 804.
30. S. Singh, V. R. Kanetkar, G. Sridhar, V. Muthuswamy and K. Raja, *J. Luminescence*, 2003, **101**, 295.
31. M. Z. Jacobson, *Energy Environ. Sci.*, 2009, **2**, 148.
32. J. Nelson, *The Physics of Solar Cells (Properties of Semiconductor Materials)*, Imperial College Press, London, 2003.
33. A. R. Jha, *Solar Cell Technology and Applications*, CRC Press, Taylor & Francis Group, Boca Raton, FL, 2009.
34. A. Shah, *Thin-Film Silicon Solar Cells*, EPFL Press, Lausanne, 2010.
35. K. Kalyanasundaram, *Dye-Sensitised Solar Cells*, EPFL Press, Lausanne, 2010.
36. A. Hagfeldt, G. Boschloo, L. Sun, L. Kloo and H. Petterson, *Chem. Rev.*, 2010, **111**, 6595.
37. M. Grätzel, *J. Photochem. Photobiol. C: Photochem. Rev.*, 2003, **4**, 145.
38. W. H. Weber and J. Lambe, *Appl. Opt.*, 1976, **15**, 2299.

39. A. Goetzberger and W. Greubel, *Appl. Phys*, 1977, **14**, 123.
40. P. F. Scudo, L. Abbondanza, R. Fusco and L. Caccianotti, *Solar Energy Mater. Solar Cells*, 2010, **94**, 1241.
41. R. Reisfeld, *Opt. Mater.*, 2010, **32**, 850.
42. J.-J. Wanègue, *Opt. Disc Syst.*, 2003, **5**, 42.
43. J.-J. Wanègue, *Opt. Disc Syst.*, 2003, **6**, 52.
44. H. Mustroph, M. Stollenwerk and V. Bressau, *Angew. Chem. Int. Ed.*, 2006, **45**, 2016.
45. M. Matsuoka, *Infrared Absorbing Dyes*, Plenum Press, New York, 1990.
46. J. Fabian, H. Nakazumi and M. Matsuoka, *Chem. Rev.*, 1992, **92**, 1197.
47. R. S. Gairns, in *Chemistry and Technology of Printing and Imaging Systems*, ed. P. Gregory, Blackie Academic, London, 1996, pp. 76–112.
48. I. M. Hutchings and G. D. Martin (ed.), *Inkjet Technology for Digital Fabrication*, John Wiley & Sons Ltd., Chichester, UK, 2013.
49. P. Gregory, *Chem. Br.*, 2000, **36**, 49.
50. K. Carr, in *Colorants for Non-Textile Applications*, ed. H. S. Freeman and A. T. Peters, Elsevier, Amsterdam, 2000, pp. 1–34.
51. M. Fryberg, *Rev. Prog. Color.*, 2005, **35**, 1.
52. Y. Fujie, N. Hanaki, T. Fujiwara, S. Tanaka, M. Noro, K. Tateishi, K. Usami, A. Hibino, N. Wachi, T. Taguchi and Y. Yabuki, *Fujifilm Res. Develop.*, 2009, **54**, 35.
53. H. Ujiie (ed.), *Digital Printing of Textiles*, Woodhead Publishing, Cambridge, UK, 2006.
54. R. Bradbury, in *Colorants for Non-Textile Applications*, ed. H. S. Freemen and A. T. Peters, Elsevier, Amsterdam, 2000, pp. 35–60.
55. R. Bradbury, in *Modern Colorants: Synthesis and Structure*, ed. H. S. Freemen and A. T. Peters, Blackie Academic, London, 1995, pp. 154–175.
56. J. A. Kiernan, *Color. Technol.*, 2006, **122**, 1.
57. R. B. Thomson, *Fluorescence Sensors and Biosensors*, CRC Press, Taylor & Francis Group, Boca Raton, FL, 2006.
58. R. W. Horobin, *Color. Technol.*, 2014, **130**(3), 155.
59. R. T. Ranasinghe and T. Brown, *Chem. Commun.*, 2005, 5487.
60. S. M. Borisov and O. S. Wolfbeis, *Chem. Rev.*, 2008, **108**, 423.
61. C. Gota, K. Okabe, T. Funatsu, Y. Harada and S. Uchiyama, *J. Am. Chem. Soc.*, 2009, **131**, 2766.
62. C. J. Gomer (ed.), *Photodynamic Therapy: Methods and Protocols*, Humana Press, New York, 2010.
63. M. P. Goldman (ed.), *Photodynamic Therapy*, Elsevier Saunders, Philadelphia, 2005.

64. R. Bonnet, *Chemical Aspects of Photodynamic Therapy*, CRC Press, Boca Raton, FL, 2000.
65. M. Wainwright, *Rev. Prog. Color.*, 2004, **34**, 95.
66. M. Wainwright, *Color. Technol.*, 2019, **126**, 115.
67. R. W. Sabnis, *Handbook of Acid–Base Indicators*, CRC Press, Taylor & Francis Group, Boca Raton, FL, 2007.
68. R. Narayanaswamy and O. S. Wolfbeis (ed.), *Optical Sensors: Industrial, Environmental and Diagnostic Applications*, Springer, Berlin, 2004.
69. R. M. Christie, Chromic materials for technical textile applications, in *Advances in the Dyeing and Finishing of Technical Textiles*, ed. by M. L. Gulrajani, Woodhead Publishing, Oxford, UK, 2013.
70. J. C. Crano and R. J. Guglielmetti (ed.), *Organic Photochromic and Thermochromic Compounds*, Kluwer Academic/Plenum Publishers, New York, 1999, vols. 1 and 2.
71. D. Aitken, S. M. Burkinshaw, J. Griffiths and A. D. Towns, *Rev. Prog. Color.*, 1996, **26**, 1.
72. S. M. Burkinshaw, J. Griffiths and A. D. Towns, *J. Mater. Chem.*, 1998, **8**, 2677.
73. D. C. MacLaren and M. A. White, *J. Mater. Chem.*, 2003, **13**, 1701.
74. R. M. Christie and I. D. Bryant, *Color. Technol.*, 2005, **121**, 187.
75. S. N. Corns, S. M. Partington and A. D. Towns, *Color. Technol.*, 2009, **125**, 249.
76. V. Lokshin, V. A. Samat and A. V. Metelitsa, *Russ. Chem. Rev.*, 2002, **71**, 893.
77. R. M. Christie, L. J. Chi, R. A. Spark, K. M. Morgan, A. F. S. Boyd and A. Lycka, *J. Photochem. Photobiol. A*, 2004, **169**, 37.
78. J. Hepworth and B. M. Heron Photochromic naphthopyrans, in *Functional Dyes*, ed. S.-H. Kim, Elsevier, Amsterdam, 2006, ch. 3, pp. 85–135.
79. P. M. S. Monk, R. J. Mortimer and D. R. Rosseinsky, *Electrochromism and Electrochromic Devices*, Cambridge University Press, Cambridge, 2007.

CHAPTER 12

Colour and the Environment

Over recent decades, society has developed an increasing concern over protection of the environment. There are a whole series of major global issues that are under constant current debate, including the impact of climate change attributed by most to greenhouse gas emissions, the destruction of the rainforests and the need to develop the use of renewable sources of energy and materials, thus aiming for 'sustainability'. There are also a large number and variety of individual environmental issues that are associated with particular geographical locations and specific industries. Manufacturing industry is continuously faced with the need to address its responsibility towards a wide range of health, safety and environmental concerns, which present many of its most significant current challenges. In particular, industry is challenged with the requirement to satisfy the demands of increasingly stringent legislation and controls introduced by governments and enforced by regulatory agencies to ensure compliance with environmental issues. There is concern worldwide over the potential adverse effects of the activities of the chemical industry on the environment. The various industry sectors have been addressing the issues, although it is an inescapable fact that the response in some parts of the world has been much faster and more intense than in others. The industries that are associated with colour represent a relatively small part of the overall chemical industry. This sector has been described, in general, as a small volume, multi-product industry and it has traditionally been highly innovative, constantly seeking to

Colour Chemistry, 2nd edition
By Robert M Christie
© R M Christie 2015
Published by the Royal Society of Chemistry, www.rsc.org

introduce new products. This has led to a requirement to address a wide range of toxicological and ecological issues, both in the manufacture of dyes and pigments and in their application. This chapter seeks to present an overview of some of the more important general issues facing the colour industry and the ways in which they are being addressed. However, the coverage is by no means comprehensive. This is due to the extensive range of individual issues, reflecting the diversity of chemical types of dyes and pigments and the synthetic procedures involved, together with the wide range of applications. It also reflects the complexities of the relevant legislative detail throughout the world. However, the reader will find a discussion of many of these specific issues that are associated with the colour chemistry topics covered in the preceding chapters of this book.

A detailed discussion of the legislation and controls, the introduction of which has been a major factor in ensuring compliance with the most important issues, is outside the scope of this chapter. This is due to the complexity of the legislative detail, the fact that it varies substantially from country to country, and because the situations are constantly evolving so that information presented would quickly become out of date. However, it is worth noting, by way of an example, the current situation in the European Union, the geographical region that has arguably led the way in terms of introducing the strictest environmental legislation. The environmental policy in the EU places a particular focus on energy, greenhouse gas emissions, ozone-depleting substances, air quality and pollution, waste, water use and conservation, sustainable consumption and production, chemicals, biodiversity and land use.[1] The principal EU policy objectives may be summarised as:

(a) protecting environmental quality;
(b) protecting human health;
(c) promoting prudent use of natural resources;
(d) promoting international measures to deal with environmental problems.

The application of these principles has culminated in the introduction of *REACH*, a European regulation concerning chemicals and their safe use.[2] This regulation, the introduction of which was initiated in 2007, although taking more than a decade to be phased in completely, applies directly in all 28 Member States of the EU, and also in Iceland, Liechtenstein and Norway as member countries of the

European Economic Area. The REACH regulation (*R*egistration, *E*valuation, *A*uthorisation of *Ch*emicals) is a complex piece of legislation that aims to protect human health and the environment from the risks arising from the use of chemicals while maintaining the free movement of goods in the EU. It requires manufacturers and importers to register the chemicals used, and to establish and make public the hazards posed by each substance, with reference to their entire life-cycle.

In the USA, the Environmental Protection Agency (EPA) states its mission as the protection of human health and the environment. The agency cites the twelve principles of 'Green Chemistry' that industry is encouraged to apply in its strategies towards product and process development.[3] These principles may be summarised as:

(a) better to prevent waste than to clean up afterwards;
(b) design synthetic methods to maximise incorporation of all materials used in the process into the final product (atom economy);
(c) design synthetic methods to use and generate substances that minimise toxicity to human health and the environment;
(d) design chemical products to provide their desired function while minimising their toxicity;
(e) minimize the use of auxiliary substances wherever possible, and make innocuous when used;
(f) minimize energy requirements of chemical processes, conducting syntheses at ambient temperature and pressure when possible;
(g) use renewable raw materials whenever practicable;
(h) minimize or avoid unnecessary derivatisation where possible;
(i) recognise that catalytic reagents are superior to stoichiometric reagents;
(j) design chemical products so that they degrade into innocuous species that do not persist in the environment;
(k) develop analytical methodologies to provide real-time, in-process monitoring and control prior to the formation of hazardous substances;
(l) select substances and their physical form for use in chemical processes to minimise the potential for chemical accidents, including release, explosion and fire.

The colour manufacturer and user has a duty to address environmental and toxicological risks from various points of view, including hazards in the workplace, exposure of the general public to the materials and the general effect on the environment.[4,5] The level of risk from exposure to chemicals is clearly of prime concern for those handling materials in large quantities in the workplace. This has required the introduction of modern work practices, enforced by legislation, to minimise the exposure. The approach towards addressing the problems associated with exposure to potentially dangerous chemical substances that has been adopted in most countries generally involves an evaluation of risk, including an assessment of the hazards presented by the various chemical species on the basis of the available toxicological data, and an assessment of exposure levels. From this evaluation, risk management strategies are developed.

Dyes and pigments are, by definition, highly visible materials. Thus, even minor releases into the environment may cause the appearance of colour, for example in open waters, which for obvious reasons attracts the critical attention of the public and local authorities. There is thus a requirement on industry to minimise environmental release of colour, even in cases where a small but visible release might be considered as toxicologically rather innocuous. A major potential source of release of colour into the environment is associated with the incomplete exhaustion of dyes on to textile fibres from aqueous dyeing processes.[6] The need to reduce the amount of residual dye in effluent from textile wet-processing, together with the wide range of other process-specific auxiliary organic and inorganic materials that may be used in the processes, has thus become a major concern in recent years. While this feature applies, in principle, to all application classes of textile dyes, the particular case of reactive dyeing of cellulosic and protein fibres is of special interest because of the problems associated with dye hydrolysis, a process that competes with the dye–fibre reaction, and means that hydrolysed dye inevitably appears in the effluent. An extensive programme of research to address these problems over many years has met with some success, leading to the development of high-fixation fibre-reactive systems and significantly improved processing conditions (Chapter 8). However, the development of a practical reactive dyeing system that is completely free of the problems associated with hydrolysis and incomplete dye fixation has so far proved elusive.

Several complementary approaches have been adopted to address the problems associated with the presence of colour and other materials in aqueous industrial effluent. It is clearly of vital importance that industrial organisations develop strategies for pollution abatement and waste minimisation, for example in the selection of materials, process optimisation to reduce the environmental load, and the practise of recycling, re-use and conservation wherever practicable. The recycling of water is becoming increasingly important, especially as the industry migrates to regions of the world where water may be a scarce resource.[7] Another complementary approach is the development of effluent treatment methods to ensure that any release into water courses is safe towards humans and aquatic life, and complies with regulations.[8] These methods can be effective but inevitably add to the cost of the overall process, and some may present complications associated with the possible toxicity of degradation products. The methods that are currently in use may be categorised broadly as chemical, physical or biological. In general, employing combinations of more than one method provides the most effective approach. Several chemical treatment methods for colour removal have been developed, of which the most successful involve oxidative degradation, for example using chlorine, Fenton's reagent (hydrogen peroxide with iron(II) salts) or ozone.[9] Ozone treatment is particularly effective but is rather expensive, mainly because of the specialist equipment required. Electrochemical and photocatalytic oxidation methods to degrade dyes in effluent have also been investigated. Physical treatments may also be used to remove colour, for example by the adsorption of dyes on to inert substrates such as activated charcoal, silica, cellulose derivatives or ion-exchange resins.[10] Adsorbents based on biomass, materials derived from dead animal and plant residues, micro-organisms and fungi, have also been extensively investigated. Biological treatment processes make use of the ability of living organisms to bind or degrade colour.[11,12] Biodegradation processes offer the attraction of the potential for the decoloration of effluent with complete mineralisation of the organic materials present to carbon dioxide, water and inorganic ions such as nitrate, sulfate and chloride. However, the principal problem with the development of such systems is that synthetic dyes are generally xenobiotic, *i.e.*, are not metabolised by the enzymes in the micro-organisms present naturally in waters. The solution to this issue may require the development of micro-organisms cultured specifically for the purpose of metabolising synthetic dyes. The problems associated with textile effluent are clearly not restricted to the dyes themselves. There are also issues associated with the presence of traces of heavy metals, the high levels of inorganic salts and other auxiliaries required by certain textile dyeing processes, and the use of reducing agents in the case of vat and sulfur dyes. Each specific process presents its own set of environmental issues that are required to be addressed by the development of new products, process improvements or effluent treatment methods (Chapter 7). The preceding discussion in this chapter has focussed on dyes. In the case of pigments, because of their insolubility and the particular way they are used, the loss into the environment is a much less significant issue than with textile dyes. In this respect, a convincing case may be made that pigments offer environmental advantage over dyes in use.

There has been a considerable amount of recent research carried out on textile dyeing from supercritical fluids, notably carbon dioxide, as an alternative to traditional aqueous dyeing methods, motivated to a considerable extent by environmental considerations.[13–15] We are familiar with CO_2 behaving either as a gas at normal temperatures and pressures, or as a solid (dry ice) when frozen. However, if the temperature and pressure are both increased to above the critical point (critical temperature 31.1 °C and critical pressure 73.8 bar), CO_2 can adopt properties that are intermediate between a gas and a liquid, the supercritical fluid state. Specifically, in this state it expands to fill its container like a gas but with a density more like a liquid. CO_2 is an inert, non-toxic and non-flammable solvent, and in these respects it has strong environmental credentials. At first sight, this last feature might appear surprising in view of its implication as a major greenhouse gas contaminant of the atmosphere. However, it is argued that this technology is capable of providing a use for CO_2 that otherwise might be released into the atmosphere, provided that it is recovered and recycled during the application process. Methods using supercritical CO_2 have been applied most extensively and successfully in the laboratory to the dyeing of polyester fibres with disperse dyes (Chapter 7), in a process that fulfils many of the requirements of environmental sustainability. In the dyeing process, the CO_2 medium is recyclable, uses a minimum input of chemicals (essentially only the dyes and no auxiliaries), presents low energy requirements and produces minimal waste. The dyed textile materials also show high technical performance. This process has now been established with considerable success on a laboratory scale. It is, however, difficult to speculate on the commercial future of the technology until a plant has been built and the process has been validated on an industrial scale. The use of supercritical CO_2 as a solvent for dye synthesis has also been established on a small scale for some azo dyes.[16]

The importance of being aware of the potential adverse effects of exposure to chemicals on our health is self-evident. Toxic effects may be categorised in a number of ways. Acute toxicity refers to the effects of short-term exposure to a substance. It is relatively reassuring that studies of textile dyes suggest that there is only limited evidence for acute oral toxicity, and that most show little or no toxicological effect.[17,18] However, there is evidence that certain reactive dyes for cellulosic fibres (Chapter 8) may cause problems associated with skin and respiratory sensitisation, contact dermatitis, rhinitis, occupational asthma and other allergic reactions, especially for workers who handle the dyes in their manufacture and application. The problems are reported to be associated with the ability of the fibre-reactive group in the dyes to react with human serum albumin, the most abundant protein in blood plasma. After fixation to the textile fibre, it is reassuring that reactive dyes are considered not to cause these toxicological problems because the reactive group is no longer present. A few disperse dyes (Chapter 7) have been implicated in causing allergic reactions when used in tight-fitting garments made from certain synthetic fibres. The problem may arise with dyed polyamide or cellulose acetate, because of the tendency of the dyes to migrate on to the skin *via* perspiration. However, the issue is much less prevalent with dyed polyester on which disperse dyes show good fastness towards perspiration. Pigments and vat dyes generally show low acute toxicity. This is generally attributed to their extremely low solubility in body fluids, which means that they are capable of passing through the digestive system without absorption into the bloodstream.

Chronic toxicity refers to the effect of regular exposure to a chemical over a prolonged period of time. Arguably, the most severe chronic toxicological effect for humans is the potential to induce cancer. There has been some concern over the potential carcinogenicity of certain azo dyes, which centres mainly on the possibility of metabolism by reductive cleavage of the azo group to give two primary aromatic amines, one of which is the diazo component from which the dye is derived synthetically (Chapter 3). Several primary aromatic amines that have been used in colour manufacture are recognised as carcinogens. In particular, it is well established that benzidine, **12.1a**, and 2-naphthylamine, **12.2**, are potent human carcinogens.[19] Epidemiological studies carried out in the first half of the twentieth century demonstrated a pronounced increase in the incidence of bladder cancer in workmen employed in the dye manufacturing industry who had been exposed to these two amines. As a consequence, responsible manufacturers in the Western world discontinued the manufacture of dyes from these amines, although the manufacture continued in certain other parts of the world.

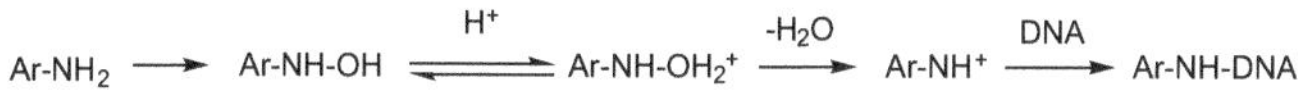

$$Ar\text{-}NH_2 \longrightarrow Ar\text{-}NH\text{-}OH \xrightarrow{H^+} Ar\text{-}NH\text{-}OH_2^+ \xrightarrow{-H_2O} Ar\text{-}NH^+ \xrightarrow{DNA} Ar\text{-}NH\text{-}DNA$$

Scheme 12.1 Mechanism proposed for the carcinogenic activity of certain primary aromatic amines.

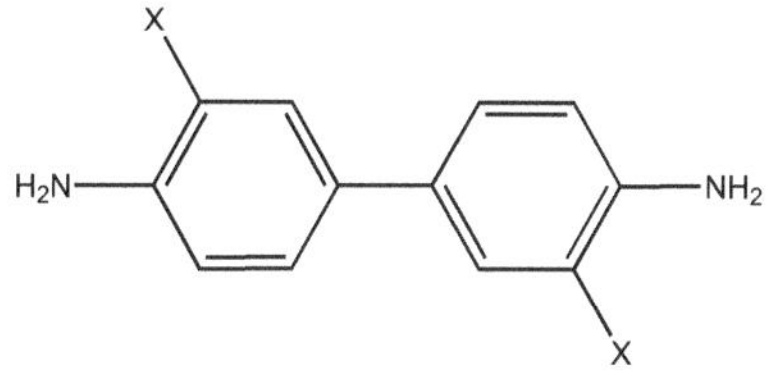

12.1a: X = H
12.1b: X = Cl

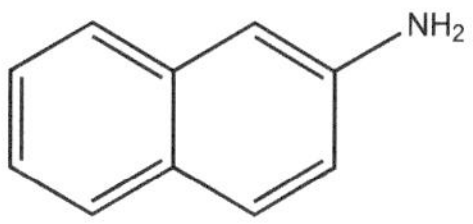

12.2

A generally-accepted mechanism that accounts for the potential carcinogenic activity of certain primary aromatic amines is shown in Scheme 12.1, although other metabolic pathways may also be involved. The initial step is N-hydroxylation, leading subsequently to the highly electrophilic nitrenium ion $(Ar\text{–}NH^+)$, which has the potential to react with nucleophilic sites on DNA. Based on this mechanism, the carcinogenic agent is a metabolite rather than the amine itself.

It has emerged subsequently that benzidine-derived azo dyes, such as Congo Red, CI Direct Red 1, **12.3**, may also present a carcinogenic risk, an effect attributed to the metabolism of the dyes to produce benzidine, **12.1a**, by enzymatic reduction. Certain European countries have examined the cancer-causing potential of azo dyes critically, by focussing on the amines that would be released if reductive cleavage of the azo group were to take place. In Germany, for example, the approach has been to ban the manufacture and importation of all dyes derived from a list of more than 20 aromatic amines believed to be animal carcinogens.[20] On the other hand, it is by no means true that all azo dyes should be considered as potential carcinogens. Studies of structure–carcinogenicity relationships in azo dyes have demonstrated, for example, that when the amine produced by reductive cleavage of the azo group contains a sulfonic acid group there is little or no toxicological risk.[21] An interesting example of this principle is CI Food Black 2, **12.4**, a first generation black dye used in inkjet printers (Chapter 11). Azo pigments are considered to present a considerably reduced risk compared with azo dyes, so that, for example, they are not included in the German legislation. 3.3′-Dichlorobenzidine (DCB), **12.1b**, is on the German list of carcinogenic amines, yet it continues to be used in the manufacture of a series of important commercial disazo pigments (diarylide yellows and oranges, see Chapter 9), since there does not appear to be strong experimental evidence that these pigments are metabolised to DCB. However, there is evidence that the diarylide pigments may cleave thermally with prolonged heating above 240 °C to give a series of decomposition products, which include DCB.[22] Since this evidenced emerged, there has been a decline in the use of these pigments in applications where high temperatures are likely to be encountered, for example in thermoplastics.

12.3

12.4

   Most foods contain natural colouring materials and, in addition, it is common to add colour to enhance the appeal of some foods to the consumer.[23,24] There are several interesting features associated with the colours present naturally in foods, and their influence on our health. There is growing evidence that certain natural colorants, for example the carotenoids present in fresh fruit and vegetables (Chapter 6), are of therapeutic value as a consequence of their anti-oxidant properties, which, it is suggested, may provide protection against cancer.[25] While there is an increasing tendency to use natural dyes as food additives, synthetic dyes are more commonly used as they offer the advantages of higher colour strength and brightness, and lower cost. Only a few synthetic dyes are permitted for use in food coloration, because of the stringent toxicological requirements, enforced by legislation that prohibits certain dyes, and regulatory agencies that impose the control mechanisms. In addition, the lists of permitted dyes are constantly being updated. In Europe, these are covered by an EU Directive introduced in the 1990s, together with a series of subsequent amendments that clarify definitions and limitations in use. In the USA, the regulating body is the Food and Drug Administration (FDA). Food dyes are also required to satisfy strict specific purity requirements. An important example of a synthetic food colorant is tartrazine, CI Food Yellow 23, a monoazo dye, **12.5**. Notably, reductive enzymatic cleavage of the azo group in this dye would lead to sulfonated aromatic amines, which are considered to present little toxicological risk. However, there is some public concern over the possibility that the use in confectionary and drinks of certain synthetic dyes (usually referred to as 'artificial' and classified by 'E-numbers') may cause hyperactivity in some children.

12.5

Another area that experiences strict legislative control in the use of colour is cosmetics, which is also the domain of the FDA in the USA. Concerns about the safety of ingredients used in hair coloration have been raised over the years in view of their extensive use in regular direct human contact (Chapter 10). In particular, the safety profile of certain precursors used in permanent hair dyeing has been extensively studied, and some original ingredients have been prohibited. In certain individuals, hair coloration may cause skin irritation or allergic reactions at varying levels of severity. International epidemiological studies on the link between hair coloration and more serious human conditions have indicated that hair coloration may play a role in the risk of certain lymphomas. The increased risk is described as only moderate and much more significant among women who had used the dyes before 1980, when the compositions would have been more likely to include potentially carcinogenic components. In parallel, the hair dye industry continues to provide reassurances over the components in current use. Nevertheless, a future change to the current situation cannot be excluded if stricter legislation, regulations and controls on the use of chemicals emerge.

The extension of the use of natural dyes into a wider range of applications including textiles, as alternatives to synthetic dyes, would appear superficially to offer attractive possibilities, in essence turning the clock back to the pre-Mauveine mid-nineteenth century when only natural dyes were available (Chapter 1). In the ongoing debate on the issues involved, the range of opinions expressed is wide. There are obvious environmental arguments that support the use of natural dyes, as this would be seen to be exploiting renewable resources, thus making a contribution towards sustainability, while at the same time presenting minimal risk to human health. There are several counter-arguments. Natural dyes are generally more expensive and show inferior colours and technical performance in application compared with synthetic dyes, to an extent that might not be acceptable to the consumer. In addition, the large-scale production and use of natural dyes might well introduce environmental problems, for example the need to cultivate large areas of land for which food production is a higher priority, and there is a likelihood that the coloration processes involved would not necessarily be free from pollution problems.[26] However, there is evidence of recent significant renewed interest in natural dye research aimed towards improved production methods, application processes and product performance, and it may well be that certain niche markets will emerge that take advantage of the positive environmental perception that is associated with natural colours.

The issue of the so-called 'heavy metals' has caused concern within the chemical industry for some time. While the term 'heavy metal' might have been derived originally on the basis of density, this property alone does not necessarily relate directly to toxicity or environmental behaviour. In fact, certain metals, for example iron, zinc, manganese, copper, chromium, molybdenum and cobalt, might technically be described as 'heavy', yet they are essential for life. The term has come to be used to represent those metals that are regarded as detrimental to the environment when a certain concentration is exceeded. The elements mercury, cadmium and lead occupy a special position because the concentration at which they begin to become detrimental to the health of organisms is very low. In addition, when absorbed in excessive amounts they may accumulate and can cause chronic health problems. Mercury is not a constituent of any significant commercial colorant. However, one important use of mercury derivatives is as catalysts in the sulfonation of anthraquinone, an essential first step in the synthesis of many anthraquinone dyes (Chapter 4). Special care is required to ensure that mercury is not released into the environment in the effluent from such processes. Inorganic pigments based on lead (lead chromates and molybdates) and cadmium (cadmium sulfides and sulfoselenides) are still important commercially (Chapter 9). Their use nowadays is restricted significantly by a series of voluntary codes of practice, reinforced by legislation in certain cases, for example in toy finishes, graphic instruments and food contact applications where ingestion is a possibility. Cadmium pigments are used mainly in the coloration of certain engineering plastic materials which require very high temperature processing conditions. At present, completely satisfactory substitutes for such applications are not available, especially in terms of thermal and chemical stability. Lead chromates continue to be used, mainly in coatings, because they remain by far the most cost-effective, high durability yellow and orange pigments. Furthermore, it is argued by the manufacturers that lead and cadmium pigments do not present a major health hazard or environmental risk, because of their extreme insolubility. Nevertheless, it seems likely that the progressive replacement of these products by more acceptable inorganic and organic pigments will continue in an increasing range of applications. Chromium(vi) is also considered to be highly toxic. In addition to its presence in lead chromate pigments, dichromates may be used in the dyeing of wool by the chrome mordanting process (Chapter 7). This process was formerly of special importance as a cost-effective means of producing deep colours, which are fast to light and washing, on wool. Despite process improvements, for example the development of dyeing auxiliaries and procedures that minimise the level of residual excess chromium in textile dye effluent, the process is little used nowadays and continues to decrease in importance. There are toxicological and environmental issues associated with several other metals used to a certain extent in colorants, for example, barium, manganese, cobalt, nickel, copper and zinc, but at present their use remains acceptable.

Certain organic chlorine compounds are of serious toxicological and environmental concern, most importantly the polychlorobiphenyls

(PCBs).[27] This group of materials, characterised by structure **12.6** in which 1–10 chlorines may be attached, consists of 209 possible congeners, due to the various numbers of chlorines and substituent patterns. PCBs were formerly used on a large scale as insulating materials in electrical equipment, hydraulic oils and heat transfer systems. In addition to their extreme toxicity, their high stability, while a technical advantage in their former applications, means that they persist strongly in the environment, and they are thus classified as persistent organic pollutants (POPs). Although their intentional production has long been discontinued, quantities remain in the environment. PCBs constitute problem waste and their disposal is problematic and expensive. The potential to form PCBs in trace amounts has been noted in the manufacture of certain colorants, including a few azo pigments and copper phthalocyanines, when certain aromatic chloro compounds are used either as reactants or solvents. In such situations, the industry is required to respond by developing processes either to eliminate PCB formation or to ensure that the levels are below the rigorous limits set by legislation, generally not exceeding 50 ppm. Some environmental agencies and activist groups have taken a more extreme view and advocated a ban on all chlorinated organic chemicals. Such a ban would have a major effect on the colorant industry because of the large number of organic dyes and pigments that contain chlorine substituents. So far, legislation to this effect has not been introduced. Indeed, any move in such a direction would be an overreaction, since there is little evidence that the mere presence of chlorine in a molecule means that it poses an environmental risk.

12.6

Environmental arguments are highly emotive. There are many reasons why the public has become sensitive to environmental issues and, to a certain extent, suspicious of industry attitudes. The industry has made errors in the past, some with severe human and environmental consequences, and has had to examine its conscience on occasions for having paid inadequate regard to health and safety, and also to broader ecological issues. However, there is a point of view that the pendulum may have swung to the opposite extreme, so that, for example, arguments based on perception, rather than solid, scientific evidence, often win the day. The need for rigorous experimentation to address toxicological and environmental concerns, while absolutely necessary, has acted as a deterrent to innovation, in view of the considerable expense required in testing before a product may be introduced into the marketplace. This has had a particular impact on the colour manufacturing industry and is a major reason why fewer and fewer new dyes and pigments, especially for traditional applications, have been introduced commercially in the last few decades. We should not lose sight of the fact that the main reason for the existence of the synthetic colour industry is that dyes and pigments enhance our environment, influencing our moods by bringing attractive colours into our lives in such a diverse range of applications. In addition, many of the recent developments towards functional dye applications, such as their use in harvesting solar energy and in the treatment of cancers and other ailments using photodynamic therapy, as described in Chapter 11, have the potential to provide immense benefit to mankind and our environment. It is indeed significant that a colour, green, has been adopted as the universal ecological symbol. Green is, after all, the predominant colour in the natural, fertile world, due mainly to the chlorophyll in foliage, and has for centuries been associated with natural forces, growth and regeneration. Green is also considered as a cool, calm, restful and safe colour.

It would be a dull world if, for example, television, movies, photography and magazines essentially only provided black and white images, and if automobiles were available in 'any colour we like provided it's black!'. The days when such a situation existed are really not so long ago in relative historical terms. Equally, it is vital that those industries involved in the manufacture and application of colour should continue to be increasingly sensitive to any potential adverse effect on the environment in its wider sense, and respond accordingly. A balanced approach will ensure protection of the environment and allow an innovative colour industry to thrive into the future.

REFERENCES

1. Towards a green economy in Europe: EU environmental policy targets and objectives 2010–2050, *EEA Report No. 8/2013*, European Environmental Agency, Copenhagen, 2013.
2. E. S. Williams, J. Panko and D. J. Paustenbach, *Crit. Rev. Toxicol.*, 2009, **39**, 553.
3. P. Anastas and N. Eghbali, *Chem. Soc. Rev.*, 2010, **39**, 301.

4. A. Reife and H. S. Freeman (ed.), *Environmental Chemistry of Dyes and Pigments*, John Wiley & Sons, Inc., New York, 1996.
5. R. Anliker, *Rev. Prog. Color.*, 1977, **8**, 60.
6. R. M. Christie (ed.), *Environmental Aspects of Textile Dyeing*, Woodhead Publishing in Textiles, Cambridge, UK, 2007.
7. K. Skelly, *Rev. Prog. Color.*, 2000, **30**, 21.
8. M. Joshi and R. Purwar, *Rev. Prog. Color.*, 2004, **34**, 58.
9. I. R. Hardin, in *Environmental Aspects of Textile Dyeing*, ed. R. M. Christie, Woodhead Publishing in Textiles, Cambridge, UK, 2007, ch. 8.
10. R. Sanghi and B. Bhattacharya, *Color. Technol.*, 2002, **118**, 256.
11. A. Kandelbauer, A. Cavaco-Paulo and G. M. Gübitz, in *Environmental Aspects of Textile Dyeing*, ed. R. M. Christie, Woodhead Publishing in Textiles, Cambridge, UK, 2007, ch. 9.
12. M. A. Rauf and S. S. Ashraf, *Chem. Eng. J.*, 2012, **209**, 520.
13. M. Banchero, *Color. Technol.*, 2013, **129**, 2.
14. E. Bach and E. Schollmeyer, in *Environmental Aspects of Textile Dyeing*, ed. R. M. Christie, Woodhead Publishing in Textiles, Cambridge, UK, 2007, ch. 5.
15. E. Bach and E. Schollmeyer, *Rev. Prog. Color.*, 2002, **32**, 88.
16. J. Hooker, D. Hinks, G. Montero and C. Conlee, *Color. Technol.*, 2002, **118**, 273.
17. P. Gregory in *Environmental Aspects of Textile Dyeing*, ed. R. M. Christie, Woodhead Publishing in Textiles, Cambridge, UK, 2007, ch. 3.
18. K. Hunger, *Rev. Prog. Color.*, 2005, **35**, 76.
19. S. Spitz, W. H. Maguigan and K. Dobriner, *Cancer*, 1950, **3**, 789.
20. German ban of the use of certain azo compounds in some consumer goods, revised version, *ETAD Information No. 6*, Basel, Switzerland, 1998.
21. P. Gregory, *Dyes Pigments*, 1986, 7, 45.
22. R. Az, B. Dewald and D. Schnaitmann, *Dyes Pigments*, 1991, **15**, 1.
23. D. B. MacDougall (ed.), *Colour in Food*, Woodhead Publishing, Cambridge, UK, 2002.
24. D. Frick, *Rev. Prog. Color.*, 2003, **33**, 15.
25. H. Nishino, M. Murakosh and T. Li, *et al.*, *Cancer Metastasis Rev.*, 2002, **21**, 275.
26. B. Glover, *Textile Chem. Colorist*, 1995, **27**(4), 17.
27. M. D. Erickson, *Analytical Chemistry of PCBs*, CRC Press, Boca Raton, FL, 2nd edn, 1997.

Subject Index